New Renewable Energy Resources

World Energy Council

CONSEIL MONDIAL DE L'ENERGIE

New Renewable Energy Resources

A Guide to the Future

KOGAN PAGE

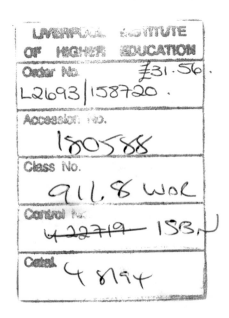

First published in 1994

Kogan Page Limited
120 Pentonville Road
London N1 9JN

British Library Cataloguing in Publication Data

A CIP record for this book is available from the British Library.

ISBN 0 7494 1263 1

Typeset by Saxon Graphics Ltd, Derby
Printed in England by Clays Ltd, St Ives plc

WORLD ENERGY COUNCIL

TECHNICAL STUDY COMMITTEE MEMBERSHIP

New Renewable Energy Resources

Chairman Prof. E. P. Volkov (Russia)
Vice Chairman Mr. J. R. Darnell (USA)

Members

Dr. M. Schneeberger (Austria)
Dr. D. L. P. Strange (Canada)
Mr. J. J. Edens (Denmark)
Dr. J. R. Frisch (France)
Prof. Dr. C. J. Winter (Germany)
Dr. M. J. Grubb (Great Britain)
Mr. J. M. Jefferson (WEC)
Dr. L. Radonyi (Hungary)
Mr. P. R. Bapat (India)
Mr. L. Y. Bronicki (Israel)
Dr. P. Chungmoo Auh (Korea)
Mr. Husam Taher (Jordan)
Prof. Dr. W. C. Turkenburg (Netherlands)
Mr. A. Oistrach (Spain)
Dr. A. A. Eberhard (South Africa)
Mr. B. Agrenius (Sweden)
Mr. Tsahn-Rarn Cheng (Taiwan, China)
Mr. J. Tillinghast; Mr. W. R. Gould (USA)

Corresponding Members

Mr. J. Speziale (Argentina)
Mr. G. M. Drew (Zimbabwe)
Dr. S. de Salvo Brito (Brazil)
Dr. T. Horigome (Japan)
Mr. Ahmed Shadzli bin Abdul Wahab (Malaysia)
Mr. P. J. Graham (New Zealand)
Mr. Suheyl Elbir (Turkey)

International Organizations

Dr. J. Christensen (UNEP)
Mr. A. A. Churchill (IBRD)
Mr. S. F. Garriba (IEA)
Mr. K. Brendow/Mrs. J. Andorfer (UNECE)
Mr. A. Radjai (UNESCWA)
Mr. G. Best (UNFAO)
Mr. A. Colling (CEC)

Secretary

Dr. A.V. Misulin (Russia)

NEW RENEWABLE ENERGY RESOURCES

Working Group Membership

Working Group 1: Solar Energy
Mr. J. R. Darnell, *Chairman* (USA)
Mr. J. T. Cole (Canada)
Dr. D. L. P. Strange (Canada)
Mr. B. Devin (France)
Dr. J-R. Frisch (France)
Prof. Dr. C. J. Winter (Germany)
Mr. B. Doron (Israel)
Mr. I. Dostrovsky (Israel)
Dr. H. Tabor (Israel)
Dr. T. Horigome (Japan)
Prof. Dr. W.C. Turkenburg (Netherlands)
Mr. G. White (New Zealand)
Mr. A. Oistrach (Spain)
Dr. M. J. Grubb (United Kingdom)
Mr. T. D. Bath (USA)
Mr. J. R. Birk (USA)
Mr. G. W. Braun (USA)
Mr. L. Coles (USA)
Mr. P. DeLaquil III (USA)
Mr. E. DeMeo (USA)
Mr. B. D. Kelly (USA)
Mr. P. C. Klimas (USA)
Mr. R. L. Lessley (USA)
Mr. C. W. Lopez (USA)
Mr. L. M. Murphy (USA)
Mr. J. N. Reeves (USA)
Mr. R. L. San Martin (USA)
Mr. M. J. Skowronski (USA)
Mr. T. Stoffel (USA)
Mr. M. M. Koltun (Russian Federation)
Mrs. J. Andorfer (UNECE)
Mr. G. Best (UNFAO)

Working Group 2: Geothermal Energy
Mr. L. Y. Bronicki, *Chairman* (Israel)
Dr. D. L. P. Strange (Canada)
Mr. G. Cuellar (El Salvador)
Mr. B. Doron (Israel)
Mr. M. Lax (Israel)
Mr. G. Allegrini (Italy)
Mr. J. T. Lumb (New Zealand)
Mr. R. DiPippo (USA)

Working Group 3: Wind Energy
Prof. W. C. Turkenburg, *Chairman* (Netherlands)
Mr. N. Meyer (Denmark)
Mr. E. Sesto (Italy)
Mr. J. P. Coelingh (Netherlands)
Mr. P. Smulders (Netherlands)
Dr. A. J. M. van Wijk (Netherlands)
Mr. L. Tallhaug (Norway)
Mr. K. Averstad (Sweden)

Working Group 4: Biomass Utilisation
Mr. P. R. Bapat, *Co-Chairman* (India)
Mr. S. de Salvo Brito (Brazil)
Mr. J. J. Edens, *Co-Chairman* (Denmark)
Prof. R. K. Dutkiewicz (South Africa)
Dr. A. A. Eberhard (South Africa)
Mr. A. T. Williams (South Africa)
Prof. D. O. Hall (United Kingdom)

Working Group 5: Hydro (Mini/Micro) Energy
Dr. D. L. P. Strange (Canada)
Mr. T. P. Tung (Canada)
Mr. G. W. Mills (New Zealand)
Mr. A. Bartle (United Kingdom)
Mr. K. Goldsmith (United Kingdom)
Mr. F. Jenkin (United Kingdom)
Mr. L. P. Mikhailov (Russian Federation)
Mr. A. A. Zolotov (Russian Federation)

Working Group 6: Ocean Energy
Dr. D. L. P. Strange, *Chairman* (Canada)
Mr. G. C. Baker (Canada)
Mr. R. H. Clark (Canada)
Mr. T. P. Tung (Canada)
Mr. G. Hagerman (USA)
Mr. L. F. Lewis (USA)

Soviet Working Group
Prof. Dr. E. P. Volkov *(Chairman)*
Dr. G. V. Tsyklauri
Dr. M. M. Koltun
Dr. F. G. Salomzoda
Dr. B. V. Tarnizhevsky
Dr. V. A. Vasilyev
Dr. A. V. Misulin
Dr. E. S. Pants'hava
Dr. E. V. Tveryanovich
Dr. D. N. Militeev
Prof. Dr. E. E. Shpilrain
Dr. V. I. Savin
Mr. E.V. Nadezhdin
Dr. R. A. Zahidov

CONTENTS

List of Figures

List of Tables

Acronyms

BOO/BOT	Build, Own, Operate/Build, Own, Transfer
BOD	Biological Oxygen Demand
BUN	Biomass User's Network
CAES	Compressed Air Energy Storage
CHP	Combined Heat and Power
CIS	Commonwealth of Independent States (former USSR)
CP	Current Policies (scenario)
EC	European Community
ED	Ecologically Driven (scenario)
EEC	European Economic Community
FAO	Food and Agriculture Office (UN)
GDP	Gross Domestic Product
GHP	Geothermal Heat Pump
HAWT	Horizontal-axis Wind Turbine
HDR	Hot Dry Rock
IEA	International Energy Agency
IEC	International Electrotechnical Committee
IPP	Independent Power Producer
LRMC	Long-range Marginal Cost
MSW	Municipal Solid Waste
MOWC	Multi-resonant Oscillating Water Column
NGO	Non-Governmental Organization
NREL	National Renewable Energy Laboratory (USA)
NUG	Non-Utility Generator
OECD	Organization for Economic Co-operation and Development
O&M	Operation and Maintenance
OTEC	Ocean Thermal Energy Conversion
OWC	Oscillating Water Column
PURPA	Public Utilities Regulatory Policy Act
R,D&D	Research, Development and Demonstration
RDF	Refuse-derived Fuel
SERI	Solar Energy Research Institute (now NREL)
toe	Tonne of Oil equivalent (= 42 MJ)
UN	United Nations
UNDP	United Nations Development Programme
UNSEGED	UN Solar Energy Group on Environment and Development
USDOE	United States Department of Energy
VAWT	Vertical-axis Wind Turbine
WECS	Wind Energy Conversion System
WMO	World Meterological Organization

Foreword

Over 80 specialists from many countries – industrialized, with economies in transition, and in various stages of development – contributed to this book. They began their work in mid-1989 under the direction of a World Energy Council (WEC) Study Committee created for this purpose.

The objective was to evaluate the possible place of "new" renewable energy resources within the world's total energy consumption over the coming decades. The main focus was to the year 2020, although in the first overview chapter a longer perspective of possibilities is outlined to the year 2100. The term "new" renewable energy resources is used to cover solar, wind, geothermal, ocean/tidal, small hydro and modern biomass. Two major forms of renewable energy are thereby excluded: "traditional" biomass from fuelwood, crop wastes and animal dung; and large hydro. Particular issues and problems arise with the provision of energy from these last two which are sufficiently different to warrant their exclusion.

This study is both a comprehensive and global overview of its subject, and a series of surveys and prospects for specific forms of new renewable energy. It is the WEC's first systematic collective study of renewable energy, and by developing two cases (current policies and environmentally driven) illustrates the range of future possibilities with and without concerted efforts to accelerate the contribution from renewable energy resources to primary energy supplies. In this respect, the book reflects a degree of realism and practical experience which is often found wanting in studies of renewable energy. This derives from the "bottom up" nature of the study which emanated from local views and expertise rather than from any "top down" academic work.

The book is also unusual in seeking to achieve a balanced perspective on environmental aspects: reflecting concerns about potential climae change *and* concerns about the adverse local environmental impacts that energy provision – including renewables – can have if not handled with sensitivity. Too often, this study stresses at various points, these local environmental impacts receive insufficient weight – either because of technological overenthusiasm or because concern with potential climate change may cause important local issues to receive insufficient scrutiny.

The report, we believe, represents a constructive, sympathetic, balanced and realistic step forward in the technical, economic and environmental understanding of the place which these energy forms could occupy in the future. Inevitably, at this stage in the development of new renewable energy, there are areas of considerable uncertainty – not least in relation to future technological developments and costs. The economics of many forms of new renewable energy forms will largely dictate their future. We have thus included as much data as we have been able to verify. It is however an area of constant change, for the most part reflecting falling costs. Inevitably, where there is uncertainty there will be gaps and weaknesses which we would have liked to fill, but cannot as yet.

Jack Darnell (USA)
Michael Jefferson (WEC)
General Editors
May, 1994

Acknowledgements

Particular acknowledgements and thanks go to:

- Professor Eduard P. Volkov (Russian Federation), Chairman of the WEC's Study on Renewable Energy Resources, 1989–93
- Mr. Jack R. Darnell (United States), Vice Chairman of the Committee, who was not only the lead author for the chapter on solar energy but prepared the Overview and had primary responsibility for producing the full report.
- To the actively participating members of the Study Committee and their respective organisations or companies, and especially to:

  Mr. A. Oistrach (Spain) – solar;
  Mr. L. Y. Bronicki (Israel) – geothermal;
  Professor W. C. Turkenburg (The Netherlands) – wind;
  Mr. J. J. Edens (Denmark) – biomass;
  Dr. D. L. P. Strange (Canada) – ocean/tidal; and
  Mr. T. P. Tung (Canada) – ocean/tidal and small hydro.

- To Dr. J. R. Frisch (France) and Mr. J. M. Jefferson (WEC) who made wide-ranging contributions.
- To the many outside the Committee who reviewed its work.

Executive Summary

This book considers the prospects for new renewable resources, from a global perspective, over the next three decades and beyond. The study is ideally read in conjunction with the final report of the World Energy Council (WEC) Commission, *Energy for Tomorrow's World,* published in 1993.

The new renewable resources investigated were: solar; wind; geothermal; modern biomass; ocean; and small hydro. Each of these areas was thoroughly researched and is the subject of a separate chapter. Recent information on large-scale hydroelectric and traditional biomass is included for added perspective on total use of renewable energy, but both fall outside the definition of new renewable energy used here.

The current use of renewable energy resources is dominated by traditional biomass, which is an important energy supply in the developing countries of Asia, Africa, and Latin America. There are major concerns about the sustainability of this biomass usage. When large-scale hydro and traditional biomass use are included, the total renewable resources contribute about 18% to mankind's total energy use today. However, the current contribution of new renewables is only about 1.9% of mankind's primary energy use.

Two alternative cases are considered up to 2020, and a qualitative assessment of the longer-term potential was also made. For the first case, referred to as "current policies", the total estimated contribution has a growth of 3.4 times compared to 1990, but still makes a relatively modest contribution of about 4% of total primary energy supply. Modern biomass will continue to be the largest contributor, with solar and wind also accounting for a notable fraction of the increase, with impact in most regions. When traditional biomass and large hydro are included, the total renewable contribution would represent about 1.8 times the current level, and would account for an estimated 21% of world total energy supply. This case provided data for Case B of the WEC Commission, includes significantly more energy conservation than a business-as-usual approach.

The second, "ecologically driven" case, openly encourages the faster and more extensive penetration of renewables into the energy markets. This assumes strong government support on an international scale. It also assumes that environmentally consistent criteria will be applied and policies pursued so that severely adverse local environmental impacts and ecosystem degradation more generally is avoided. Penetration will also be assisted through the recognition of the external costs of conventional energy use that are not sufficiently considered in microeconomic decision making today, as well as greater energy efficiency efforts and other positive measures. This case concludes that the contribution from new renewables might reach as much as 12% by 2020, and the corresponding value for total renewables is nearly 30%, with much of the renewable use in the developing countries. Although these levels, nearly a 10-fold increase compared to 1990, are technically possible, they could be achieved only with a comprehensive and sustained new effort of the entire international community that includes the realignment of priorities and economic policy. For the WEC Commission, this scenario corresponds generally to Case C. This case incorporates a strong "minimum regrets" strategy which targets lower total energy consumption and less reliance on fossil fuels as a hedge against concerns about global climate change.

However, to be logically consistent this case will also have to take due account of the adverse local environmental impacts which new renewable energy developments may have – potentially significant with such forms as modern biomass, tidal barrages, and wind power. There is a need to balance concern about potential climate change and such issues as loss of bio-diversity, competing requirements for land, destruction of natural habitats and visual intrusion. These local issues are likely to prove a severe constraint on new renewable forms of energy achieving their full technical potential.

Achieving the level of penetration of the current policies scenario will be relatively easy, but it will involve the active effort of a large number of people over the entire period. The challenge is to approach the projections of the second scenario, if this is accepted as a positive and necessary contribution to overall world progress. This will take fundamental changes in the way decisions about energy supply are made by the public and private sector around the world. It will also involve investment in the long-term future, and greater sharing of technology, know-how and financial resources of the developed countries with the developing world.

Extensive penetration of renewables into the energy markets will not happen quickly for several reasons, such as the time lag for technology development and the deployment of a large manufacturing industry, and a limitation on total investment capital available. Although there is broad public support for the concept of renewable energy, there is also the barrier posed by existing investments and interests in fossil fuel and nuclear energy provision and use. But as the 21st century progresses, renewables will continue to penetrate the energy markets, and will make a major contribution to the earth's long-term future.

The work of the Study Committee on Renewable Energy Resources led to a number of important conclusions, a few of which are highlighted below:

- Besides the traditional use of biomass resources in the developing countries, there are significant numbers of economically attractive applications for new renewable energy systems today.
- Costs for energy from new renewable energy systems are expected to be reduced over the next few decades, in contrast with the predictions for fossil-fueled systems, which will experience increasing costs over time. The apparent economic advantages that currently

exist for fossil fuels are expected to decline or reverse with time, and renewable options are likely to become the economic choice in an increasing number of regions and applications.

- Many of the environmental advantages of renewable energy systems are real, and renewables are seen as one part of a total solution to our environmental problems, present and future. However, just because they are called "renewable" does not assure environmental acceptability, and especially concentrated or large-scale applications need to be applied with sensitivity to their possible environmental impacts. Modern biomass, wind, tidal, and small hydro utilisation can have significantly adverse local impacts.
- Research and development of the emerging renewable technologies needs to be continued and expanded, so that these options can be rapidly moved to maturity.
- One of the benefits of the ecologically driven scenario compared to a continuation of current trends is a projected reduction in annual global, energy-related CO_2 emissions of 25% in 2020, of which the growth in new renewables contributes about one-third.
- Although the aggregate contribution of distributed small renewable energy systems by 2020 appears small compared to global energy use, these applications are very important because they can make a tremendous difference in the lives of hundreds of millions of people.
- Short-term planning based on current economic decision-making patterns is a major obstacle to renewables, which tend to have higher capital costs than fossil-fueled alternatives, as a lifetime of fuel is paid for in advance. Closer focus on long-term energy and development needs is required.

This Committee makes a number of recommendations:

- Increase the priority for funding of renewable energy projects among financing institutions and for bilateral aid programmes, especially those programmes that will lead to sustainable development based on indigenous resources. These programmes will principally increase the flow of technology and financial resources to the developing countries.
- Encourage private sector investment and involvement in the programmes for training and implementation of new renewables. Promote the establishment of joint-venture manufacturing plants for local production of renewable energy systems.
- Establish or designate a single organisation to give international focus and leadership to the increased use of renewable energy.
- Establish regional centres of excellence for renewable energy, to provide training, technology support, and resource databases appropriate to the regional needs.
- Increase R&D funds for renewables, eliminate subsidies on conventional energy, and reduce barriers that exist for the introduction of new and different (especially small and distributed) energy systems.

- Review and update economic decision-making methodology to include the external impacts of the options under consideration. This is an appropriate area for future study by the WEC.

The accelerated introduction of renewable energy resources into the world's energy markets is one necessary and beneficial element in an integrated development strategy. Renewables are distributed so widely that they are indigenous to almost every country on earth in significant quantities. Renewables also have the potential to make a major contribution to economic development by distributing economic activity over a wider area, thereby slowing the migration of rural people to the cities, and alleviating many social problems. But unless due recognition is given to the barriers which exist in our present energy system and the investment already sunk in it, the technology and financing required to develop renewable energy in bulk, and the local environmental impacts which need to be avoided or minimised, then unrealistic assumptions will be made about the pace and scale of new renewable energy development.

CHAPTER 1

Overview, Conclusions and Recommendations

INTRODUCTION

The conventional energy resources, fossil fuels with electricity contributions from nuclear fission and large-scale hydroelectric energy, dominate the near-term global energy supply picture. Other energy resources have not yet developed to any great degree. Among these others are those referred to as the new and renewable energy resources. As a group, they are difficult to compare to the conventional energy resources, since they are widely distributed and relatively diffuse. They are now only partially developed to commercial status, and ask somewhat different compromises from the end-user. These supplies are essentially limitless, but they are perceived as expensive and unreliable. Most are or can be made environmentally benign. There will be a need to apply strict environmental standards to renewable energy developments, as part of a consistent process toward total energy provision and use. However, capital costs have fallen remarkably (for example, photovoltaic modules by a factor of 10 from 1970–1990), and can be expected to fall further. Reliability can be expected to rise. As the balance of the environment and the energy economy is debated now and in the future, the potential and necessary contribution of these renewable resources is a key area for discussion.

Objective/Goal

This book examines the growth opportunities for, and constraints to, the full utilization of renewable energy resources, from a global perspective, over the next three decades and beyond. By assessing the current status and prospects for these vast resources, as well as two scenarios for future growth, the book also develops recommendations for future strategy at the national and international levels. The book covers both developed and developing countries. The original terms of reference are given in Annex 1 and the results were used as input to the World Energy Council (WEC) Commission's Report, *Energy for Tomorrow's World.*[1]

Methodology

The new renewable resources were divided in six separate areas for investigation:

- solar;
- wind;
- geothermal;
- biomass (modern);
- ocean; and
- small hydro.

Each of these areas was thoroughly researched and forms a separate chapter in this book. The treatment of the subject matter was intended to be parallel, but the chapters include significant variation in approach, which is indicative of the differences among the renewable energy resources. Other resources are not investigated in this book, namely large-scale hydroelectric and fusion energy, because there is a significant amount of information already available and they fall outside the immediate terms of reference. Data on large-scale hydroelectric energy and recent assessments of traditional biomass are included in this summary for additional perspective on the total picture of renewable energy and to assist in comparison with recent studies by others. Moreover, it was also felt that the widespread concern about the environmental impacts of large hydro, and grave doubts about the sustainability of traditional biomass in some regions of the world, placed both these forms of energy in a different context to new renewables.

SOLAR ENERGY

The earth receives solar energy as radiation from the sun, and the amount greatly exceeds mankind's use. This resource has a familiar daily variation, along with seasonal variations, and is significantly affected by weather. It has a relatively low intensity, with a peak of about $1kWt/m^2$ at sea level. Every country has access to this resource, to different degrees. The applications of solar energy are quite diverse, including direct thermal (both active and passive systems), electric power generation using thermodynamic cycles, and direct conversion to electricity with photovoltaic (PV) systems. Storage is relatively inexpensive for thermal systems, providing some decoupling of the resource from the time of end use.

A large amount of R&D work has been done in the past 20 years, with a vast amount of data generated on technology and applications, and some dramatic gains in cost effectiveness. A few of these applications are now fully commercial, but further improvement in cost, achievable through mass production and technical development, is needed for widespread application. Millions of low-temperature thermal collectors have been installed, over 350 MWe of thermal power is operational, and the PV manufacturing capacity has reached 50 MWe (peak) per year.

Solar energy is environmentally benign, although concentrated, large-scale use can lead to localized environmental impacts, and solar can be integrated well with different types of cultural settings. A number of institutional and economic constraints must be overcome for extensive use, but with adequate support, the contribution of this resource is likely to become significant in the next several decades.

WIND ENERGY

Wind is the result of thermal heating of the earth by the sun, and has global patterns of a semi-continuous nature. It is significantly affected by topography and weather, with seasonal, daily and hourly variations. Much of the wind resource is located along coastlines (including offshore) and in mountain regions, but significant resources are also found on plains. Estimates of global availability, assuming 0.33 MWe/km^2 average for land with greater than 5 m/sec average wind, indicate that the maximum technical potential is about twice the current global electricity production.

Wind energy can be converted into mechanical energy by wind turbines, either horizontal- or vertical-axis types, and used for pumping water or generating electricity, or even thermal energy through friction. Wind turbine technology has been under intensive development for the past 20 years, and currently there is about 2000 MWe of installed capacity, mostly in the United States and Europe. The trend in size is increasing, with proven machines in the 250–500 kWe range now available. Installed costs have fallen sharply in the past decade, and wind energy is nearly competitive for non-firm energy supply in many high-resource areas. Wind pumps can be used throughout many rural areas where groundwater is near the surface. Improvement goals include increased long-term reliability, lower maintenance costs, and good efficiency over a wider range of wind conditions. Several constraints to wider application of wind energy include lack of detailed understanding of the very site-specific resource, impacts such as visual intrusion and aesthetics, noise, bird mortality and telecommunication interference, and availability of capital resources.

Each chapter discusses the extent and distribution of the subject resource, the technologies ready now or under development to tap the resource for useful purposes, the economics of resource utilization both now and projected over the next three decades, the many constraints to widespread use, the opportunities for co-operation in further utilization, and projections of the contribution under two general scenarios. The first case represents a continuation of current trends and policies into the future, and no special emphasis on these renewable resources. The second case represents a shift toward the accelerated development of renewable energy supplies, as part of a global ecology initiative, and is intended to represent the highest penetration of global energy markets that is reasonable by 2020. Such an acceleration would require a major switch of emphasis and resources to new renewables, propelled by government support. Each chapter also discusses the ultimate potential in the longer term, beyond the arbitrary 2020 horizon.

A questionnaire was developed for gathering national perspectives on the forecast for penetration of renewables. A request for data was sent to WEC member committees but responses were not sufficient to complete the study reported in this book, and they were used as partial input to the work of the individual renewable resource investigations.

GEOTHERMAL ENERGY

Geothermal energy refers to heat stored beneath the surface of the earth. The quantity is much larger than man's current use, but the intensity is very low except along the boundaries of the tectonic plates, in areas also known for volcanism and earthquakes. It is renewable if the heat extracted is not more than the replenishment from the centre of the earth, and if the water that brings the heat to the surface is reinjected.

The low temperature resources (less than 100°C) have been used from ancient times for bathing and space heating, and recently for greenhouses and process heat. Dry steam (about 240°C) and high temperature water (90 to 350°C) have been in commercial use for electric power production. In the last decade, considerable progress has been made in using moderate temperature water (as low as 100°C) for power generation through binary cycles. Commercial use of geothermal energy has been developed in at least 20 countries, with estimated 1991 electric generation capacities of more than 5900 MWe and direct use equivalent to 5.6 Mtoe.

Although concentrated in certain areas, geothermal energy reserves of significant quantity exist in all regions of the world. To tap these resources, it is necessary to continue development of exploration, extraction and conversion technologies, and to manage the environmental issues such as small quantities of dissolved gases including H_2S and CO_2 and disposal/reinjection of concentrated brines. In the long run, development of effective means for extracting useful energy from hot dry rock, the largest fraction of this resource, as well as geopressured and magma resources will greatly expand the potential contribution of geothermal energy.

BIOMASS ENERGY

Biomass encompasses a range of products derived from photosynthesis, and is essentially chemical storage of solar energy. It also represents a renewable storage for carbon in the biosphere. It is distributed world-wide, and is available in some form to every country on earth, although varying significantly in capacity per hectare. It is a source of thermal energy, often collected by individuals outside the commercial energy markets (traditional biomass) and, for example, contributing over 90% of total energy use in Nepal and Malawi. Biomass can also be converted into various carbon-based fuels parallel to petroleum-derived products (modern biomass). In 1990, biomass (mostly traditional) contributed an estimated 12% of the total global energy supply, with major use in countries such as India, China and Brazil. The energy potential from biomass significantly exceeds world energy use.

The diverse forms of biomass include fuelwood, crops specifically grown for energy, agricultural and forestry residues, food and timber processing residues, municipal solid waste, sewage, and aquatic flora. Each has its own related technologies, but most forms are low in energy density and high in water content, relatively costly to gather and transport, and are used in other contexts such as fertilizers.

Economics of biomass are very difficult to generalize, because value is often set by competing uses. Large scale commercial usage would have significant impacts on other markets such as food and paper. The costs and energy investment of delivery, water and fertilizer must be considered, as well as the overall impact on land use and population patterns. On the other hand, energy production from waste biomass can represent a better solution to a disposal problem. Biomass is often most appropriate for dispersed applications, with low population density, and typical facilities are small. It is the only renewable resource that is likely to significantly impact the transport fuels sector energy use by 2020 without electric vehicles. However, for large-scale use of bio-energy, we need to guard against reduction of bio-diversity, protect areas of great natural beauty and environmental sensitivity, and control effluents and emissions.

Although this book presents quantitative projections of the use of these resources in the future, the reader is cautioned that it is often too easy to focus on the numbers or percentages, and miss the important messages and issues. These resources are not directly similar to current conventional energy systems, and involve different commitments from the user, especially in distributed applications. For example, for renewable systems such as solar, ocean and wind, capital must be available for the user to pay, in initial costs, the equivalent of the fuel cost for the life of the facility. This represents opportunity costs and risks related to future "fuel" availability that are much different from those experienced with fossil fuel systems.

OCEAN ENERGY

Ocean energy includes several diverse, low-intensity phenomena that can be tapped for useful purposes, including thermal gradients, tides and waves. These resources are located in a wide number of coastal areas, with thermal gradients generally distributed in the tropical regions. Although these types of resources have very different characteristics, they have in common the low density and the engineering design challenges of ocean conditions. Ocean thermal is by far the largest ocean energy resource, more than 10 times current global energy use.

All technologies for tapping ocean energy are very immature. Tidal energy is most advanced since it uses dams and turbines very similar to low-head hydroelectric plants. However, there may be some potentially severe adverse impacts on local ecology at many of the sites with high technical potential, and effective mitigation or avoidance strategies are necessary for development to proceed. Wave energy uses a variety of mechanical means to capture the oscillation of the surface waters. Ocean thermal uses large heat exchangers and pumps to extract energy to power a very low efficiency thermodynamic cycle. Current use of ocean energy is limited to a few demonstration facilities, such as the 240 MWe tidal plant in France.

Ocean energy systems are expected to have high capital costs and must run reliably over an extended period in a hostile environment. Their action can change the local aquatic habitat, especially near river mouths. Ocean thermal and wave energy plants can produce fresh water as a by-product, and therefore are attractive to remote coastal areas without fresh water resources. Transport of energy from ocean energy plants to serve loads is one of the major issues to be resolved for widespread application.

SMALL HYDROPOWER

Hydropower is well known throughout the world, and currently contributes about 6% of the total world energy production. This is due mostly to large facilities on major rivers, since these have been the focus of past development. However, to reach the maximum potential use of natural resources, smaller water flows must also be tapped. These facilities of less than 10 MWe capacity are the focus of this work. There are currently several thousand such facilities identified with combined capacity of more than 19,000 MWe. These systems can involve new dams or diversions, of the repowering or irrigation or flood control barriers.

Such small facilities depend on special local situations for economic viability. They will prosper if they are as modular and standardized as possible, have low barrier/diversion construction costs, and serve a local market with few other options. Various turbine designs are used, depending on the available head. Improvements are possible in turbine designs for small applications, low cost construction techniques, and training of the local people on operation. Small systems also can lead to advantageous situations for local agriculture and water supply.

Small hydro facilities have lower environmental impact than similar very large projects, but great care will be required to mitigate impacts, especially near scenic and environmentally sensitive areas. They are less disruptive of the local land use and environment than larger facilities, but typically provide less storage and must be operated nearly continuously to maintain adequate stream flow.

IMPORTANT FEATURES OF THE RENEWABLE RESOURCES

Each of the next six chapters presents an in-depth analysis of one segment of the renewable resources. It is important to understand the basics of these resources, the status of the technology to use them, the current experience (both positive and negative), and the situations where they will play a significant role. This overview presents an abbreviated discussion of each renewable resource (see text panels), with highlights of the major issues that they face as emerging energy vectors. Table 1.1 also presents some highlights of these features. The reader is encouraged to study the corresponding chapters, since without this complete perspective, the projections of future potential become only a numbers game.

Table 1.1 *Summary of important characteristics of new renewable energy sources*

		Solar	*Wind*	*Geothermal*	*Biomass*	*Ocean*	*Small hydro*
Resource	Magnitude	Extremely large	Large	Very large	Very large	Very large	Large
	Distribution	World-wide	Coastal, mountains, plains	Tectonic boundaries	World-wide	Coastal, tropical	World-wide, mountains
	Variation	Daily, seasonal, weather-dependent	Highly variable	Constant	Seasonal, climate-dependent	Seasonal, tidal	Seasonal
	Intensity	Low 1 kW/sq m peak	Low average 0.8 MW/sq km	Low average Up to 600C	Moderate to low	Low	Moderate to low
Technology	Options	Low to high temp. thermal systems, photovoltaics, passive systems, bioconversion	Horizontal and vertical-axis wind turbines, wind pumps, sail power	Steam and binary thermodynamic cycles, total flow turbines, geopressured, magma	Combustion, fermentation, digestion, gasification, liquefaction	Low temp. thermo-dynamic cycles, mechanical wave oscillators, tidal dams	Low to high head turbines and dams and turbines
	Status	Developmental, some commercial	Many commercial, more developmental	Many commercial, some developmental	Some commercial, more developmental	Developmental	Mostly commercial
	Capacity factor	<25% w/o storage, intermediate	Variable, most 15–30%	High, base load	As needed with short-term storage	Intermittent to base load	Intermittent to base load
	Key improvements	Materials, cost, efficiency, resource data	Materials, design, siting, resource data	Exploration, extraction, hot dry rock use	Technology, agriculture and forestry mgmt.	Technology, materials, and cost	Turbines, cost, design, resource data
Environmental characteristics		Very clean Visual impact, local climate, PV manufacturing	Very clean Visual impact, noise, bird mortality	Clean Dissolved gases, brine disposal	Clean Impacts on fauna and other flora, toxic residues	Very clean Impact on local aquatic environ., visual impact	Very clean Impact on local aquatic environ., land use

In order to place the new renewables into context, it should be noted that the current use of all renewable energy resources is dominated by biomass, which is a major energy supply in the developing countries of Asia and Africa, with some additional contribution in North and Latin America. When large-scale hydro use is included, the renewable resources contribute 1559 Mtoe (million tons of oil equivalent) or about 18% to mankind's total energy use today. Without traditional biomass and large hydro, the current contribution of all other renewables is 164 Mtoe or about 2%. To build on such a small base and make a significant contribution in 3 decades is a major challenge, considering that nuclear energy achieved a 5% share in 30 years of commercial development with sustained government-supported R&D.

CONSTRAINTS FOR EXPANDED RENEWABLES USE

The expanded use of new renewables will continue throughout the next 30 years and beyond. However, extensive penetration of renewables into the energy markets is not likely to happen quickly. There is a natural time lag between success of the prototypes and demonstrations, leading to a few initial commercial applications, and then incorporating learning into the next generation, and so on. There is also a need to deploy a significant manufacturing industry which requires time for planning, financing, and training. Experience has shown that too rapid a growth, through massive tax and other incentives, encourages marginal companies and technologies into the market, with disappointing results compared to expectations of long-term reliable operation.

The constraints are numerous and complex. The diversity of the renewable resources makes summarizing them difficult, and the technical chapters contain much more complete data on the constraints for each renewable resource. A few common threads can be found, including:

- **Resource Definition.** The magnitude of each renewable resource is dependent on local conditions. To optimize the use of these resources, better data is needed, especially on their variation. The comprehensive understanding of local conditions throughout the world will take extensive effort, but each country has indigenous energy reserves that should be understood as part of comprehensive national or regional energy planning.
- **Technology Development.** Many (but not all) of the technologies necessary to make efficient use of the renewables are quite immature, and/or relatively costly. Renewables have a relatively low energy density in their raw form, and this either leads to high cost for concentration (if possible), or for structure to capture the energy in a useful form. Therefore, material costs are often a high percentage of the capital cost, even for relatively simple designs. Further development, covering the range from basic R&D on advanced materials to learning from a wide range of field demonstration facili-

ties, will need to take place for users to invest the equivalent of future fuel costs with confidence. The areas for development include not only the design, but also the manufacturing, installation, operation, and maintenance of these systems. The success of renewables at higher market penetrations will also depend to a certain extent on the development and application of both improved energy storage and transportation systems. There is a need to minimize any localized environmental impacts, which will mean avoiding large-scale, concentrated use in environmentally sensitive areas.

- **Economic and Institutional Policies.** Renewable energy systems typically have higher capital costs than fossil-fuelled systems, since all the fuel equivalent over the useful lifetime is purchased at the beginning of the system life. Emphasis on life-cycle costs and reduction of the risks of high capital investments will be necessary for the success of renewables. Renewable energy resources have been encouraged in some locations and special situations, and as a result we have a reasonable set of data on significant use in regional settings, such as California and Brazil. But widespread use will involve recognition that energy systems have significant impacts beyond their physical envelope, and recognition and quantification of these impacts are not normally taken into account when energy systems are compared. Also, there is significant infrastructure and incentives that have been instituted over the past decades, which favour continued or expanded use of conventional fuels. Indeed, one of the greatest practical barriers to our rapid introduction of "new" renewable forms of energy is the huge existing level of investment/infrastructure in traditional fossil fuels. It is unrealistic to expect involved investors, companies, suppliers or customers to walk away from such investments, abandon the infrastructures or switch to other forms of energy until these have become price competitive. Many renewable energy companies are small and poorly capitalized, and have limited capability to endure the longer payback times that are typical of renewable energy use. Production tax credits or investment tax credits are examples of mechanisms that have been used successfully to stimulate renewable resource use.
- **Environment.** Much of the recent enthusiasm for developing renewable energy has been propelled by fears of climate change induced by anthropogenic activity, notably fossil fuel combustion. This is readily understandable. However, one unfortunate consequence has been that the emphasis given to potential climate change has far outweighed the emphasis which should be given to the local adverse environmental impacts which some forms and scales of new renewable energy can have. This book seeks to maintain a fair balance between global and local environmental concerns. As a result the need for particular care in developing modern biomass and wind energy is called for, and tidal power developments involving barrages and headponds are not favoured.

- **Public Education.** The energy user takes for granted the supply of conventional energy unless this supply is disrupted. The average user does not understand the systems he depends on, and that is one of the reasons that achievement of substantial gains in energy conservation or efficiency is very difficult. The widespread use of renewables will require even greater understanding of energy systems, and application of that knowledge in making compromises in many aspects of daily life. For example, local benefit or economy needs to be balanced against impact on regional or global environment.
- **International Co-operation and Sharing.** The developed countries of the North have most of the technology and the financial resources for utilizing renewable resources. They have also seen both the advantages and disadvantages of an energy-intensive economy based primarily on fossil fuels. The opportunities for using renewables exist in these countries, but exist to a greater extent in the developing countries, which in general have neither the technical nor financial resources to make significant progress in this area. The mounting pressure from population growth in the developing world, and their worsening environmental and social problems, point to a need for action while there is still time, and for international co-operation with a different set of priorities and techniques.

These constraints are likely to limit significantly the capabilities of the new renewables to play a much larger role in the next decade or two. The use of the new renewables is likely to advance slowly with a continuation of current policies – too slowly to respond to a possible threat to global environmental balance. Unless ways are found to recognize and develop plans to overcome the constraints, we will become inexorably more dependent on fossil fuels in the next few decades. In that event, our options to deal with the threat of global environmental changes will be limited.

EXPECTED CONTRIBUTION BY 2020

For comparison with the total picture of future energy supplies, reference is made to two baseline WEC scenarios, presented in Montreal in 1989 at the WEC Congress[2]. These are referred to as the "M" for moderate growth, and "L" for low growth, scenarios. Comparisons to other baselines can be easily carried out using data in this book. However, it is our purpose to promote dialogue on the potential of the renewable resources, and these scenarios represent the 1989 WEC baselines at the time this work was begun. The projections by the WEC in 1989, for the M and L scenarios, indicated the following contribution from renewable sources:

	Contribution (Mtoe)			*Percent of total*		
	1985	*2020–M*	*2020–L*	*1985*	*2020–M*	*2020–L*
Large hydroelectric	445	1043	848	5.8	7.7	7.3
Traditional biomass	880	1055	1310	11.5	7.8	11.3
Other renewables	19	365	170	0.2	2.7	1.5
Sub-total	1344	2463	2328	17.5	18.2	20.1

The current study took a more extensive look at the new renewables sector, previously called "new energies" or "other renewables". Two very different cases or "scenarios" were considered for this evaluation. The first case, referred to as "current policies", is intended to be a baseline or reference for what can probably be expected in the future without significant modification of the trends that are now apparent. It is based on several major assumptions, including a slow increase in price of conventional fossil fuels, continued development of renewable technologies at a modestly increasing pace as their economic market opportunities expand, increased pressure over time on energy efficiency and environmental protection, and sustained economic development. This is not business-as-usual from the perspective of the conventional energy industry, because the new renewables will make increasing gains within the energy market-place, and these gains will accelerate toward the 2020 horizon. It implies a modest growth rate of the industries supporting the new renewables, as well as increasing acceptance by the users in a variety of applications and regional settings.

For the WEC Commission, this scenario compares approximately to Case B, which is an update of the reference scenario "M" from Montreal. And, as discussed in the final report of the WEC Commission, it also implies a significant amount of energy conservation compared to extrapolation of current per-capita energy use, and is therefore not a business-as-usual approach. The Commission incorporated this Committee's current policies numbers into its report (Case B) as the minimum evolution of the new renewables.

The results of the evaluation of this current policies scenario indicate that the new renewables contribution by 2020 is likely to be higher than previously estimated by the WEC, with a growth of 3.4 times compared to 1990. This is still a relatively modest contribution of 540 Mtoe or about 4% of the global total expected in 2020. Modern biomass will continue to be the largest contributor, with most use in the developing countries. Solar and wind will also account for a notable fraction of the increase, with impact in most regions. When traditional biomass and large hydro are included, the total renewable contribution is expected to represent about 1.8 times the current level, and will account for an estimated 2844 Mtoe or 21% of the global total energy supply. While the total global energy consumption in the WEC Commission's Case B is assumed to rise by a factor of about 1.5 during the next 30 years, total renewables share will rise somewhat faster (18% in 1990 vs 21% in 2020) and new renewables share will almost double (2% vs 4%). These

Table 1.2 *Summary of regional and global projections for current policies scenario*

Values are Mtoe	*Solar*	*Wind*	*Geo-thermal*	*Modern biomass*[3]	*Ocean*	*Small hydro*	*"New" total*	*% of global*[1]	*Trad. biomass*[3]	*Large hydro*[1]	*Renew. total*[2]	*% of global*[1]
1990												
North America	3	1	4	19	0	4	31	0.4%	38	127	196	2.2%
Latin America	1	0	1	46	0	1	49	0.6%	125	80	254	2.9%
Western Europe	1	0	2	10	0	7	20	0.2%	20	99	139	1.6%
CIS & East Europe	2	0	1	10	0	2	15	0.2%	30	55	100	1.1%
Mid East/No. Africa	1	0	0	0	0	0	1	0.0%	21	5	27	0.3%
Sub-Saharan Africa	1	0	0	5	0	0	6	0.1%	141	9	156	1.8%
Pacific/China	1	0	4	23	0	4	32	0.4%	351	70	453	5.1%
Central/South Asia	2	0	0	8	0	0	10	0.1%	204	20	234	2.7%
TOTAL	12	1	12	121	0	18	164	1.9%	930	465	1559	17.7%
	0.8%	0.1%	0.8%	7.8%	0.0%	1.2%	10.5%		59.6%	29.8%	100.0%	
2000												
North America	5	3	7	26	0	5	46	0.5%	40	143	229	2.3%
Latin America	2	0	2	54	0	2	60	0.6%	145	126	331	3.2%
Western Europe	1	1	3	14	0	8	28	0.3%	20	119	167	1.6%
CIS & East Europe	2	0	1	12	0	3	18	0.2%	32	54	104	1.0%
Mid East/No. Africa	2	0	0	0	0	0	2	0.0%	28	7	37	0.4%
Sub-Saharan Africa	1	0	0	7	0	0	8	0.1%	198	12	218	2.1%
Pacific/China	3	1	5	29	0	6	44	0.4%	377	111	532	5.2%
Central/South Asia	2	1	0	10	0	1	14	0.1%	240	28	282	2.8%
TOTAL	18	6	18	152	0	25	219	2.2%	1080	600	1899	18.7%
	0.9%	0.3%	1.0%	8.0%	0.0%	1.3%	11.6%		56.9%	31.6%	100.0%	

Table 1.2 *Contd.*

Values are Mtoe	*Solar*	*Wind*	*Geo-thermal*	*Modern biomass*[3]	*Ocean*	*Small hydro*	*"New" total*	*% of global*[1]	*Trad. biomass*[3]	*Large hydro*[1]	*Renew. total*[2]	*% of global*[1]
2010												
North America	11	14	10	40	0	6	81	0.7%	43	153	277	2.3%
Latin America	5	0	3	63	1	3	75	0.6%	163	196	434	3.7%
Western Europe	3	7	4	19	0	10	44	0.4%	20	132	196	1.7%
CIS & East Europe	5	1	2	18	1	3	30	0.3%	34	58	122	1.0%
Mid East/No. Africa	4	0	0	0	0	0	5	0.0%	32	12	48	0.4%
Sub-Saharan Africa	2	0	0	10	0	0	12	0.1%	250	21	283	2.4%
Pacific/China	7	4	7	36	1	11	66	0.6%	398	147	610	5.2%
Central/South Asia	5	2	0	12	0	1	20	0.2%	268	49	337	2.9%
TOTAL	42	29	26	198	3	34	332	2.8%	1208	768	2308	19.5%
	1.8%	1.3%	1.1%	8.6%	0.1%	1.5%	14.4%		52.3%	33.3%	100.0%	
2020												
North America	28	42	15	55	1	6	147	1.1%	46	165	358	2.7%
Latin America	13	1	5	72	3	5	100	0.7%	179	286	565	4.2%
Western Europe	9	18	6	24	4	12	74	0.6%	20	148	242	1.8%
CIS & East Europe	11	4	3	23	1	4	46	0.3%	36	64	146	1.1%
Mid East/No. Africa	9	1	0	0	0	0	10	0.1%	38	18	66	0.5%
Sub-Saharan Africa	4	1	0	12	0	1	18	0.1%	299	32	349	2.6%
Pacific/China	22	13	11	43	5	18	112	0.8%	414	192	718	5.4%
Central/South Asia	13	5	0	14	0	2	34	0.3%	291	76	401	3.0%
TOTAL	109	85	40	243	14	48	540	4.0%	1323	981	2844	21.3%
	3.8%	3.0%	1.4%	8.5%	0.5%	1.7%	19.0%		46.5%	34.5%	100.0%	

Notes:
1. Based on WEC Commission Case B (with interpolation where necessary) – global total in 2020 is approximately 13,340 Mtoe.
2. Renewable total adds traditional biomass and large hydro to new renewables.
3. Biomass is divided into traditional and modern categories.

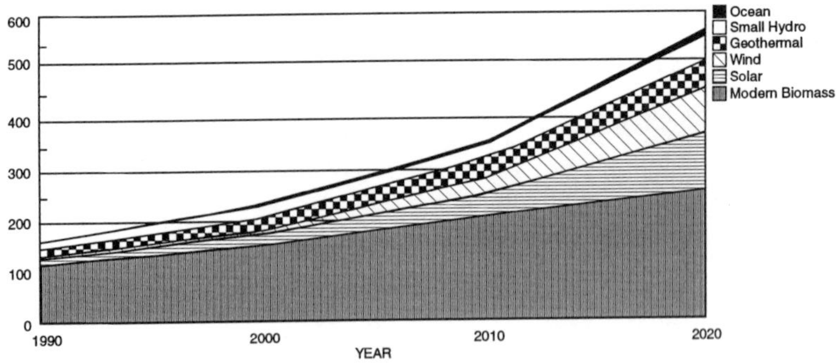

Figure 1.1 *Penetration of new renewables by type – current policies scenario*

Note:
Large hydro and traditional biomass are excluded.

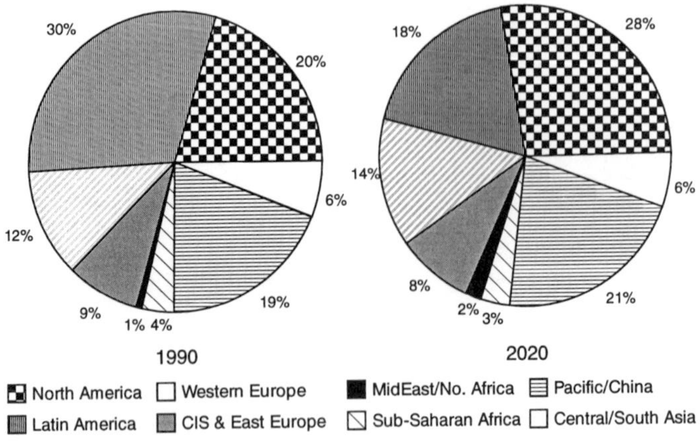

Figure 1.2 *New renewable contribution by region – current policies scenario*

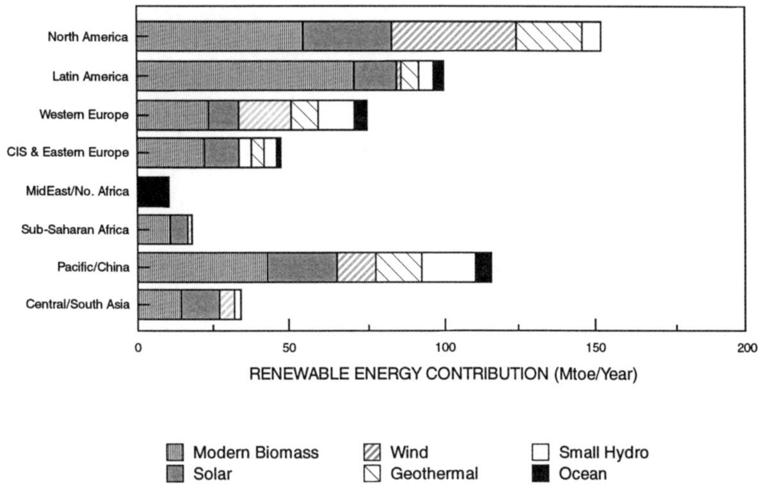

Figure 1.3 *New renewable distribution by region in 2020 – current policies scenario*

Note:
Large hydro and traditional biomass are excluded.

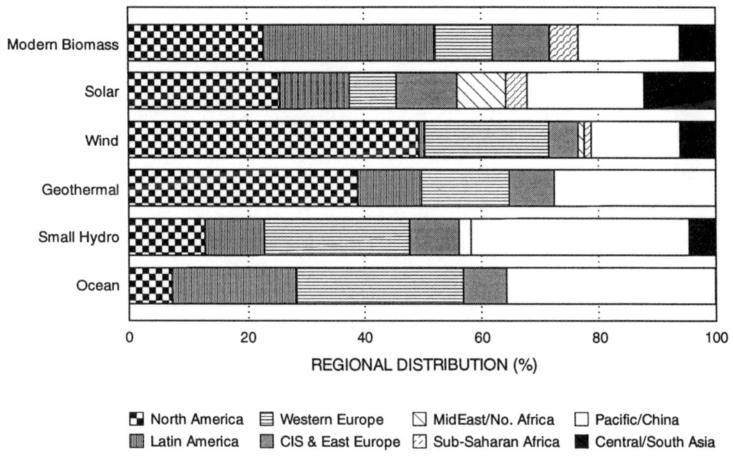

Figure 1.4 *Resource breakdown for new renewables in 2020 – current policies scenario*

Note:
Large hydro and traditional biomass are excluded.

quantitative results, presented in Table 1.2, are further illustrated in Figures 1.1 through 1.4.

With the exception of biomass, this penetration is primarily in the electric sector, with very little impact on the transport sector, although in some countries, notably Brazil, Zimbabwe, and Kenya, biomass-derived fuels have reached national importance. Analysis of this forecast reveals that this study has taken more credit than past WEC reports for the pace of development in areas such as wind and photovoltaics, and anticipated more joint venture sharing of the technologies around the world. However, it must be reiterated that even this modest contribution will not be achieved without continued progress in the development, high-volume manufacture, and government leadership inspired by public activism toward promotion of expanded use of renewables.

Penetration figures do not include the effects of competition between the renewable forms for the same markets. The study did not attempt to quantify this effect, since it is highly dependent on local situations. There will be some limited competition, especially for investment capital, and it is reasonable to expect a minor reduction in the penetration percentages. But in general, the choice between these resources will not be made solely by economic competition based on average costs. Each resource has its own role to play, and for example solar and geothermal are more complementary than mutually exclusive. Wind and ocean (tidal and wave) may directly compete in some regions for a share of the electric market, but their cycles of variation are different, and on average they may not be directly competing. Renewables are expected to compete primarily with fossil fuels, but hybrid systems that share the advantages of both renewables and fossil fuels hold near-term promise.

OPPORTUNITIES FOR ACCELERATED PENETRATION OF RENEWABLES

The current policies scenario discussed above leads to the conclusion that the new renewables will continue to play a minor role in the energy supply picture until at least 2020. This will result in greater dependence on fossil fuels, with increased pressure on the global environment. Depletion of oil and natural gas along with political instability of supply regions will bring major increases in the use of coal. The build up of carbon dioxide in the atmosphere will continue, and if global climate change resulting from fossil fuel burning is eventually proven to require corrective action, then there will be no option but forced major cutbacks in fossil fuel use until emergency measures (requiring decades) allow nuclear and/or renewables to substitute for sizeable fractions of the fossil fuels likely to be used at that time.

The accelerated introduction and expanded use of renewable energy could be one component of a more optimistic view of the future. Through sustained effort and dedication to preserving our energy options, a more flexible future can be achieved, but it will still take time to increase our dependence on renewables substantially. But by 2020, the optimistic view concludes that we can achieve a higher penetration than the expected result of the current policies approach.

The major growth in energy demand between 1990 and 2020 is focused in the developing countries, which have sizeable rates of population growth. Renewables are well suited to distributed applications, providing the opportunity to enhance economic growth in the rural areas and reduce the pressure of migration of these people toward the already overcrowded major cities. Potential renewable supplies are very large, and their distribution allows for use of some types in every country and region. The environmental advantages of the renewables are real, as long as the same sensitivity to environmental impact is observed for large scale installations as is currently being used in developed countries. Another important asset of the renewables is that the public opinion is very much in favour of expanding our use of renewables where it makes sense.

There are some interesting examples of the encouragement of renewable resources that have led to success in terms of penetration into regional or national markets, including:

- The efforts of the UN Geothermal Programme to explore geothermal resources in many areas, which played a major role in the achievement of significant penetration of the electricity supply market, eg 30% in El Salvador and 15% in the Philippines.
- The ProAlcool programme in Brazil has led to production of 62% of the national transport fuels, as part of a national profile where fossil fuels contribute only 25%. Many lessons, both positive and negative, have been learned about dependence on renewable energy to such a large degree.
- In the US, the implementation of the Public Utilities Regulatory Policy Act (PURPA) in 1978 opened the electricity supply markets to non-utility generation sources, and combined with renewable energy investment tax credits, led to an additional 8000 MWe from biomass, 1500 MWe of wind, over 730 MWe of small hydro and 350 MWe of solar thermal, and increased use of geothermal as well as many types of waste fuels. Recently, a new programme of incentive payments of 1.5 US cents/kWh was enacted for power generation facilities using solar, wind, biomass or non-steam geothermal resources.
- From 1979 to 1989, Denmark encouraged its local wind energy industry with decreasing subsidies as it matured, producing 350 MWe of installed capacity in a relatively small country, and a thriving manufacturing industry that exports technology to many other countries.
- Since 1980, Austria has substituted up to 10% of primary energy from biomass with 11,000 district heating systems, supported by national and local government grants, with goals of increased energy security and increased economic activity in rural areas.
- In Israel, government policy led to massive solar hot water installations in about 70% of all residences. These 890,000 systems with 2.5

million m^2 of collectors are estimated to save some 0.5 Mtoe/year of imported fossil fuel.

- A government-supported programme in Indonesia has helped finance over 10,000 individual home photovoltaic/battery systems that result in better lighting, access to television and improved educational opportunities for the people, who pay the same cost as they did buying kerosene for lamps. These villages are beyond the near-term economic reach of the national electric grid.

POSSIBLE ACCELERATION OF CONTRIBUTION FROM RENEWABLES BY 2020

The world now has a major opportunity to follow a long-term path towards a sustainable balance of energy supplies between fossil fuels, nuclear energy and the renewables. Therefore, the study reported in this book also evaluated a second case for the future which seeks to accelerate the penetration of new renewables into the global energy markets to the maximum extent possible over the next 30 years. It represents a complete shift in policy towards this goal, along with the maximum practical use of energy efficiency to keep total energy use restrained and limit the growth of fossil fuel use for environmental reasons. From another perspective, it represents part of a "minimum regrets" strategy, which maintains sustainable options while curbing fossil fuel use to levels below those which would otherwise be reached. Developmental lead times must however be remembered in any strategic change situation.

This "ecologically driven scenario" openly encourages the faster and more extensive penetration of new renewables into the energy markets. The external costs of renewable energy are arguably much lower than those for conventional energy use, when consistently high environmental standards are applied to all energy forms. These external costs are not being considered in microeconomic decision making today, and if applied in conjunction with greater energy efficiency and other positive measures, it is clear that higher penetration could be achieved.

The ecologically driven scenario indicates the contribution from new renewables may reach as much as 1343 Mtoe or 12% by 2020, and the corresponding value for total renewables is about 3278 Mtoe or 30%, with much renewable use still in the developing countries. The breakdown is given in Table 1.3 and further illustrated in Figures 1.5 through 1.8. The study concludes that these levels, more than an eight-fold increase compared to 1990, are technically possible, but could be achieved only with a comprehensive and sustained new effort of the entire international community that includes the realignment of government priorities and economic policy.

The WEC Commission took this Committee's higher case data as a maximum evolution, requiring major national and international policy support consistent with its environmentally led Case C. For its Cases A and B1, which feature higher total energy consumption, the Commission

Table 1.3 *Summary of regional and global projections for ecologically driven scenario*

Values are Mtoe	*Solar*	*Wind*	*Geo-thermal*	*Modern biomass*[3]	*Ocean*	*Small hydro*	*"New" total*	*% of global*[1]	*Trad. biomass*[3]	*Large hydro*[1]	*Renew. total*[2]	*% of global*[1]
1990												
North America	3	1	4	19	0	4	31	0.4%	38	127	196	2.2%
Latin America	1	0	1	46	0	1	49	0.6%	125	80	254	2.9%
Western Europe	1	0	2	10	0	7	20	0.2%	20	99	139	1.6%
CIS & East Europe	2	0	1	10	0	2	15	0.2%	30	55	100	1.1%
Mid East/No. Africa	1	0	0	0	0	0	1	0.0%	21	5	27	0.3%
Sub-Saharan Africa	1	0	0	5	0	0	6	0.1%	141	9	156	1.8%
Pacific/China	1	0	4	23	0	4	32	0.4%	351	70	453	5.1%
Central/South Asia	2	0	0	8	0	0	10	0.1%	204	20	234	2.7%
TOTAL	12	1	12	121	0	18	164	1.9%	930	465	1559	17.7%
	0.8%	0.1%	0.8%	7.8%	0.0%	1.2%	10.5%		59.6%	29.8%	100.0%	
2000												
North America	6	3	10	29	0	7	55	0.6%	38	133	226	2.3%
Latin America	2	0	3	75	0	2	82	0.8%	140	142	364	3.8%
Western Europe	2	2	4	14	0	11	33	0.3%	19	107	158	1.6%
CIS & East Europe	3	0	2	12	0	3	20	0.2%	29	56	106	1.1%
Mid East/No. Africa	2	0	0	4	0	0	6	0.1%	25	9	40	0.4%
Sub-Saharan Africa	1	0	0	15	0	0	16	0.2%	190	15	222	2.3%
Pacific/China	3	1	7	49	0	7	67	0.7%	359	102	529	5.4%
Central/South Asia	3	1	0	20	0	1	25	0.3%	226	37	287	3.0%
TOTAL	22	7	26	218	1	31	304	3.1%	1026	602	1932	19.9%
	1.1%	0.4%	1.4%	11.3%	0.0%	1.6%	15.8%		53.1%	31.1%	100.0%	

Table 1.3 *Contd.*

Values are Mtoe	*Solar*	*Wind*	*Geo-thermal*	*Modern biomass*[3]	*Ocean*	*Small hydro*	*"New" total*	*% of global*[1]	*Trad. biomass*[3]	*Large hydro*[1]	*Renew. total*[2]	*% of global*[1]
2010												
North America	24	24	19	46	1	10	124	1.2%	37	139	300	2.8%
Latin America	8	0	5	125	4	4	147	1.4%	144	205	495	4.6%
Western Europe	9	10	6	24	4	15	68	0.6%	17	114	199	1.9%
CIS & East Europe	9	2	5	22	1	5	44	0.4%	26	58	128	1.2%
Mid East/No. Africa	6	0	0	7	0	0	14	0.1%	27	13	53	0.5%
Sub-Saharan Africa	6	0	1	29	2	0	38	0.4%	222	22	282	2.6%
Pacific/China	17	8	13	79	4	12	133	1.2%	355	135	622	5.8%
Central/South Asia	14	3	0	40	2	1	59	0.6%	234	53	347	3.2%
TOTAL	92	48	49	372	18	48	626	5.9%	1062	738	2427	22.7%
	3.8%	2.0%	2.0%	15.3%	0.7%	2.0%	25.8%		43.8%	30.4%	100.0%	
2020												
North America	85	94	37	68	9	13	306	2.8%	36	145	487	4.4%
Latin America	33	5	9	186	11	6	250	2.3%	144	267	661	6.0%
Western Europe	26	36	11	34	5	20	132	1.2%	15	122	269	2.4%
CIS & East Europe	29	27	9	32	4	6	107	1.0%	23	59	189	1.7%
Mid East/No. Africa	18	5	1	11	0	0	36	0.3%	27	17	80	0.7%
Sub-Saharan Africa	23	4	1	48	6	1	83	0.7%	239	28	350	3.2%
Pacific/China	77	36	22	114	12	20	281	2.5%	344	167	792	7.1%
Central/South Asia	64	8	1	68	7	2	150	1.3%	232	70	452	4.1%
TOTAL	355	215	91	561	54	69	1344	12.1%	1060	875	3279	29.6%
	10.8%	6.5%	2.8%	17.1%	1.6%	2.1%	41.0%		32.3%	26.7%	100.0%	

Notes:
1. Based on WEC Commission Case B (with interpolation where necessary) – global total in 2020 is approximately 11,050 Mtoe.
2. Renewable total adds traditional biomass and large hydro to new renewables.

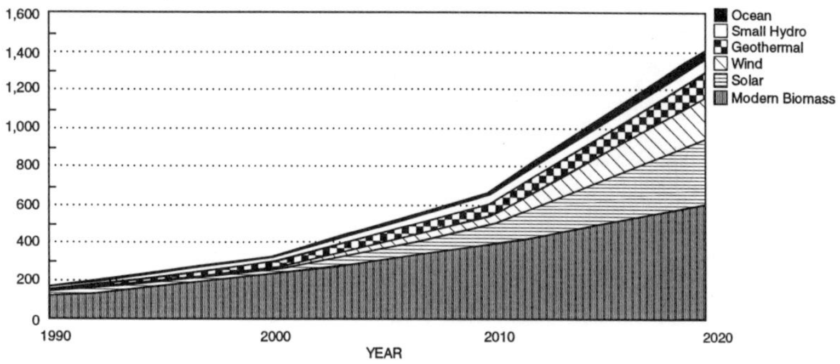

Figure 1.5 *Penetration of new renewables by type – ecologically driven scenario*

Note:
Large hydro and traditional biomass are excluded.

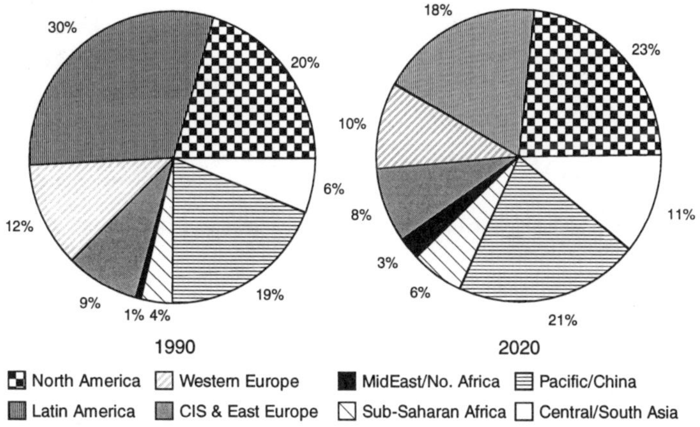

Figure 1.6 *New renewable contribution by region – ecologically driven scenario*

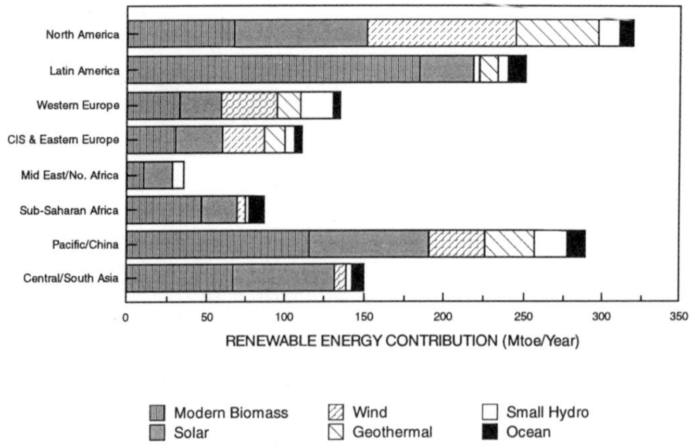

Figure 1.7 *New renewable distribution by region in 2020 – ecologically driven scenario*

Note:
Large hydro and traditional biomass are excluded.

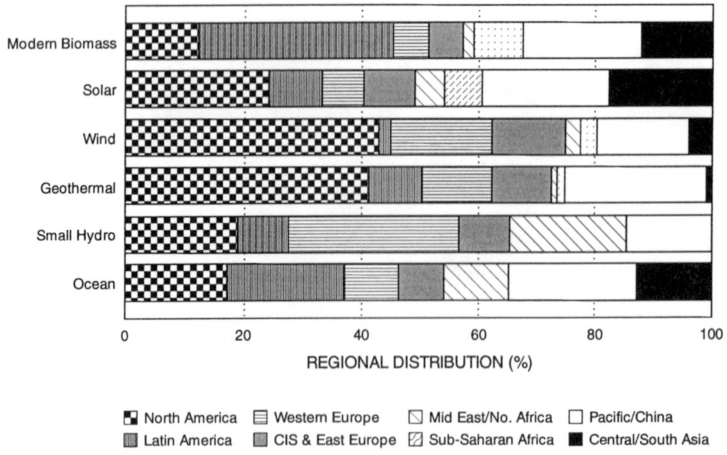

Figure 1.8 *Resource breakdown for new renewables in 2020 – ecologically driven scenario*

Note:
Large hydro and traditional biomass are excluded.

adopted an intermediate value (somewhat closer to the current policies percentage although about 60% higher in absolute terms in 2020) for the contribution of new renewables. As stated above, the WEC Commission took this Committee's lower case (current policies) data as the minimum evolution of new renewables.

Pursuit of the ecologically driven scenario is projected in 2020 to increase the energy contribution from all renewable sources by only 16%, due to a lesser contribution from traditional biomass on a per capita basis, and more constraints on large-scale hydro development. But the contribution from new renewables is projected to be 2.5 times higher than for the first case. Modern biomass, solar and wind have the largest contributions, with a lesser contribution from geothermal, small hydro and ocean.

The modern biomass growth is largely in Latin America and Asia. The solar growth is more widely distributed, with the largest impact in North America and Asia. Wind growth is largest in North America, with lesser contributions in Europe and across Asia. Geothermal growth is primarily in the North America and Pacific/China regions, while ocean appears most promising for Latin America and the Pacific. Overall, the largest renewable growth appears in the developing countries of the South, along with a ten-fold increase in North America. The wide geographic distribution of contributions by indigenous renewable resources is a very important finding of this study.

Implicit in the ecologically driven scenario is a major energy conservation effort, which results in 17% less total energy demand in 2020 than for the current policies case. This allows the contribution of new renewables to reach a higher percentage of the total global energy supply by 2020 than would be indicated by comparing the gross energy contribution of the two scenarios. One of the benefits of the ecologically driven scenario compared to a continuation of current trends is a reduction in annual global, energy-related CO_2 emissions. Energy efficiency is the largest single factor in the reduction, accounting for almost two-thirds of the change, but the growth in new renewables accounts for more than one-third of the total difference. There would also be corresponding reductions in other air emissions and solid waste.

Achieving the level of penetration of the first scenario will be relatively easy, but it will involve the active effort of a large number of people over the entire period. The challenge is really to approach the projections of the second case, which represents the limits of what may be possible to achieve within the next 30 years, if this is accepted as a positive and necessary contribution to overall world progress. This would take fundamental changes in the way decisions about energy supply are made by the public and private sector around the world. It would also involve investment in the future, and sharing of technology and know-how from the developed countries to the developing world. There have been hopeful signs in the last several years that the level of international co-operation necessary could be achieved in the next several decades. One example is world-wide action to prevent ozone depletion following the Montreal Protocol, and another is the UN Climate Convention.

A higher contribution of renewables will not happen simply because of pioneers experimenting with new technology, private sector companies taking all the high risks and long paybacks of establishing new markets, or political rhetoric. It will require a sustained effort by people aware of the resources around them, and careful to conserve what they have. It could contribute to a reversal of the mass population movement from the rural agricultural areas to the major urban areas, and a willingness to adapt to the diversity of the renewables.

Above all, changes of this magnitude are likely only under conditions of sustained government leadership and intervention, within a global planning framework, to adapt market forces to work in the direction of balanced and sustainable energy systems. It would require an integrated approach with other efforts necessary to resolve the many problems facing the world, such as adequate food and healthy water supplies, environmental protection and the developmental progress for most of the earth's population. It is important to recognize that a sustainable energy supply will eventually be necessary for the earth, and renewable energy resources will play a significant role in that long-term future.

LONGER-TERM PROSPECTS FOR RENEWABLE ENERGY USE

Application of renewables to a significant extent in the world's energy markets will take place in the longer-term future, *but how quickly is the most important issue*. Unless energy decision-making strategy is modified to address long-term impacts, renewables will make slow penetration into energy markets. By 2020, they will still be a minor contributor to the energy supply, and significant penetration is likely to occur later in the century if oil and natural gas shortages take hold.

One of the significant insights that was gained in the process of the Committee's investigation of the future of the renewable resources is the potential long-term impact of the acceleration caused by pursuing the ecologically driven scenario. To illustrate this critical point, a hypothetical extension of the two scenarios has been constructed, looking beyond the 2020 horizon toward the end of the 21st century. No one can predict the results of the two paths, but conceptually the analysis can be extended to see one possible impact of proactive promotion of the new renewables.

The concept is illustrated in Figures 1.9 and 1.10, which in a series of charts illustrate important parameters including population, per capita energy use, total energy use and the projected growth of renewable energy use, out to the year 2100. Population is assumed to stabilize at about 12 billion and 10 billion respectively, for the current policies and ecologically driven cases, by the end of the next century. Per capita

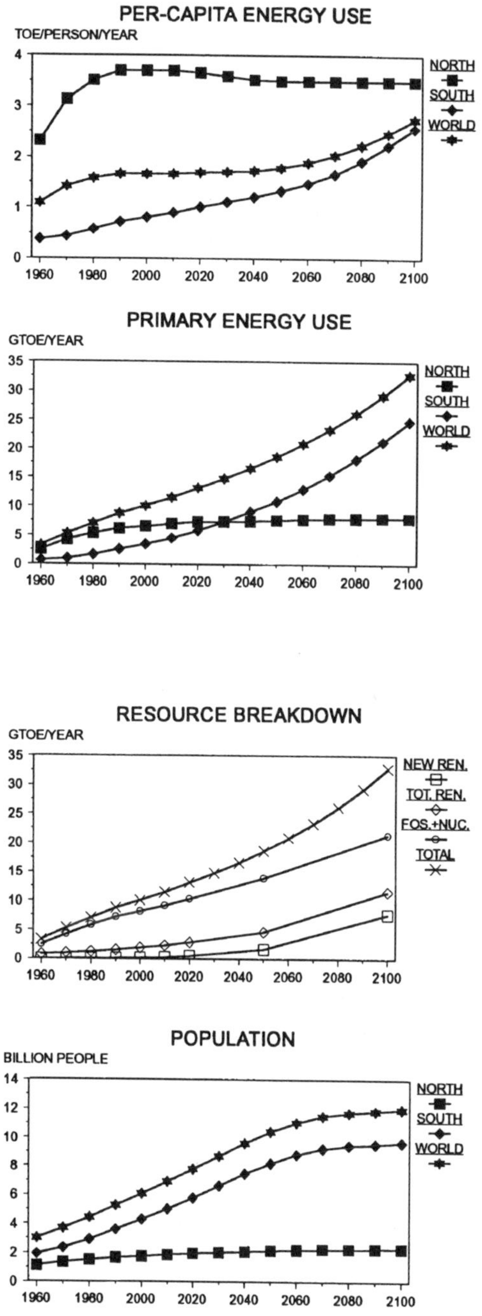

Figure 1.9 *Long-range current policies case*

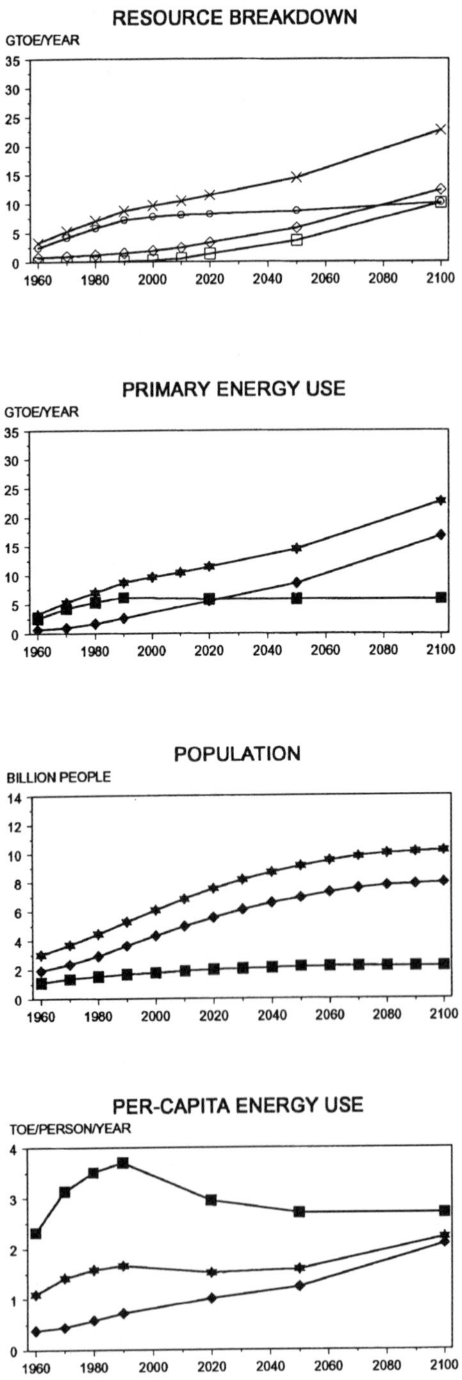

Figure 1.10 *Long-range ecologically driven case*

energy use is lower for the ecologically driven case since one of the major tenets of this approach is aggressive energy conservation and increased efficiency. This per capita data, combined with the assumed population growth, leads to total primary energy consumption in 2100 of about 33 Gtoe and 23 Gtoe for the two cases, respectively.

This Committee's work on the long-range perspectives differs from that of the WEC Commission Report, *Energy for Tomorrow's World* (1992). This was to be expected, since the groups worked independently. However, the implied global primary energy consumption in 2100 compares closely with the Commission's 33 Gtoe for its Case B and 20 Gtoe for its Case C (lower than the Case A value of 42 Gtoe). The Commission assumes that world population reaches 12 billion in 2100, with per capita consumption in the developing countries between 1.5 toe annually to as high as 3.15 toe. The per capita values used by this Committee are in a much narrower range, and the ecologically driven case is seen to be consistent with a lower world population of 10 billion in 2100. In both cases, the per capita consumption in the developing world (South) is found to continue to lag behind the North even in 2100, and this has implications on the pace of development that can be achieved.

But the fourth graph for each case displays the key impact of pursuing accelerated promotion of conservation and renewables in the next three decades. The current policies case is still dominated by the combination of fossil fuels and/or nuclear by the year 2100. The global resources of petroleum and natural gas may be constrained by that time. Carbon dioxide production from energy use will be several times higher than in 1990 unless there is heavy dependence on either fission breeder reactors with fuel reprocessing or fusion energy is introduced commercially in mid-century. New renewables are postulated for the current policies case to supply about 25% of primary energy in 2100.

The major difference in the ecologically driven case is that the dominance of fossil and nuclear can be contained over the next 30 years, by limiting the total growth in energy use, and renewables can supply on the order of 50% of total energy by 2100. These proportions are identical with those of the Commission's Cases A and B in the first, and Case C in the second instance, respectively. Some would argue that renewables can supply a larger percentage because the resources are so large, but if renewables make a major contribution and overall energy growth is low, we will still have supplies of petroleum and natural gas by 2100, and a good mix of energy supplies has many advantages. The contribution from new renewables is projected to reach 5 to 15 Gtoe by 2100, and this range is based on evaluation of market growth for renewables, assuming lower annual growth rates as the size of the penetration increases. Global energy use above 33 Gtoe by 2100 could lead to a higher absolute contribution from renewables. Again there is a good fit with the WEC Commission's assumptions, though they have a narrower range of 8.5 to 10 Gtoe for the contribution of new renewables by 2100.

Much of the contribution will probably come from solar and wind, which are intermittent resources, with significant modern biomass contribution, and lesser contributions from other new renewable resources. Large hydroelectric will still be a major contributor, but traditional biomass use will probably be much less, and its usage will be at much higher efficiency.

There is no proof or evidence at this time that different outcomes to the two cases are not equally probable. Several other analyses of long term energy issues have been published recently, giving alternate perspectives on the likely outcome[3–6]. However, the above discussion points out a widely held opinion that unless action is taken in the near future, the future generations of the earth will be living with heavy dependence on fossil fuels and/or nuclear energy, regardless of the consequences. There is a chance to keep another option available, should it be needed after enough is learned about the atmosphere to be certain of the impacts of man's emissions due to energy use and deforestation. This is the essence of the "minimum regrets" strategy.

It is clear that renewables will play a significant role in the long-term energy future. The penetration of renewables can be accelerated by a deliberate act of choice and policy. Even if current policies guide the next several decades, renewable penetration will continue to increase at a slower rate, and reach a significant level before the end of the next century. Natural limits on the use of renewables will be reached locally and regionally, due to the intermittent nature and relatively low capacity factor of some renewables, or to environmental limits on land use or species protection. But, provided due care is taken to avoid renewable energy developments which significantly reduce bio-diversity, natural habitats, species and numbers within a species, and sites of historic significance or natural beauty, those limits are not approached in the current policies scenario by 2020, and are reached in only a few situations in the ecologically driven scenario by that time.

ADDITIONAL PERSPECTIVE ON ECONOMIC ISSUES

There is clearly a concern among major segments of the energy community that use of renewables will be significantly more expensive than use of fossil fuels. Numerous papers and reports on the relative economics of renewables have been published over the last two decades, and the views expressed cover a wide range of practicalities and speculation. A simple and complete analysis to the question "can renewables compete economically?" does not exist. This book does not attempt to provide a simple, direct answer, but to highlight the issues that affect the conclusion in various situations.

There are several major elements to an economic analysis of energy alternatives. First, the total investment cost must be measured, including plant facilities, land, transmission facilities and financing assumptions, and this investment is typically higher for renewables than for fossil fuels. Then the total cost of operation and maintenance of an energy facility must be considered, including clean up and disposal of waste produced. Fuel cost is applicable to some options, and this includes the transportation to the site and assumptions on future price escalation. But the most controversial part is the external impacts of construction and operation of the facility. A brief examination of the degrees of freedom of such an analysis will demonstrate that the answer can be manipulated to favour the intended result of the analyst. Complex studies are done by electric utilities for such cases, and the results vary considerably from region to region.

A range of examples provides more insight into the future opportunities for renewables than a single "average" analysis. One reason for this statement is that an average analysis would not explain the current use of renewables, which is more dependent on local situations at the margins of the market than on broad averages of market factors. Another factor is that the conventional energy markets have significant variations. The cost of energy, even from fossil fuels, can vary greatly around the world, depending mainly on government policies and fuel transportation and/or distribution costs. Consider the following range of situations:

1. grid-connected electricity in an industrialized country;
2. direct thermal energy supply to an industrial facility;
3. rural electrification in a developing country compared to stand-alone diesel generator compared to stand-alone renewable generator/storage;
4. rural direct thermal use on a small scale, with either renewable or fossil source.

It might appear that these are arranged in the order from simple to most complex for the purposes of economic comparison of renewable and conventional global energy economics, but the discussion of external costs makes each of these situations very complex. These external costs include environmental effects (eg, forest damage and human health), derivative economic effects (eg, employment and local economic value), and subsidies (eg, public-supported R&D and existing infrastructure). Economic approaches to including these costs may involve using market pricing, cost of replacement or restoration, indirect or derivative market effects, or public surveys. These impacts have an inherently high uncertainty, but the range of published numbers indicates that they may have substantial impacts on the comparison of energy systems. Increasingly, local government regulation is forcing arbitrary values for these external factors to be applied to energy decision making, especially in the developed countries.

Even without the consideration of external costs, there are a number of situations where renewable energy systems are economical today. The following several examples, included in Chapters 2–7, are summarized here to help understanding of the current penetration of these energy resources:

- Specific comparisons were made of the long-run marginal cost for wind energy and conventional fossil-fueled generation for a number of developing countries. For example, the results indicate that the economic potential for wind use is high in Romania and Chile, but unlikely for Colombia and Kenya. Overall, wind is marginally competitive in the best areas today, and this situation will improve to regionally competitive by 2005 under ecologically driven conditions.

- Comparisons of solar thermal electric plants with fossil alternatives indicate that further developments in technology and higher volume production should allow solar to compete with peaking power plants starting about 2000, and compete in a large range of high resource situations by 2020. With policies to accelerate development and deployment, the attainment of this competitive position might improve by as much as 15 years.
- A specific comparison of a 25 MWe biomass gasifier coupled with combined cycle power plant in north-east Brazil shows that it will be competitive with expansion of the large hydro-based local utility grid. A demonstration project, supported by an up to US$30 million grant from the Global Environmental Facility, is planned for operation in 1997 with fuel provided from a eucalyptus plantation.
- For a village of 200 people in India, beyond the national electric grid, a biomass gasifier was installed to displace fuel from a diesel generator that powered village lighting, water pumping and milling equipment. The investment was lower than the extension of the grid, and resulted in more reliable electric power of higher consistent quality. An analysis of oil displacement only indicated a simple payback of about 10 years, which will decrease as the energy demand in the village grows. Co-operative labour from the villagers was used to grow and harvest biomass on marginal land, reducing the economic drain from oil purchases.
- Separate analyses have been conducted on the feasibility of replacing grid extension with stand-alone hybrid photovoltaic systems in both California and South Africa. The results depend on the local labour cost, the distance from the grid, and the size of the peak load. For California, the results for a single-family home show PV more economical for a distance greater than 0.5 km, and for South Africa, the economical distance is about 1 km for a small village, and somewhat further for a larger village.

Although there are many constraints that must be overcome for renewable energy to reach its long-term contribution potential, it is likely that the capital investment required and its sources will be the most controversial. It is relatively easy to calculate the investment required to generate the predicted penetration of renewables for the two cases, and these investments are large. These costs include the R&D expenditures, and the equivalent of a lifetime of fuel costs as part of the initial payment for some of the renewables. Table 1.4 summarizes the investment needed to support the two scenarios discussed above. A transmission allowance is included for those resources which are often located remotely from the users.

Table 1.4 *Estimated cumulative total renewable investment (US$ billion, 1990)*

	Current policies			*Ecologically driven*		
	2000	*2010*	*2020*	*2000*	*2010*	*2020*
Solar	52	134	313	65	265	1205
Wind	14	62	181	16	98	412
Geothermal	15	20	30	20	50	80
Biomass	50	100	150	66	140	260
Ocean	1	10	55	2	50	150
Small hydro	21	50	100	36	88	150
Sub-total	153	376	829	205	691	2257
Transmission	10	23	55	15	47	141
Total	163	399	884	220	738	2398

This total investment can be compared to an estimate of the total investment necessary to deliver the same amount of energy using conventional fossil fuels. With an average of US$100 million per Mtoe for fuels and US$1000 million per Mtoe for electric generation, this investment for the ecologically driven case is about US$700 billion. But to this should be added an allowance for electric transmission and distribution facilities for comparison with distributed renewable applications, as well as an allowance for the typical energy losses expected in rural distribution in developing countries. With such an allowance, the total capital investment for fossil-based systems is expected to be about US$930 billion.

Although this comparison indicates that the renewable approach will require significantly greater capital formation than the fossil equivalent, it is not appropriate to base economic comparisons solely on these numbers since fuel costs are not included for either the fossil or biomass facilities. Another aspect is that these prices control the penetration of energy efficiency into the markets, and government action to encourage efficiency, such as the high gasoline prices in Europe, can have a significant effect on market size and technology share. This is also true of the current subsidies for conventional energy in many developing countries.

The projected investment in renewables for the ecologically driven scenario can also be compared to total GDP as a measure of its financial feasibility. Using the 1989 WEC "M" Scenario as a basis, the combined current annual GDP for the world is about US$21 $\times 10^{12}$, and the projected value by 2020 is an estimated US$36 $\times 10^{12}$. Of these values, about 5% is spent on energy. For the aggressive renewables scenario, the total cumulative investment between now and 2020 amounts to about 0.25% of the cumulative global GDP for that period. In annual terms, the projected share in 2020 is about 0.5%. Keeping in mind the intent of the ecologically driven scenario, this brief comparison indicates that it appears possible to achieve the projected use of renewables. The view of the WEC Commission is that somewhat higher cumulative total renewable investment would be necessary to achieve these results, but the overall conclusion on feasibility is similar.

Table 1.5 *Examples of external costs of airborne emissions in US states*

Emission species	*Evaluation factor (1989 US$/kg)*
SO_2	0.90–30.18
NO_x	0.11–30.45
VOC	1.30–19.25
CO	0.99
Particulates	0.35–5.83
CO_2	0.002–0.119
CH_4	0.24
N_20	4.36–4.55

Notes:

1. Values are compiled from US state regulatory agencies, and are used in making decisions for new electric generating plants, but are not paid by individual projects.
2. Narrow ranges indicate only a few states are applying this criteria.
3. Values are very dependent on local conditions and political will.
4. High end of the ranges are generally significantly higher than actual costs of emission controls, up to the capability of best available current technology.
5. For coal-fired power plants, these factors generally increase fuel costs the equivalent of 2 to 5 US cents/kWh, and 1 to 2 US cents/kWh for natural-gas-fired plants.

But for meaningful insight, it is necessary to compare these costs with the associated benefits in relation to other options, and the impact of external costs of energy systems must be included in a comprehensive manner. Several analyses of this type have been published, but the results have not achieved general agreement in the energy industry. Some examples related only to airborne emissions are included in Table 1.5. The ranges in values being considered are very large, and the assumptions used are very diverse. Those based on political considerations can change frequently, and cause large variations in the economic signals sent to the energy market. A more rational and well-documented set of external costs is needed for stability of the markets. Because such an analysis involves broader expertise than currently included in the WEC Renewables Study Committee, it is deferred to future WEC investigation by an appropriate study group.

The currently prevailing market system excludes many important side-effects and longer-term ramifications. The possible contribution from renewable energy sources, with the diversity and choice it provides, is one possible casualty of short-term focus for decision making. The high capital cost of renewable technologies makes it very attractive to use public sector financing, through government-backed bonds, for example, rather than private sector financing. The long-term payback characteristic of these technologies does not match well the objectives of many private lenders. However, involvement of the private sector in a major way is necessary for success.

At present, new renewables face a severe "break-in" barrier or commercialization hurdle due to existing asset and investment values of fossil fuel systems, and their present inability to achieve economies of large-scale production.

CONCLUSIONS

The work of the Study Committee on Renewable Energy Resources led to a number of important conclusions. Those specific to an individual resource are stated in the relevant subsequent chapter, and the most significant and generally applicable are given below, divided into four categories: general; international; regional; and national and local.

General

- Besides the traditional use of biomass resources in the developing countries, there are significant numbers of economically attractive applications for renewable energy systems today. Some are mature technologies in conventional applications, such as geothermal power generation, but many can best be classed as special situations, where local conditions such as remoteness from conventional energy supplies, very severe environmental conditions, small size, or special incentives have created conditions favourable for the application of renewables.
- For many new applications that are economic by life-cycle-cost measures, there is still a major hurdle because the high capital investment is a disadvantage when initial cost is the major criteria used for energy system choices. It is even more difficult to displace fuel from existing fossil-fueled facilities with renewable energy, because of the remaining value of the infrastructure.
- Recent penetration of renewables for many applications has been slowed due to low prices for conventional fossil fuels. To expand more rapidly the use of renewables, the economic decision-making methodology would have to be updated to include the external costs of energy use, which would be generally more favourable to renewables than to fossil fuels.
- Costs for energy from renewable energy systems are expected to be reduced over the next few decades, due primarily to higher production volume, but incremental improvements are expected in efficiency, materials, reliability and application. This is in contrast with the predictions for fossil-fueled systems, which will experience some cost-saving technological advancements, but increased pressure on emission control and rising fuel costs will lead to increasing costs over time. As a result, the apparent economic advantages that presently exists for fossil fuels will decline or reverse, and renewable options are likely to become the economic choice in an increasing

number of regions and applications by 2020 and beyond. This will result in a gradual penetration of renewables without major facilitation of this process.

- The penetration up to 2020 was examined in detail in this study, but there is much evidence that the penetration of renewables will not peak at this time, but will continue to increase in the following decades. There may be some structural limits to the penetration of individual resources that can be reached after that, but the renewable resources derived from the sun are large enough theoretically to satisfy the entire energy needs of the world in the next century and beyond. Other finite resources will reach their natural limits, even if coal will be available for several centuries, but the renewables offer very long-term promise for a permanent solution to a major fraction of the energy supply problem.
- The environmental advantages of renewable energy systems are real, and renewables are seen as one part of a total solution to our environmental problems, present and future, as well as a major component of sustainable development. However, just because they are called "renewable" does not assure environmental acceptability.

  Environmentally unacceptable features identified in individual chapters here include: risk of loss of bio-diversity, natural habitats, species and ecosystems generally; visual intrusion; and noise. These unacceptable features could severely constrain modern biomass, tidal power using barrages and wind power developments below their technical potential. Recognition of ways to incorporate environmental/ecological considerations successfully into development could (and should) raise the trajectory of modern biomass development considerably closer to its technical potential. This is not the case with tidal barrages and only to some extent the case for wind farms. Large-scale schemes for all forms of renewable energy risk having severe local environmental impacts.
- Many of the renewable technologies are modular, which is beneficial for mass production of standard sizes/models, and contributes to a relatively short planning and implementation cycle. But to be successful, the standard models must be well matched to the resource and user characteristics in the target market. Otherwise, unwise investment will lead to less effective use of both economic and energy resources.
- The renewable energy resources, relevant technologies and applications have been the subject of intensive investigation for at least the past 20 years. A significant amount of data and experience has been developed, and the emphasis is shifting from technical to economic demonstration, but the renewables should still be considered in the childhood stage of their development.
- Research and development of the emerging renewable technologies needs to be continued and expanded, if these options are to be rapidly moved to maturity. Only a small portion of the energy R&D funds are

now devoted to renewables, although the percentage is rising again in some regions after declining significantly during the 1980s.

- Electricity production and use is continuously increasing, because it is clean and convenient at the point of use. This trend is not an advantage for overall primary energy efficiency, especially when the power generation plants are based on thermodynamic cycles with built-in efficiency limitations. Several of the renewables, which do not use this type of thermal cycle, can produce electric power without the usual efficiency loss or unwanted waste heat.
- Overall, biomass is the dominant resource today, and will continue to be the largest new renewable supply until at least 2020. However, the relative contribution of modern biomass to total new renewable contribution will decline from about 75% today to between 40 and 45% by 2020. Solar and wind will fill in much of the difference, each projected to be similar in magnitude to the other three renewable categories combined. Longer term, solar is expected to dominate.

International

- Efficient operation of energy markets will not solve all of the world's energy problems. There needs to be guidelines given to the market to help the prioritization of society's needs if truly sustainable energy systems are to be achieved.
- There is a definite trend towards diversity in the energy markets, as they are decentralized and deregulated. This has been seen dramatically in the US, as individuals take advantage of specific local opportunities and unique situations. When small-scale systems are given the opportunity to compete with large systems, the resulting mix is very diverse and contributes to the reliability of the system to unforeseen circumstances. Increased diversity also includes hybrid systems using both fossil and renewable resources, which may have economic advantages over renewable-only systems in the short term.
- Renewables have the capability to address all energy-use sectors, but each individual resource probably does not. All can address electricity generation, but biomass is likely to have the largest impact on transport fuels. Solar and biomass can best address the needs of the individual in residential applications. Process heat can be provided by biomass, solar, or geothermal. Current trends toward urban lifestyles present a challenge for renewable use in many applications, except for the production of transportable fuels and electricity generation.
- Based on evaluation of the Committee's two global cases for future energy use, it is important to note that new renewables are projected to account for a significant fraction of the future increase even for the current policies case, and represent more than half of the total increase in energy supply for the next 30 years for the ecologically driven case.

- The difference between the results of the two scenarios are quite impressive. The ecologically driven approach is projected to increase the modern biomass, geothermal, and wind contributions by at least a factor of two compared to the continuation of current policies. For solar and ocean, it is more than a factor of three. The total contribution of new renewables rises from about 4% to about 12% by 2020. The 2020 projected contribution for total renewables is about 21% and 30%, respectively, compared to about 18% today. It would take a major effort, such as the ecologically driven scenario, to achieve more than a very gradual increase in the renewable percentage contribution.

Regional

- Technology transfer is needed from the developed countries, which hold most of the technological expertise, to the developing countries, where many of the opportunities are. However, only proven technologies should be transferred, and these should be appropriate for the end use situation. Joint completion of development is desirable with co-operation on the specific design features and desirable financing methods.
- The regional contribution from renewables for the two scenarios examined points out the major differences in the regional opportunities, as shown in Figure 1.11. For the current policies scenario, moderate penetrations of the developed countries are projected, along with very significant contributions for many of the developing countries, with heavy use of traditional biomass. The Commonwealth of Independent States (CIS) and Eastern Europe show the lowest penetration of total renewables in 2020, closely followed by the MidEast/North Africa and Western Europe regions. But for the ecologically driven case, North America reached 20% penetration,

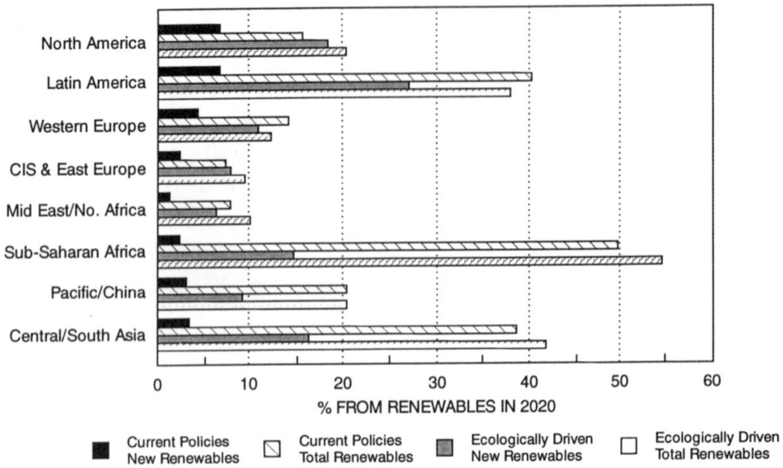

Figure 1.11 *Regional penetration of renewables in 2020*

with more than 10% for the other developed countries. The contribution is about 40% in the developing countries, with Sub-Saharan Africa at about 50%. Renewables will play a major role in several regions, and could play a significant international role for this higher case. It is also interesting to note that higher contribution from new renewables combined with lower use of "old" renewables and lower total energy consumption results in about the same total renewable percentage for the two cases in several regions for 2020.

- For larger penetration of intermittent renewables, there are still technological issues of energy transportation and storage that need to be resolved. Storage and/or back-up systems are usually needed for reliable operation of renewable applications not connected to an electric distribution system. These technologies need to be included in the R&D priorities.
- Although the aggregate contribution of distributed small renewable energy systems by 2020 appears minor compared to global energy use, it is very important because these systems can make a tremendous difference in the lives of millions of people.
- Traditional biomass will still be a very large part of the total contribution from renewables by 2020, but it is not expected to increase greatly from current levels. This use will continue to cause major environmental stresses in developing countries unless sustainable harvests are used, and will make a lower contribution for the ecologically driven scenario than the current policies projection.
- Today's contribution by renewables is poorly recognized and largely underestimated.

National and Local

- The costs for installations of renewables are usually very site-specific, because the resource can be extremely localized. It is therefore very difficult to generalize on the cost at which renewables will compete economically with conventional resources. Better information on the site-specific distribution of the resources needs to be developed, so that the specific economics of local applications can be accurately assessed.
- The site-specific impacts of renewables include risk of severely adverse local environmental impacts, and damage at sites of national significance (eg on historic, landscape, and ecological grounds). There needs to be a balance between concern for potential climate change and the ensuing abatement and mitigation policies, and the need to avoid severe local environmental damage or destruction of the national heritage.
- Renewable resources are by nature indigenous to the locality, and their use can ease foreign payment issues, as well as the environmental concerns for international transportation of fossil fuels. Use of renewables will also increase the geographic diversity of energy use, allowing more energy development in areas remote from conventional supplies.

- Short-term planning and economic decision making is a major obstacle to renewables. There is a need for more permanent, inter-governmental focus on the long range problems such as renewable energy development. Major initiatives are beginning in non-governmental organizations, including the United Nations and the World Bank, to provide leadership.
- Many types of renewables are well adapted to rural energy use patterns. A number of countries are now heavily dependent on renewable use. More countries can benefit from this situation, with a balanced energy development plan. Within governments, especially in the developing countries, there is a need for effective co-ordination of sectoral plans dealing with energy, agriculture, forestry, and rural development. It is possible to "leap-frog" the problems due to massive dependence on fossil fuels, and proceed directly on a path to sustainable development. But this approach will need the trained action by large numbers of people who understand the vision and their local conditions, and are sensitive to the combined environmental impact of their actions. It will also need the availability of significant capital resources.

RECOMMENDATIONS

This is the first major study of new renewable resources by a WEC Study Committee. The Committee arrived at international, regional and national/local recommendations.

International

- Increase the priority for funding of renewable energy projects among financing institutions and for bilateral aid programmes, especially those programmes that will lead to sustainable development based on indigenous resources.
- Establish or designate a single organization to give international focus and leadership to increased use of renewable energy. This organization should co-ordinate with other international organizations that have complementary objectives, while maintaining momentum towards greater renewable energy use.
- Develop international treaties, programmes and criteria for preserving ecosystems and reversing degradation.
- Review and update economic decision-making methodology to include the external impacts of the options under consideration. This is a very appropriate area for future study by the WEC, since initial investigations have pointed out some major opportunities, but a consistent and well-debated basis should be established for use by the world-wide energy community.

Regional

- Include renewable energy where relevant as one of the key elements in the overall strategy for sustainable development. Integrate programmes for food, water, renewable energy and social development.
- Establish regional centres of excellence for renewable energy, to provide training, technology support, and resource databases appropriate to the regional needs. Develop and implement regional demonstration programmes as showcases of the best elements of renewable technologies.
- Gather, review and publicize success stories involving renewable energy, to give realistic examples of what has been done and is possible.

National/Local

- Create an inventory of resources for the area, using a multi-disciplinary group including energy and environmental experts, and make assessments of the potential and desirable contribution of different renewable resources.
- Establish targets based on tangible objectives for implementation of renewables. Select at least one resource and pursue it in advance of conventional economic thinking to gain experience with the different characteristics of renewables within the overall energy systems. For example, look at the Non-Fossil-Fuel Obligation for UK utilities. Require that government facilities use appropriate renewable energy resources.
- Introduce programmes, criteria, assessments and appraisal techniques to avoid schemes which impose significant local environmental damage and which ensure the protection and reclamation of ecosystems, species and landscapes. Ensure, by approaches which balance environmental and energy provision criteria, that renewable energy development is optimized, and where environmental criteria cannot be met that projects do not go ahead.
- Pursue programmes of education, specialized training, and public information on the benefits and characteristics of the renewable resources within the local context.
- Increase R&D funds for renewables, eliminate subsidies on conventional energy, and reduce barriers that exist for the introduction of new and different (including small and distributed) energy systems.
- Encourage private sector investment and involvement in the programmes for training and implementation of renewables. Promote the establishment of joint-venture manufacturing plants for local production of renewable energy systems.
- Strengthen national capabilities in the assessment, planning and implementation of energy programmes, stressing the need for inter-

sectoral co-operation and multi-disciplinary strategy, especially for rural areas of developing countries.

- Enhance understanding of renewables within overall energy systems, making it possible to incorporate larger percentages of renewables within the operational limits of electric utilities and other energy systems.

Renewable energy can reach its potential contribution in an accelerated manner only if there is a willingness to evaluate the benefits from a global perspective, and enable action at the local level. A longer-term view must transcend the short-term focus of many political and economic systems, as well as efficient market economies which currently ignore some important global and longer-term considerations.

Approaching the beginning of a century of development of the various new renewable forms of energy, it is vitally important not to be driven by unsound technological imperatives, or loss of balance in appreciation of all environmental impacts, or unsound economics. In a world concerned with sustainable development, tough and consistent environmental criteria will need to be applied to the provision and use of all forms of energy.

If this is done then, steadily, over the next century, "new" renewable energy will make a growing – and eventually massive – contribution to the world's energy supply.

REFERENCES

1. WEC Commission. *Energy for Tomorrow's World – the Realities, the Real Options, and the Agenda for Achievement.* London, UK, Kogan Page, 1993 (subsequently published in US, French, Spanish, Portuguese, Arabic and other editions).
2. WEC Conservation and Studies Committee. *Global Energy Perspectives 2000–2020.* London, UK: 14th Congress of the World Energy Conference, Montreal, Canada, September, 1989.
3. T. B. Johansson, et al (editors). *Renewable Energy: Sources for Fuels and Electricity.* Island Press, 1993.
4. Dennis Anderson. *The Energy Industry and Global Warming: New Roles for International Aid.* London, UK, Overseas Development Institute, 1992.
5. S. W. Gouse, et al. Potential World Development through 2100: The Impacts on Energy Demand, Resources and the Environment. London, UK, *World Energy Council Journal*, December 1992, pp 18–32.
6. Nebojša Nakićenović, et al. *Long-term Strategies for Mitigating Global Warming: Towards New Earth.* Laxenburg, Austria: International Institute for Applied Systems Analysis, March 1992. (To be published in *Energy – The International Journal.*)

CHAPTER

2

Solar Energy

NATURE OF THE SOLAR RESOURCE

Solar energy results from the process of continuous nuclear fusion in the Sun. Solar irradiance at the average Earth orbit has an intensity of 1.367 kW/m^2[1]. The Earth's perimeter is 40,000 km and so the intersected power amounts to 174,000 TW. The peak intensity at sea level is typically 1 kW/m^2, and the 24 hours annual average at a location on the earth's surface is 0.20 kW/m^2. This amounts to 102,000 TW to drive our environment, on which we depend for our sustaining life support, including all other forms of renewable energy (excepting geothermal energy) which are treated separately in this book.

Although the total solar resource is over 10,000 times the current energy use of the humankind its low power density and its geographical and time variation represent major challenges which may limit its contribution to the total energy mix. Before evaluation of the possible future contribution of this resource it is therefore necessary to understand its major characteristics on the planes of interest for solar collectors.

Day, Night, and the Seasons

Day and night are caused by the Earth's rotation and the seasons are caused by the 23°27' tilt of the rotation axis relative to the axis of the Earth's orbit around the Sun. The Sun's elevation is always highest at noon, but while for the tropical latitudes (at less than 23°27' from the equator) the noon Sun is directly overhead twice a year, in higher latitudes the Sun always appears toward the equator. In the arctic and antarctic regions (at more than 90° – 23°27' from the equator) the Sun sinks below the horizon for extended periods in the winter, and stays above the horizon for extended periods in the summer.

Beam, Diffuse and Reflected Irradiance

Solar irradiance on the horizontal plane or "insolation" has two components – direct or beam, and diffuse irradiance. Beam irradiance comes directly from the solar disk. Diffuse irradiance results from scattering in the atmosphere and can be modelled as consisting of three components – circumsolar (sky brightening around the solar disk), horizon (sky brightening or darkening near the horizon), and rest of the sky diffuse irradiance. A non-horizontal plane also receives reflected irradiance from the ground. The sum of the three components (direct, diffuse, and reflected) is the total, or global, irradiance.

Beam irradiance can be concentrated by lenses or reflectors. If the concentration ratio is high, very high power densities can be reached but the diffuse irradiance is lost. If concentration is low, part of the circumsolar diffuse irradiance can also be concentrated. Diffuse irradiance varies from as little as 10% of the total for clear sky conditions and high solar elevations to 100% when the solar disk is not visible due to clouds. Concentrating collectors will thus normally collect significantly less energy than the non-concentrating types. Reflected irradiance is usually very small but can be up to 40% of the total irradiance on a vertical plane when there is snow on the ground.

Clear Sky Conditions

For clear sky conditions – ie very low humidity and airborne dust and no pollution – solar irradiance at any given time only depends on the site latitude, distance above sea level (height) and ground albedo. Maximum values, increasing with ground height and reflectivity, are always obtained at noon. For lower solar elevations increased atmospheric absorption and scattering reduce the global irradiance while increasing its diffuse fraction[2].

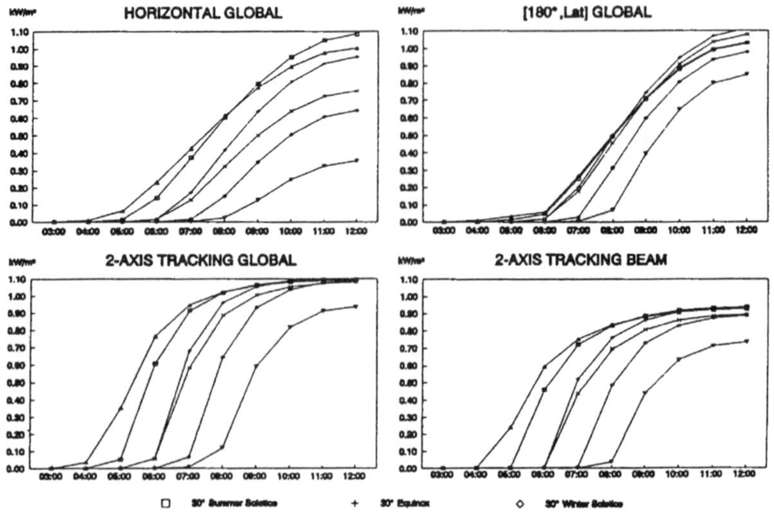

Figure 2.1 *Clear sky hourly solar irradiance at 30° and 45° latitudes*

The upper left graph of Figure 2.1 shows the significant seasonal variation of the horizontal irradiance. This is what drives weather and solar collectors laid out on the horizontal plane, such as solar ponds. But non-concentrating collectors are usually fixed on the latitude-tilted plane (towards the equator), as shown in the upper right graph, leading to a fairly constant output throughout the year. A steeper tilt allows for more winter output, but less annual collection. The shape of the solar irradiance curves on the normal plane is much more square, allowing for maximum energy collection. This is achieved by a collector that tracks the Sun in two axes. If tracking is only done in one axis, with the polar or East–West axis fixed, slightly less energy is collected.

The difference between the normal global and the normal beam irradiance is also worth noting. It is higher than calculated with traditional solar engineering algorithms because it has been computed with Surya[3], a new simple yet powerful hourly solar model that treats solar irradiance according to the recent findings of a higher and highly anisotropic diffuse fraction, leading to more normal global and less normal beam irradiance than estimated with the traditional models[4].

Table 2.1 shows the sea level clear sky 24 hours average seasonal and yearly solar irradiance at intervals of 15° latitude. By using power units instead of energy units a direct comparison with the hourly irradiance values can be obtained. This table shows that the facts mentioned above for the hourly distribution – ie significant seasonal latitude variation away

Table 2.1 *Clear sky seasonal solar irradiance (kW/m²)*

	Horizontal global				*[180° Lat] global*			
Lat	*Summer*	*Equinox*	*Winter*	*Year*	*Summer*	*Equinox*	*Winter*	*Year*
0	0.303	0.344	0.324	0.330	0.303	0.344	0.324	0.330
15	0.351	0.331	0.251	0.317	0.317	0.345	0.306	0.330
30	0.378	0.292	0.166	0.283	0.328	0.341	0.271	0.321
45	0.380	0.230	0.078	0.231	0.336	0.327	0.202	0.300
60	0.366	0.152	0.022	0.174	0.341	0.298	0.078	0.257
75	0.364	0.091	0.002	0.141	0.345	0.260	0.003	0.211
90	0.449	0.071		0.151	0.495	0.115		0.198

	2-axis tracking global				*2-axis tracking beam*			
Lat	*Summer*	*Equinox*	*Winter*	*Year*	*Summer*	*Equinox*	*Winter*	*Year*
0	0.462	0.483	0.493	0.481	0.379	0.399	0.405	0.397
15	0.504	0.479	0.444	0.479	0.418	0.396	0.362	0.395
30	0.542	0.468	0.369	0.465	0.448	0.383	0.292	0.379
45	0.587	0.444	0.253	0.435	0.482	0.357	0.192	0.349
60	0.655	0.393	0.089	0.387	0.526	0.306	0.037	0.294
75	0.847	0.332	0.004	0.380	0.676	0.196		0.260
90	1.099	0.193		0.436	0.743	0.001		0.263

from the tropics for all planes of interest except the latitude-tilted plane facing the equator, and maximum irradiance on the normal plane with a significant difference between the normal global and the normal beam irradiance – hold true for the all day figures. But it also shows that the latitude dependence of the annual average is much more significant for the horizontal plane than for the other planes of interest for solar energy technology. The annual solar irradiance even increases in the arctic due to the increased snow cover that causes higher values for the diffuse and reflected components.

Actual Sky Conditions

Clear skies are seldom found in many regions and clouds have a significant impact on the total solar energy available and its specific variability. Figure 2.2 gives a generalised map of the insolation resources across the Earth, although this view greatly oversimplifies the local weather patterns[5]. Dense cloud cover reduces the global insolation to one-third of the clear sky value and the beam irradiance to zero, and in many areas clouds pass frequently and at random intervals over the land. This may be especially noticeable in humid climates near mountains in the trade wind zones, such as the leeward side of tropical islands.

To show the weather effect, compared to the clear sky conditions, the annual average solar irradiance on the latitude-tilted plane and the 2-axis tracking plane has been computed with the Surya model for a 108-site database of monthly average horizontal irradiance. The data and the modelled results are shown in Figures 2.3 and 2.4, and in Table 2.2. It can be seen that, in all the planes of interest for solar energy technology, the latitude dependence is fairly weak and clouds play a much more important role. With the help of this data the reader may easily interpolate in order to estimate roughly the annual irradiance on a solar collector in her or his region. The resulting figure, which usually varies less than 10% from year to year, will have a direct impact on the cost of solar energy.

Wavelength Distribution

The Sun's energy has a wavelength distribution which can be modelled as a black-body radiation, with a temperature of 5,800°K. The distribution covers the ultraviolet, visible and infrared regions, which are differently affected by atmospheric attenuation. Most of the visible radiation reaches the ground. But ozone in the upper atmosphere absorbs much of the ultraviolet radiation. Recent thinning of the ozone layer, especially in the antarctic and arctic regions, is allowing more ultraviolet radiation to reach the Earth's surface. Part of the incoming infrared radiation is absorbed by carbon dioxide, water vapour and other gases, and the longer wavelength infrared radiation emitted from the Earth's surface at night mostly escapes to outer space. But the accumulation of these green-

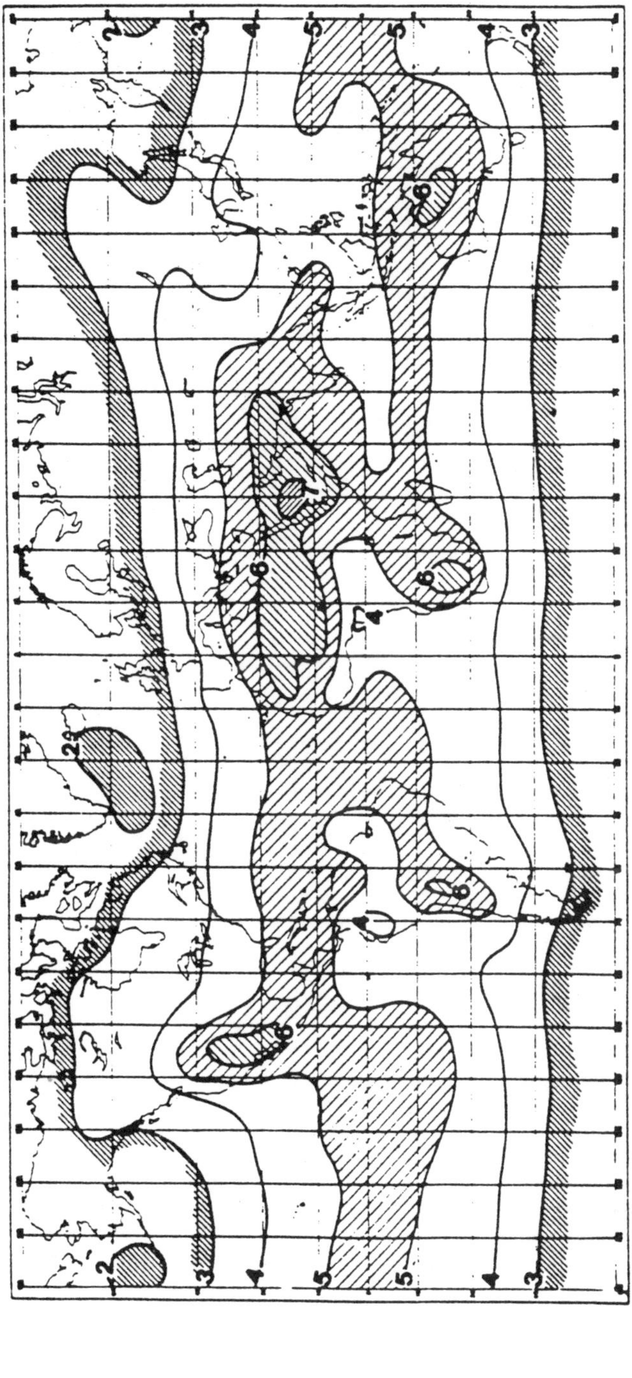

Figure 2.2 *Map of annual average insolation (1 kWh/m²/day)*

Source: World Meteorological Organisation (reproduced with permission).

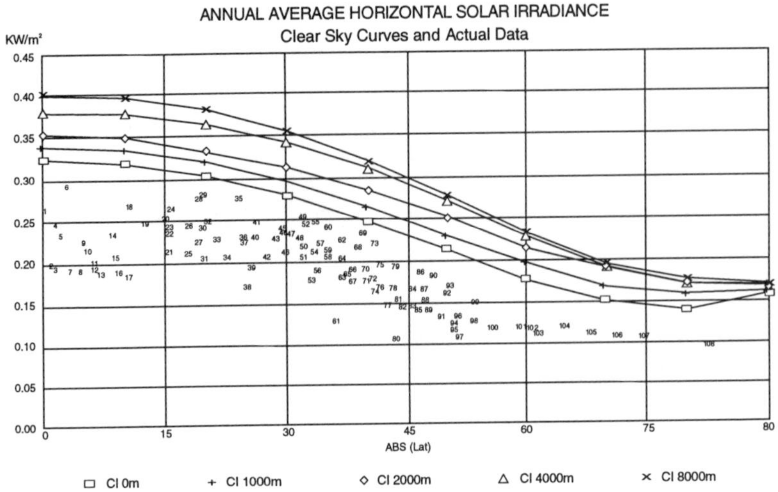

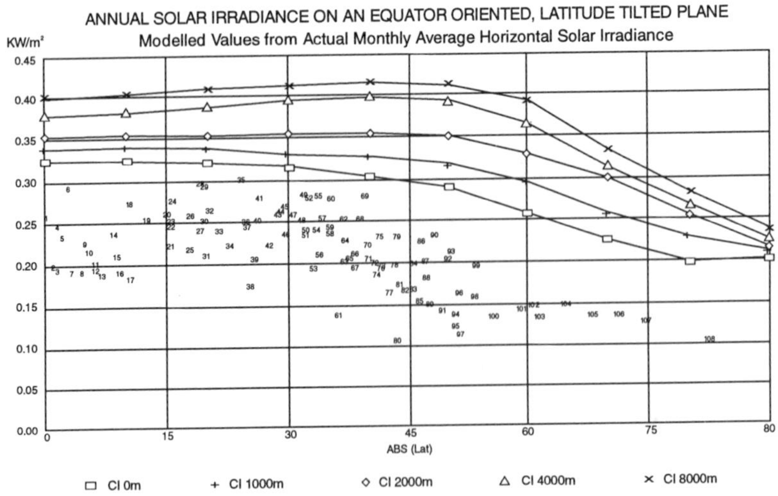

Figure 2.3 *Horizontal and latitude-tilted plane annual average solar irradiance*

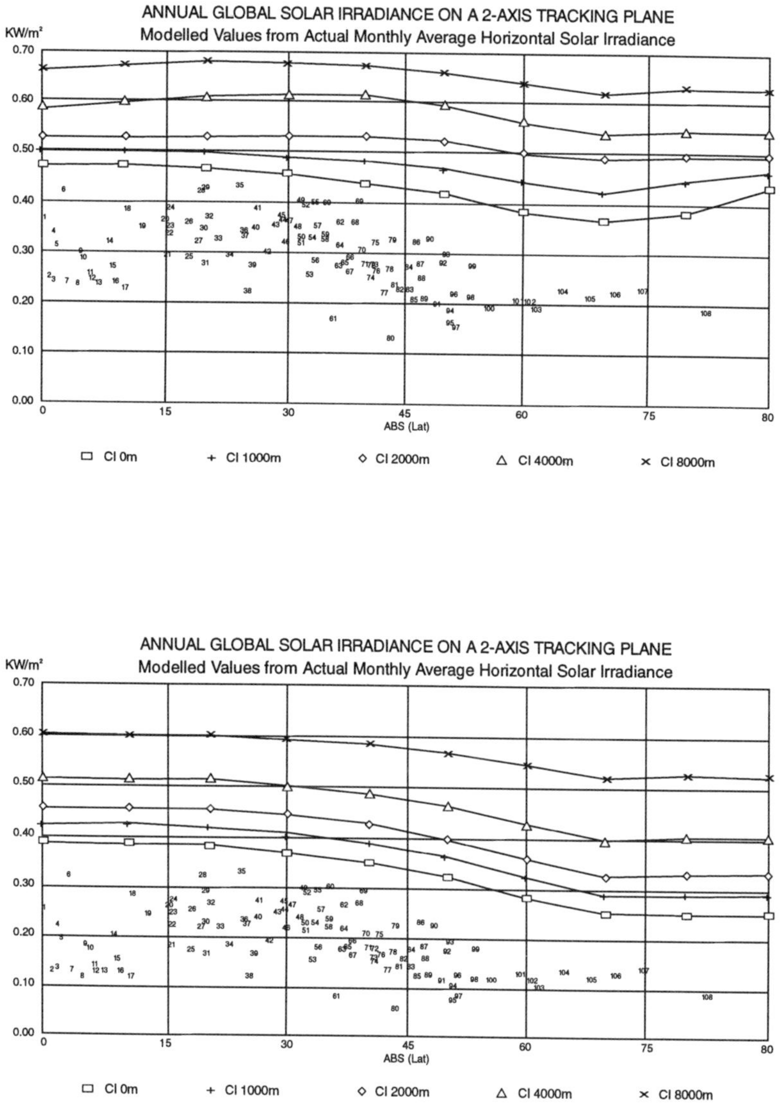

Figure 2.4 *2-axis tracking plane annual average solar irradiance*

Table 2.2 *Annual average solar irradiance (kW/m²)*

N	*Site*	*Lat*	*Long*	*h*	*Htot*	*ref*	*CI*	*df*	*ELTot*	*2ATot*	*2ADir*
1	Kisumu	0.1	33.8	1134	0.262	0.20	77%	35%	0.262	0.364	0.244
2	Yangambi	0.8	24.5	487	0.199	0.20	59%	53%	0.199	0.250	0.134
3	Singapore	1.3	104.0	10	0.195	0.15	60%	52%	0.195	0.245	0.134
4	Kiribati	1.4	172.9	0	0.247	0.20	75%	37%	0.248	0.339	0.222
5	Lodwar	-1.9	35.6	500	0.235	0.20	70%	42%	0.236	0.316	0.196
6	Canton Isl.	-2.8	-171.7	4	0.294	0.20	89%	23%	0.294	0.422	0.321
7	Kuala Lumpur	3.1	101.6	19	0.192	0.15	59%	53%	0.192	0.240	0.129
8	Cameron HL	4.5	101.4	1689	0.191	0.20	55%	56%	0.191	0.236	0.120
9	Juba	4.9	31.6	460	0.226	0.20	68%	44%	0.228	0.301	0.181
10	Penang	5.3	100.3	3	0.217	0.15	67%	44%	0.219	0.287	0.174
11	Kota Bahru	6.2	102.3	5	0.201	0.20	61%	51%	0.201	0.255	0.141
12	Zanzibar	-6.2	39.3	0	0.195	0.20	59%	53%	0.195	0.245	0.130
13	Ibadan	7.0	4.0	50	0.188	0.15	58%	53%	0.190	0.237	0.129
14	Trivandrum	8.5	76.9	61	0.235	0.20	72%	42%	0.239	0.319	0.198
15	La Chorrera	8.9	-79.8	94	0.208	0.20	64%	48%	0.211	0.272	0.156
16	Lago Gatún	9.2	-79.9	90	0.189	0.20	58%	53%	0.192	0.241	0.130
17	Caracas	10.4	-66.8	1225	0.183	0.15	55%	56%	0.185	0.228	0.118
18	Pto. Cabello	10.5	-68.0	2	0.268	0.20	83%	30%	0.275	0.385	0.275
19	Darwin	-12.4	130.5	30	0.247	0.15	77%	33%	0.256	0.351	0.244
20	Dakar	15.0	-17.0	0	0.253	0.15	80%	31%	0.263	0.365	0.262
21	Lusaka	-15.3	27.5	1154	0.214	0.15	65%	44%	0.224	0.295	0.182
22	Goa	15.5	73.8	55	0.236	0.20	74%	37%	0.250	0.337	0.225
23	Sana'a	15.5	44.2	2210	0.243	0.15	72%	34%	0.254	0.351	0.246
24	Khartum	15.6	32.5	380	0.265	0.25	82%	31%	0.278	0.389	0.273
25	Nandi, Fiji	-17.8	177.5	0	0.212	0.20	67%	45%	0.220	0.290	0.174
26	Salisbury	-17.8	31.0	1471	0.245	0.15	75%	34%	0.260	0.359	0.251
27	Bombay	19.0	72.8	0	0.225	0.15	73%	38%	0.242	0.325	0.217
28	Dongola	19.2	30.5	225	0.278	0.20	89%	24%	0.297	0.423	0.324
29	Mauna Loa	19.5	-155.6	3399	0.281	0.37	80%	34%	0.295	0.425	0.287
30	Port Sudan	19.6	37.2	5	0.242	0.25	77%	37%	0.254	0.349	0.230
31	Hilo, Hawai	19.7	-155.1	12	0.205	0.20	67%	46%	0.212	0.279	0.167
32	Bulawayo	-20.2	28.6	1343	0.248	0.15	77%	31%	0.266	0.370	0.265
33	Honolulu	21.3	-155.8	0	0.227	0.15	75%	37%	0.241	0.328	0.219
34	Calcutta	22.5	88.5	6	0.206	0.15	69%	45%	0.224	0.294	0.181
35	Aswan	24.0	32.8	200	0.277	0.25	91%	23%	0.302	0.433	0.331
36	Riyadh	24.5	46.0	550	0.230	0.15	76%	35%	0.249	0.343	0.234
37	Karachi	24.6	67.0	0	0.223	0.15	76%	36%	0.246	0.334	0.228
38	Cananéia	-25.0	-48.0	10	0.170	0.15	57%	54%	0.174	0.221	0.121
39	Miami	25.6	-80.5	2	0.194	0.15	66%	46%	0.208	0.272	0.165
40	Pretoria	-26.0	27.5	1418	0.230	0.15	74%	36%	0.254	0.348	0.238
41	Jodhpur	26.3	73.0	224	0.248	0.25	83%	32%	0.280	0.388	0.275
42	Brisbane	-27.4	153.1	0	0.206	0.15	71%	41%	0.223	0.300	0.192
43	Delhi	28.6	77.2	216	0.228	0.15	79%	34%	0.260	0.356	0.249
44	Maseru	-29.3	27.0	1571	0.234	0.15	77%	34%	0.262	0.364	0.255
45	Kuwait	29.3	48.0	20	0.238	0.20	84%	30%	0.268	0.375	0.270
46	Gainsville	29.7	-82.4	59	0.212	0.15	75%	37%	0.236	0.321	0.216
47	Cairo	30.1	31.1	112	0.233	0.15	83%	30%	0.259	0.362	0.262
48	Mersa Matruh	31.3	27.2	20	0.228	0.25	81%	33%	0.253	0.352	0.242
49	El Paso	31.8	-106.4	1194	0.250	0.25	86%	28%	0.284	0.405	0.295
50	Perth	-31.9	116.0	0	0.217	0.15	78%	34%	0.239	0.331	0.230
51	Lahore	32.0	74.0	150	0.206	0.15	74%	38%	0.234	0.318	0.213
52	Tucson	32.2	-111.0	744	0.245	0.25	86%	28%	0.282	0.400	0.290
53	Charleston	32.9	-79.9	18	0.177	0.15	65%	43%	0.195	0.259	0.166
54	Baghdad	33.2	44.2	34	0.211	0.15	78%	34%	0.239	0.330	0.228
55	Phoenix	33.4	-112.1	347	0.246	0.25	88%	27%	0.282	0.401	0.295
56	Atlanta	33.7	-84.4	310	0.189	0.15	69%	43%	0.211	0.283	0.179
57	Los Angeles	33.9	-118.4	38	0.220	0.15	82%	31%	0.254	0.353	0.253
58	Canberra	-34.9	149.0	177	0.206	0.15	76%	36%	0.236	0.325	0.220
59	Sta. María	34.9	-120.4	71	0.211	0.15	79%	32%	0.242	0.336	0.238
60	Albuquerque	35.1	-106.7	1620	0.240	0.17	85%	26%	0.280	0.400	0.300
61	Tokyo	36.0	139.5	0	0.127	0.15	48%	64%	0.138	0.167	0.082

Table 2.2 *Contd.*

N	*Site*	*Lat*	*Long*	*h*	*Htot*	*ref*	*Cl*	*df*	*ELTot*	*2ATot*	*2ADir*
62	Fresno	36.8	-119.8	110	0.225	0.23	85%	28%	0.254	0.362	0.266
63	Marmaris	36.8	28.0	0	0.181	0.15	70%	42%	0.201	0.273	0.176
64	PSA, Tabernas	36.8	-2.5	500	0.198	0.20	74%	35%	0.228	0.315	0.218
65	Sevilla	37.4	-6.0	30	0.183	0.15	71%	41%	0.206	0.280	0.181
66	Athens	38.0	23.6	21	0.187	0.15	73%	39%	0.213	0.291	0.193
67	Columbia	38.0	-92.0	248	0.175	0.15	68%	44%	0.194	0.262	0.165
68	Lisboa	38.8	-9.1	0	0.216	0.15	86%	27%	0.254	0.361	0.269
69	Ely, Nevada	39.3	-114.9	1914	0.232	0.25	85%	28%	0.282	0.404	0.295
70	Palma	39.6	2.7	28	0.188	0.15	76%	37%	0.223	0.306	0.207
71	Mahón	39.9	4.2	87	0.176	0.15	71%	42%	0.206	0.279	0.179
72	Madrid	40.4	-3.6	667	0.178	0.18	70%	42%	0.201	0.275	0.178
73	Salt Lake City	40.7	-111.9	1306	0.220	0.22	84%	28%	0.260	0.374	0.275
74	New York	40.7	-74.2	57	0.163	0.20	66%	47%	0.184	0.246	0.149
75	Laramie	41.3	-105.6	2172	0.195	0.30	72%	43%	0.234	0.323	0.204
76	Barcelona	41.4	2.1	95	0.167	0.15	69%	44%	0.194	0.262	0.166
77	Boston	42.4	-71.4	194	0.146	0.15	61%	46%	0.166	0.221	0.141
78	Madison	43.0	-90.0	271	0.166	0.15	69%	43%	0.197	0.267	0.171
79	Macerata	43.3	13.5	338	0.192	0.24	79%	35%	0.233	0.326	0.222
80	Bilbao	43.3	-2.9	32	0.104	0.15	44%	70%	0.108	0.130	0.058
81	Toronto	43.8	-79.6	135	0.153	0.19	65%	47%	0.174	0.235	0.144
82	Genova	44.4	9.0	55	0.144	0.05	64%	41%	0.167	0.227	0.155
83	Montreal	45.4	-73.6	60	0.145	0.17	63%	46%	0.169	0.227	0.144
84	Ottawa	45.5	-75.6	115	0.164	0.18	72%	42%	0.198	0.272	0.177
85	Genève	46.2	6.0	380	0.139	0.20	61%	51%	0.153	0.206	0.122
86	Richland	46.6	-119.6	223	0.185	0.18	82%	31%	0.225	0.322	0.232
87	Bismark	47.0	-102.0	502	0.165	0.19	73%	39%	0.202	0.281	0.191
88	Jassy	47.2	27.6	102	0.154	0.18	70%	44%	0.182	0.250	0.160
89	Seattle	47.5	-122.0	122	0.139	0.18	63%	49%	0.151	0.208	0.126
90	Glasgow	48.2	-106.6	699	0.180	0.22	80%	35%	0.236	0.332	0.231
91	Vancouver	49.3	-123.2	94	0.131	0.20	61%	51%	0.143	0.196	0.116
92	Winnipeg	49.9	-97.2	250	0.159	0.31	73%	45%	0.204	0.281	0.174
93	Suffield	50.3	-111.2	777	0.168	0.32	76%	41%	0.215	0.301	0.196
94	Bonn	50.8	7.1	120	0.122	0.20	59%	55%	0.137	0.184	0.105
95	Brüxelles	50.9	4.2	105	0.116	0.30	54%	63%	0.123	0.161	0.080
96	Mussonee	51.3	-80.7	34	0.131	0.26	63%	52%	0.163	0.218	0.130
97	London	51.5	-0.5	70	0.106	0.20	52%	58%	0.114	0.152	0.084
98	Goose Bay	53.3	-60.5	10	0.125	0.31	62%	54%	0.157	0.210	0.120
99	Edmonton	53.6	-113.5	677	0.149	0.25	73%	42%	0.196	0.275	0.179
100	Copenhagen	55.7	12.3	30	0.116	0.22	62%	49%	0.134	0.189	0.118
101	Stockholm	59.4	18.0	12	0.117	0.25	68%	47%	0.141	0.206	0.130
102	Bethel	60.8	-161.8	49	0.115	0.43	65%	53%	0.147	0.206	0.119
103	Palmer	61.6	-149.1	69	0.109	0.43	62%	57%	0.134	0.189	0.105
104	Fairbanks	64.8	-147.9	138	0.118	0.47	71%	48%	0.149	0.226	0.137
105	Aklavik	68.2	-135.0	122	0.109	0.52	69%	52%	0.134	0.211	0.123
106	Barrow	71.3	-156.8	16	0.106	0.52	72%	50%	0.135	0.221	0.132
107	Resolute	74.7	-95.0	64	0.103	0.57	73%	49%	0.127	0.228	0.141
108	Alert	82.5	-64.0	0	0.093	0.68	65%	67%	0.103	0.189	0.093

Notes:

1. All solar irradiances are the 24-hour annual average, in kW/m^2.
   H refers to the horizontal plane.
   EL refers to the equator-oriented, latitude-tilted plane.
   2A refers to the 2-axis tracking, or normal, plane.
   Tot refers to the total, or global, irradiance.
   Dir refers to the direct, or beam, irradiance.
2. ref = ground reflectivity, or albedo.
   Cl = sky clearness (ratio of actual to clear sky horizontal irradiance).
   df = diffuse fraction (ratio of diffuse to global horizontal irradiance).
   Lat = latitude (degrees).
   Long = longitude (degrees).
   h = site height (m).

house gases in the upper atmosphere may result in an increased absorption leading to global warming and cloudier weather. While ozone depletion has little effect for solar collectors, the greenhouse effect would increase the diffuse fraction and solar concentrators might be seriously hampered.

In conclusion, most of the reliable data available on the solar resource has been gathered with ground-based solar measurement instruments. This method is expensive and data is not available in many regions. Approximate methods, based on weather records, are used to predict the resource in these areas. The lack of good resource data is therefore a constraint for solar energy development in many areas of the world. Satellite measurements are expected to increase greatly the availability of reliable solar data world-wide.

CURRENT SOLAR TECHNOLOGY – DESCRIPTION AND STATUS

Because solar energy is an integral part of everyday life on earth, man has attempted to harness its power for useful purposes since the dawn of the Technological Age. A variety of technical approaches have been developed, many schemes have been demonstrated with a wide variation of success, and new commercial industries have been established. Although this section is not intended to be a complete discussion of these technology options, since there are many excellent references on this subject (eg *WEC Committee on Solar Power Report, 1989*[1] and *Solar Power Plants*[6]), it must include a summary review of the basics, their current status and momentum, and the important lessons learned in the process. This section presents such an overview of the technology options and their development status, and establishes a baseline for subsequent discussion.

The discussion is arranged by technology category, and includes discussion of the relevant applications. Table 2.3 presents an overview of the important technical characteristics of each category, as well as the status of the current production industry[7–11]. For each technology type, a specific example of a complete system was used to generate calculated cost and performance numbers, and ranges were given where necessary to show the amount of variation expected. Some values were estimated based on the best available data.

Solar Thermal Systems

This category refers to systems based on low-temperature thermal collectors that tap the solar resource for thermal end uses.

Table 2.3 *Summary of current solar system characteristics*

Solar technology category	*Specific technology type*	*Key physical characteristics of typical modular unit*	*Basic materials of construction*	*Development status*	*World-wide annual manufacturing capacity*	*Annual industry growth (%/yr)*	*World-wide installed & operating quantity*	*Rationale for current quantity*
Low temperature thermal	Metal flat-plate	1–3 sq m absorber/tubes/glazing	Aluminium/copper	Commercial	1,000,000 sq m	10%	>10,000,000 sq m	Incentives
	Plastic flat-plate	1–3 sq m integral absorber/glazing	Polymer/composite	Commercial	1,000,000 sq m	15%	>10,000,000 sq m	Incentives
	Low concentration	Evacuated tube/internal pipe	Glass/copper	Demonstrated	Low demand	N/A	Unknown	Incentives
	Greenhouse/crop drying	Glazing/frame	Wood/metal/glass	Commercial	Not estimated	Unknown	Unknown	Climate
Solar architecture	Glazing, mass storage	Integral with building design	Glass/masonry	Commercial	Not estimated	Unknown	Unknown	Pioneers
Thermal-electric	Solar pond	10 m deep/plastic lined/earth retained	Earth/plastic	Demonstrated	Low demand	N/A	>5 MWe	R&D
	Parabolic trough	80x concentrator reflector/vacuum tube	Metal/glass	Commercial	100 MWe	N/A	>350 MWe	Incentives
	Dish/Stirling engine	10 m dia reflector/focal point engine	Metal/glass	Prototype	N/A	N/A	>100 kWe	R&D
	Central receiver	>50 sq m hellostats/central tower	Metal/glass	Pilot plant	N/A	N/A	>15 MWe	R&D
Concentrating PV	Fresnel lens	Single crystal	Silicon/other	Demonstrated	<1 MWe	N/A	>10 MWe	Incentives
Non-concentrating PV	2nd generation – ingot	Single crystal	Silicon	Commercial	20 MWe	12%	120 MWe	Remote and consumer markets
	2nd generation – block	Poly or single crystal	Silicon	Commercial	15 MWe	25%	60 MWe	
	2nd generation – ribbon	Poly or single crystal	Silicon	Commercial	<1 MWe	20%	2 MWe	
	3rd generation – thin film	Amorphous or microcrystal	Silicon/other	Commercial	15 MWe	12%	80 MWe	

Flat-plate Collectors

These are the most widely used solar collector type. Their basic element is a plate which is heated by the global solar irradiance and transmits the heat to a circulating absorber fluid which is usually water or air. The plate is always dark in colour and may have a selective coating to maximize solar absorption. Rubber, plastic and metal plates provide for an increasing range of temperature outputs. The plate can be the only component of the collector but, to reach higher temperatures, it is normally sealed in an insulating box with a high performance glazing cover to achieve the greenhouse effect – visible irradiance enters through the cover but part of the longer wavelength infrared irradiance emitted by the heated plate is trapped inside.

Low-concentration Collectors

These concentrate the beam and part of the circumsolar diffuse irradiance. With their improved geometry and thermal insulation they achieve higher temperatures than flat-plate collectors. One example of these collectors uses evacuated glass tubes to house the absorber, with sealed pipe connections. They often include metal or glass reflectors to increase the concentration ratio. Planar reflectors may achieve a ratio of about 2, while compound parabolic, non-imaging reflectors may achieve a ratio of up to 10. With higher concentration ratios, higher peak temperatures and efficiencies are possible, but the annual collection may be lower because an increasing part of the circumsolar diffuse irradiance is lost.

The solar collectors are usually fixed on a plane oriented to the equator and tilted to an angle close to the latitude of the site to maximize the annual collection, although other tilts are possible to maximize the seasonal performance as mentioned in the previous section. Monthly tilt adjustments may be warranted for concentrating collectors.

System Applications

The system usually includes storage to let the solar heat be used overnight. If the circulating fluid is liquid the storage is an insulated tank, and if it is air, thermal mass of rock or concrete can be used but this is bulky and phase-change materials offer a better solution. But even with these more advanced materials it is still not practical to store heat for prolonged periods and therefore conventional energy backup is included in most systems to supplement the solar heat.

Single collector systems can use natural circulation based on the thermosyphon effect between the collector and storage, but larger systems require forced circulation with a pump or fan. These systems have often been designed for operation at high and constant speed flow, controlled by on-off switches driven by the temperature difference between collector and storage. But new systems operate at low and variable flow, proportional to the solar irradiance, to provide increased efficiency and reduced costs. This is the natural mode of operation of the small thermosyphon systems, and it is achieved in large systems with variable speed pumps or fans driven by a small photovoltaic module.

Solar thermal systems are mainly used commercially for water heating. Either swimming pool or service water for single family homes, apartment buildings, and hotels or other buildings in the commercial and services sectors can be easily heated by solar energy with low investment payback times. With an average annual efficiency of 40%, only 2 m^2 of collector area is required to provide 80% of the water heating demands for a family in Mediterranean weather. Larger – but still modest – areas are necessary in less sunny regions.

These systems can also provide a large fraction of the building space heating demand, but much larger areas are required resulting in impaired economics and building aesthetics. Commercial success has only been achieved aiming at lower solar heating fractions, with no-storage systems used for preheating building ventilation air. Energy is thus saved in three ways – by heating the air, by reducing heat loss through the walls, and by mixing the air in the building for more constant temperature. Newer, better insulated buildings have lower cost/performance ratio than typical structures in areas such as Canada[12].

The heat supplied by solar thermal collectors can be used to drive desiccant or absorption heat pump circuits to provide building space cooling. This is easier to implement with the higher temperatures of concentrating collectors, but their higher prices and the added price and complexity of the refrigeration equipment has meant that a commercial market has not yet developed. Research is now aiming at improving the efficiency of the cooling systems to allow the use of lower-cost flat-plate collectors.

Process Heat for Industry

This is another possible application for solar thermal systems. Concentrating collectors offer a wider scope of potential application in this field than flat-plate collectors due to their higher temperatures. But industry tends to be very energy intensive and the lack of space for solar collectors is very often an insurmountable constraint. Just as with the space heating applications, a limited commercial success has been achieved only with no-storage systems designed to supply a small part of the load.

Solar thermal systems can also provide process heat for agriculture. Greenhouses can increase crop yields significantly and extend the growing season in cold climates. Solar drying leads to a higher quality product which can endure longer storage before spoilage, and may attract a higher market value. Both greenhouses and drying facilities may operate simply by direct solar exposure, but adding flat-plate air collectors leads to much higher solar fractions together with more weather-independent operation. These systems offer a significant technical potential, but the limited amount of capital available in many rural areas is a major constraint for market development.

With the advent of the low and variable flow operation flat-plate collectors, system control has probably already reached full maturity. But

R&D on advanced materials still continues so that polymer and composite fibre collectors can achieve the higher efficiencies that are currently possible only with metal collectors. The new materials are also more amenable to greater cost reductions through mass production so that lower prices will be possible if a sizeable market develops.

The major constraint for market development is the investment needed for the solar system. Even with the low current conventional energy prices, investment payback times of only a few years are common. Although this is a very short time from the point of view of an energy utility, it is very often much too long for potential individual customers.

Solar Architecture

Solar architecture makes the building structure itself into a solar energy collection, distribution and storage system able to provide thermal comfort and daylighting for the occupants. Space heating is achieved by letting solar energy get into the building through large windows or sunspaces, or through thermal collectors integrated into the roof or facade. Space cooling is achieved by shading, ventilation and by evaporative, radiative, or ground fresh-air cooling. Daylighting is achieved by channelling sunlight deep into the building.

Most of the early work in solar architecture was aimed at space heating of private homes relying entirely on natural heat transfer effects. The term "passive solar" was coined and is still widely used. But many designs today rely on electronic controls, or even on "active solar" collectors, to achieve a better heat distribution and a higher solar fraction.

Since building techniques vary widely from region to region, and as the relative needs of heating, cooling, and daylighting are strongly affected by climate, appropriate solar architecture design tends to be very site-specific. It can nevertheless be stated that the incremental cost to take advantage of purely passive designs is minimal in areas where high mass walls and partitions, such as concrete or bricks, are the common building practice. The cost of the large sun-facing high performance glazings (with proper shading and cross ventilation facilities to avoid summer overheating) is partly compensated for by the reduced surface of the external walls, and part of the building thermal mass can be easily transformed into vented heat storage where the winter solar heat is stored for night use and the cool summer night air is stored for day use.

Good designs can thus cut the annual heating load by 80% in cold and sunny sites, and around 50% for more cloudy weather[13]. To achieve the same solar heating fractions in lightweight buildings, solar thermal collectors and dedicated storage are needed, resulting in higher costs. Proper shading and ventilation strategies in high thermal mass buildings can cut 80% of the annual cooling load if nights are cool enough, but otherwise more sophisticated technologies based on water evaporation or underground cooling tunnels are also needed. The daylighting effect provided by the large, high performance glazings results in little energy savings for private homes but can be very important for offices, where the reduced need for electric lighting also results in less cooling loads.

Simple solar architecture, based only on high performance glazings and good thermal insulation of the building envelope, is used commercially by some architects and is often considered as energy conservation. This is fair because existing buildings can be retrofitted by such measures, resulting in energy savings of about 25%. But the higher solar fractions mentioned above can only be achieved in new and carefully designed buildings, with thermal and daylighting computer modelling being an essential tool to optimize the building performance. Several computer software tools have been developed but a validated, comprehensive and easy to use program is not yet available. This is the main hurdle for a wider dissemination of solar architecture designs, together with the lack of general awareness that such designs can lead to buildings with very little energy requirements that are also more pleasant places in which to live and work[15].

Buildings represent very significant energy use in societies that have already gone beyond the industrial age. The hurdles to solar architecture should therefore be overcome, because it offers a real potential to enhance the environment. And this potential can even increase if the advanced technology developments that are currently taking place are successful and lead to commercial products. Transparent insulating materials, solid-state phase-change storage, sunlight pipes, and electrochromic, thermochromic or holographic switching glazing (to control heat and light as desired and independently in both directions) may find significant use in the future – eg zero energy buildings in developed countries.

Solar Thermal-electric Systems

This category refers to systems based on thermal collectors that tap the solar resource mainly or solely for electricity generation through a thermodynamic cycle. It can be done by low-temperature collectors, but it has been done mainly by higher temperature linear or circular concentrating collectors.

A low boiling point organic fluid is necessary if a thermodynamic cycle is to be driven by a low-temperature solar collector. But the low efficiency inherent to the low-temperature thermodynamic cycle has precluded the commercial use of flat-plate or even low-concentration collectors. For electric generation, low temperature solar energy has only shown promise with "solar ponds", which act as a combined nonconcentrating collector and storage tank. In the salt-gradient type, the salt concentration increases with depth, which overcomes the natural buoyancy of the heated water and allows higher temperatures to develop at the bottom. These systems may be of interest in high insolation areas where a natural pond already exists, or where land, water and salt are plentiful and inexpensive. Demonstrations have been built in Israel and a few other countries, and R&D continues aimed mainly at improving maintenance of desired pond characteristics under conditions such as wind-driven waves and wind-carried contaminants.

Parabolic Trough

The parabolic trough is the major type of linear concentrating system. The troughs consist of long rows of concentrators that are parabolic in cross section. A reflective inner lining on the trough focuses solar energy on to a black pipe that runs along the focus of the parabola. The troughs are usually mounted on a single-axis tracking system that follows the sun's elevation or azimuthal movement. A fluid, such as a heat transfer oil, is circulated through the focal pipe to collect the energy and transport it to supply either a process heat application or a power generating turbine cycle. The systems are commercially available from several suppliers, but Luz International is the largest and best known. The higher concentration of these systems compared to the low temperature systems discussed previously leads to higher temperatures and better efficiencies, but requires tracking of the sun, and can only use the direct insolation component. R&D on low-cost reflective materials and system reliability, as well as higher production volume, are expected to lower energy costs for these systems.

Since its founding in 1979, Luz has developed this technology, using more than US$1 billion of private capital and solar tax credits, and now has more than 350 MWe of operating power plants in Southern California. Several generations of design have evolved, and current systems heat oil to 440°C (735°F) in the focal pipe that is surrounded by a vacuum-insulated glass tube. The parabola-shaped glass mirrors use computerized controls to track the sun. An oil-heated steam generator and a high-performance but conventional Rankine steam cycle are used, with a natural gas-fired superheater and backup heat supply to assure maximum capacity payments for the electricity. Luz was developing a direct steam-cooled collector, and planning facilities in a number of other countries, prior to filing for bankruptcy in 1991.

By concentrating in two dimensions rather than one, circular concentrators can achieve higher concentrations and temperatures than the linear collectors. The peak energy density of these systems is comparable to conventional combustion systems and is able to serve many of the same applications.

Parabolic Dish

A modular version known as the parabolic dish is a paraboloid of revolution and has a single point of focus. To be effective this system must aim directly at the sun at all times and requires a two-axis tracking mechanism. Thermal energy can be removed from the focal area by an appropriate working fluid and either routed to a remote thermodynamic cycle or used in a small engine (about 25 kWe) mounted behind the focal point. Stirling engines have been under development for this application, and Brayton and Rankine engines have also been evaluated. Full scale prototypes of these dish-electric systems have been built and tested. The dish-Stirling combination has achieved a sunlight to electric conversion efficiency of almost 30% under actual field conditions. R&D on Stirling engines, focal-point heat exchangers, and low-cost reflective surfaces is needed to improve long-term efficiency and cost effectiveness of this system.

The "central receiver" is the equivalent of a very large parabolic dish. A field of individually focused mirrors called heliostats reflects and concentrates solar energy on a tower-mounted heat exchanger called the receiver. A computer controls each heliostat so that it bisects the angle between the sun and the receiver on a continuing basis. The size and temperature of these systems can easily be comparable to industrial and power boilers, up to about 200 MWe equivalent with a 50% annual capacity factor, and conventional power generation equipment can be used. A 10 MWe pilot plant that utilized water/steam as the working fluid has been built and tested, and several other smaller facilities have been built. Prototype components for second-generation systems based on molten sodium/potassium-nitrate salt as the working fluid have been built and tested. R&D continues on advanced receivers and storage materials as well as reflective surfaces. A plan to retrofit the second-generation (salt) technology to the 10 MWe pilot plant is now underway.

Solar Chimney

A related system, much simpler but also with much lower efficiency, is the solar chimney. The circular field of heliostats is replaced by a circular area of land covered with glazing, and the central receiver tower is replaced by a chimney that houses a wind turbine. Solar heated air below the glazing is drawn up through the chimney driving the electricity generating turbine. A 100 kWe prototype has been built in Spain.

Thermal energy generated by solar ponds can also be used for building space heating and cooling through district lines, and the higher temperatures reached by concentrators can also be used to provide process heat for industry with thermal storage in tanks. If lack of space was not a constraint, these systems could provide up to 80% of the annual thermal load in very sunny sites and perhaps 50% in less favourable weather areas, with backup conventional energy to supply the rest of the load. But thermal energy prices are lower than electricity prices and a commercial market has never developed. The drop in oil prices that followed the oil crisis has even avoided commercial success for the systems designed solely to generate electricity. The US parabolic trough deployment lasted while tax credits lasted, and the circular concentrator systems have not yet gone beyond the demonstration stage. While energy prices remain low, the very high photon flux that linear and circular concentrators can provide may find use in detoxification applications (the "Thermochemical and Photochemical Systems" section below).

Photovoltaic Systems

Photovoltaic (PV) systems, developed originally for space applications, transform light directly into electricity. Their basic principle is the photoelectric effect, first explained by Einstein, whereby light makes electrons emerge from matter. Photovoltaic devices – solar cells – are flat crystals made of thin layers of semiconductor material with different electronic properties resulting in strong built-in electric fields. When light enters the crystal, photo-generated electrons are separated by these fields and an electric potential develops between the top and bottom faces of the cell. This results in a direct current if the circuit is completed.

To protect them from the environment, PV cells are linked together and encapsulated in modules. Modules mounted on a plane with the proper orientation and tilt for maximum yearly or seasonal collection form the PV panel or array. Either a single module panel or huge array fields are possible, with a wide range of DC voltages which are further turned into any desired DC or AC form by solid-state electronic power conditioning. The PV system typically includes electrochemical storage batteries for stand-alone applications.

PV development for terrestrial applications began at the time of the first oil crisis along two radically different paths. One is aimed at concentration technologies where cost reduction can be achieved by replacing PV area with lens area, and the other one is aimed at cost reduction of the PV modules by high-volume industrial fabrication.

The main goal of concentration technologies R&D is achieving higher efficiencies. Silicon point-contact cells have already reached a 30% peak efficiency. Even higher figures can be achieved by stacking silicon and different gallium-based materials (or other semiconductor materials like indium phosphide) to form a multijunction cell, where every layer collects a different part of the solar frequency spectrum. The record so far with this approach is 37%. The cells are encapsuled in high (around 100 suns) or very high (up to 1000 suns) concentration modules, made usually with Fresnel lenses. Cell efficiency declines with increasing cell temperature, and high concentrators need at least passive cooling while very high concentrators need active cooling which can also provide a source of low temperature thermal energy. Extremely precise two-axis tracking mounts are required to maintain focus on the cells, and it has also been proposed to place them on a tower receiving the beam irradiance reflected by a heliostat field.

This would be very similar to the central receiver thermal-electric system, but all high-concentration PV systems are actually similar to this system in two ways: mechanical complexity makes them only valid for central power stations and inability to use diffuse irradiance limits potential sites for high efficiency to very high insolation areas. And just as with the solar thermal-electric technologies, a real commercial market has not yet developed. Low-concentration PV, based on non-imaging, luminescent or holographic concentrators, is also being pursued while the mainstream of the PV industry efforts are based in non-concentrating technologies to supply the world-wide stand-alone systems market that accepts higher electricity costs.

Second Generation Technologies

These are a direct descendant from the first generation space technology. Their ultimate goal is the replacement of all the original processes by high-volume industrial equivalents. Starting also from sand, or other silicon rich material, the multiple purification steps leading to semiconductor silicon would be replaced by a simpler process leading to solar silicon. Batch ingot growth from a molten bath has been replaced by continuous ingot growth, which is thereafter cut sidelong into a rectangular cross section block and then sliced into rectangular wafers. It has also been replaced by direct block solidification which avoids the sidelong cut, or by ribbon growth which leads directly to thinner rectangular wafers by simple cutting. Batch wafer thermal diffusion for junction and back-surface field formation is replaced by conveyor thermal diffusion, or even higher speed ion implantation and laser annealing. All wafer and cell surface repair treatments done by chemicals are to be replaced by their plasma equivalents. And, finally, manual cell encapsulation into modules is replaced by automated encapsulation.

By careful design, all the above processes can also result in high peak efficiencies. The higher values are obtained by single crystal cell modules, with 20% already achieved. But higher speed crystal growth is the leading criteria for lower cost, and this is the path chosen for the main block and ribbon technologies. It leads to polycrystalline cells with lower peak efficiencies due to losses at grain boundaries, but hydrogen plasma passivation allows for values very close to the single crystal figures and 17% efficient modules have already been achieved.

The above processes cannot be considered yet as mature second generation technology, mainly because semiconductor-grade silicon purchased from the electronic industry is still widely used. This leads to an energy payback time of almost 10 years, while solar silicon could lead to an energy payback time as low as 9 months. This has been demonstrated in the laboratory, but PV demand has not yet reached the needed size to warrant a market for the huge production that solar silicon factories would have. Increased demand would also foster application of ion implantation and plasma equipment, which have similarly been demonstrated but are still not widely used.

Third Generation Technologies

These technologies based on the deposition of thin film layers of semiconductor material on a metal, glass or plastic substrate are a radical departure from the second generation technologies. Here is where most PV R&D funds are allocated. A variety of different materials are under development, with hydrogenated amorphous silicon being the candidate that offers the greatest potential if its degradation problem with initial exposure to sunlight can be solved. Other promising thin film materials include copper indium diselenide and cadmium telluride. With very little material being used, energy payback time is only a few months, even with very low module efficiency. Area costs are also very low, but maximum peak module efficiencies achieved are only 10%, and therefore power costs are similar to the second generation modules. The potential for further area cost reductions is good because thin film deposition lends itself to high volume manufacturing better than second generation technologies. And the potential for further efficiency increases is even greater through the multijunction approach, either with amorphous silicon only or in combination with other materials.

Thermochemical and Photochemical Systems

This category refers to systems that tap the solar resource to induce chemical reactions, either for improving the quality of existing products so that they can be used or for the synthesis of entirely new ones. Thermochemical refers to the use of heat to drive reactions, whereas photochemical refers to the use of the photons directly, such as the ultraviolet portion of the solar spectrum.

The most basic element that can be improved by solar energy is water, which is only available as brackish or sea water in many areas of the world. For large communities where conventional energy supply is available, water desalination by thermal evaporation is an established technology and parabolic troughs can be used to reduce energy demand. For small isolated communities solar stills have been used for many years as the sole source of fresh water. A sloping transparent cover over a shallow pond induces a severe greenhouse effect, water evaporates and condenses on the cover, and is collected when it runs off. But solar stills have very low efficiency and high maintenance costs, and as PV costs are dropping PV driven reverse osmosis or electrodialysis desalination, often together with ultraviolet radiation units to kill bacteria, are becoming attractive for this application.

The use of solar energy for cooking food has been the focus of large government-sponsored programmes in India, China and several other areas, with about 1 million units manufactured. Double-glazed sealed box solar cookers, particularly if they include reflective concentrators to achieve higher temperatures, can cook many types of food adequately but only during sunny, daylight hours. These systems were intended to displace the vanishing fuelwood resources, but they require modification of food preparation and cooking habits, and efforts to introduce them in rural areas have been rather unsuccessful. Programmes based on improved charcoal or biogas cookstoves have probably a much better chance of commercial success.

Solar thermal concentrators can provide temperatures to match any chemical process. A main application envisaged is the steam reforming of methane reaction leading to synthesis gas, which is a major feedstock for the production of ammonia, methanol, most monomers for plastic, synthetic fuels and hydrogen. Other proposed applications include coal gasification and refinement of heavy oil, tar sands and oil shales. But, as has already been mentioned, while energy prices remain low commercial deployment of these systems will not be possible. Since solar concentrators can even achieve temperatures higher than possible with conventional means, some central receiver demonstration facilities built during the 1980s are actually used commercially for testing materials that must endure very high temperatures.

But the main near-term application for the very high photon flux that linear and circular concentrators can provide will probably be photocatalytic water detoxification. Hazardous chemicals can thus be broken down into carbon dioxide, water and easily neutralized acid, and this may be extremely valuable for cleaning up existing contaminated water stocks. Current catalysts can only use ultraviolet irradiance, and R&D is therefore aimed at improved catalysts able to use a larger fraction of the solar spectrum in order to increase system efficiency and reduce cost.

Hydrogen Generation

Hydrogen generation from solar energy and water deserves particular attention since this would be a fuel that is inexhaustible and also environmentally benign. When hydrogen is burned, either directly for thermal or mechanical end-uses, or in fuel cells for electricity generation, only water is released. Since hydrogen can be used both for transport and long-term energy storage, with a higher density than natural gas if used in liquid form, it could displace our dependence on fossil fuels. Concentrated sunlight could drive thermochemical reactions or high-temperature electrolysis for the generation of solar hydrogen. Solar hydrogen can also be obtained from photoelectrochemical systems, that yield hydrogen and oxygen directly from water. One recent development in this field that may deserve particular attention mimics the role of chlorophyll in photosynthesis by means of titanium dioxide particles coated with a ruthenium-based photosensitive dye[14]. However, both this and all other current photoelectrochemical research needs much further development to enhance device efficiency and solve degradation problems which still plague the solid-liquid interface. Water electrolysis driven by PV electricity has already been demonstrated and is probably the simplest way to obtain solar hydrogen.

It can be envisioned that countries in the Northern Hemisphere will have exploited their national solar energy potentials relatively sooner than the South, and that after that time they may want to import renewable energy. Solar-generated hydrogen might be the simplest way to do so, especially when motivated by the need to lessen pollution levels in urban areas with high environmental stresses (although this need is not limited to developed countries). This is at least the rationale of the only large-scale demonstration of hydrogen generation from water and a renewable energy resource, a joint Canadian and German project[16] where water electrolysis is driven by hydroelectricity. Hydrogen generated in

Canada would be shipped to Europe where it will be stored and used in different ways. At present only the aerospace industry makes use of stored energy in hydrogen. But R&D is taking place throughout the world on applications in other sectors, such as ultra low-emission vehicles, hydrogen-oxygen steam generators for peaking power, intercontinental liquid hydrogen transportation similar to LNG, hydrogen-oxygen catalysis and fuel cells.

Solar Technology Status

Table 2.4 summarizes the current technology status of the main solar systems that have been developed or are currently under development. Annual system efficiencies and energy costs have been calculated assuming deployment in the Albuquerque area, a desert environment at 35° latitude and 1600 m elevation in New Mexico (US) with 0.24 kW/m^2 annual insolation, 85% sky clearness and 26% diffuse fraction.

Latitude tilt has been assumed for the thermal systems, leading to 0.28 kW/m^2 annual global irradiance for the non-concentrating collectors and 0.23 kW/m^2 for the low-concentration collectors assuming that they can use the beam and 70% of the circumsolar diffuse irradiance. Solar ponds use the 0.24 kW/m^2 annual global horizontal irradiance, and parabolic troughs tracking the Sun in elevation around a fixed East–West axis receive 0.20 kW/m^2 annual beam irradiance. Two-axis tracking has been assumed for the other systems, leading to 0.30 kW/m^2 annual beam irradiance for the concentrating PV and other thermal-electric systems and 0.40 kW/m^2 annual global irradiance for the non-concentrating PV systems.

Since maintenance costs for most solar systems are very small, the cost of solar energy for other sites can be ratioed by the annual irradiance on the collector, taken directly or intrapolated from Table 2.2. But this would only be a rough approximation because system efficiency depends both on solar irradiance and site temperature.

It is important to note some critical differences between the different types of systems. Concentrating systems (which use only direct insolation) and those systems with high thermal inertia are more significantly impacted by variable weather, such as intermittent clouds, than are fixed low-temperature thermal or PV systems. Under varying insolation conditions, the thermal systems can suffer an additional efficiency loss because the heat losses are constant regardless of the insolation, and performance suffers much more at part load than for PV systems.

Thermal systems that incorporate thermodynamic cycles to convert thermal to electrical energy suffer the same energy losses as conventional power plants with cycle efficiencies in the 20–40% range for concentrating, high-temperature systems. Also, cooling water may be very limited in the desert areas where solar insolation is highest. PV systems suffer only relatively minor losses for the conversion from direct to alternating current, and require water only for occasional washing of the modules.

Table 2.4 *Summary of current solar technology status*

Solar technology category	*Specific technology type*	*Typical system size*	*System efficiency: sunlight to* Thermal	AC elect	*Land required (1000 sq m/system)*	*Thermal output temperature (C)*	*Typical useful lifetime (years)*	*Installed system capital cost (1990 US$)*	*Annual op. & maint. cost (1990 US$)*	*Cost of energy (no storage)* Thermal[3] ($/kWht)	AC Elect ($/kWhe)	*Coupled storage option*[1] [1]	*Percent change in energy cost to increase annual capacity factor by 50%*
Low temperature thermal	Metal flat-plate	<100 sq m	50–70%	N/A	0.1–0.4	50–70	15	300–1000/sq m	[2]	0.013–0.04	N/A	Tank	+10%
	Plastic flat-plate	<100 sq m	55–70%	N/A	0.1–0.4	30–35	10	75–250/sq m	[2]	0.005–0.3	N/A	Pool	N/A
	Low concentration	<100 sq m	35–45%	N/A	0.1–0.4	100–170	15	1000–2000/sq m	[2]	0.4–0.10	N/A	Tank	+5%
	Greenhouse/crop drying	<100 sq m	20–45%	N/A	<0.2	30–45	10–20	various	[2]	N/A	N/A	Mass	N/A
Solar architecture	Glazing, mass storage	<1000 sq m	N/A	N/A	N/A	N/A	30–50	various	various	N/A	N/A	Mass	N/A
Thermal-electric	Solar pond	<100,000 sq m	10–15%	2–4%	<100	70–90	20–30	4500–5000/kWe	20/kWe	0.015–0.04	0.24–0.27	Pond	–20%
	Parabolic trough	80 MWe	40–50%	14%	1700	<380	20	2500–3000/kWe	40–50/kWe	0.035–0.045	0.13–0.16	Tank	–10%
	Dish/stirling engine	5 MWe	75–80%	24%	80	<700	20	5000–6000/kWe	150–200/kWe	N/A	0.23–0.28	Battery	+10%
	Central receiver	>50 MWe	40–50%	15%	1300	<700	20	3000–4000/kWe	40–50/kWe	0.035–0.045	0.16–0.18	Tank	–20%
Concentrating PV	Fresnel lens	<5 MWe	N/A	13–15%	130–150	N/A	20	8000–10000/kWe	12–20/kWe	N/A	0.34–0.43	Battery	+20%
Non-concentrating PV[5]	2nd generation – ingot	<5 MWe	N/A	11–12%	85–160	N/A	20	6000–8500/KWe	10–15/kWe	N/A	0.28–0.31	Battery	+20%
	2nd generation – block	<5 MWe	N/A	9–10%	100–200	N/A	20	6000–9000/kWe	10–15/kWe	N/A	0.29–0.32	Battery	+20%
	2nd generation – ribbon	<5 MWe	N/A	10–11%	90–180	N/A	20	6000–9000/kWe	10–15/kWe	N/A	0.29–0.32	Battery	+20%
	3rd generation – thin film	<5 MWe	N/A	4–5%	225–380	N/A	20	5000–7500/kWe	25–35/kWe	N/A	0.25–0.28	Battery	+20%

Notes: 1. Other columns (to the left) apply to systems without energy storage.
2. Annual costs cover a wide range, but are generally low.
3. Temperature varies with system type.
4. Annual system efficiencies and resulting energy costs refer to Albuquerque, NM, USA site. Refer to the "Solar Technology Status" section for application to other sites.
5. For non-concentrating PV, the wide range in cost is due to the difference between 2-axis tracking and fixed supports. The tracking systems have higher cost but also higher energy capture.

On the other hand, thermal systems have an advantage compared to fixed or concentrating PV systems for applications with capacity factors higher than about 25% (and non-concentrating, 2-axis tracking systems greater than 35%). Thermal energy storage, using insulated tanks, adds only a small incremental capital cost for the thermal systems and results in a reduction in energy cost. The equivalent battery system to store electrical energy for PV systems is a major additional cost and adds an efficiency penalty. In Table 2.4, this difference is illustrated in the last column, which shows the change in energy cost resulting from increasing the capacity factor from no storage (in the other parts of the table) to 50% on an annual basis. The difference between the direct electric and thermal-electric systems is due both to the much lower cost/kwh of tanks compared to batteries and the better utilization factor on the equipment of the thermodynamic cycle for the thermal systems with storage. However, thermal storage can be maintained only a relatively short time, while batteries can hold a charge for an extended period.

It is important to keep in mind, however, that solar systems have a direct relationship between design capacity factor and capital cost, unlike conventional systems which can simply burn more fuel with the same plant equipment to increase the capacity factor. Although back-up heaters based on fossil fuels can be used to maintain operation of thermal systems during periods of low insolation, their use is not included in Table 2.4. In the current market these fossil heaters usually represent a small incremental capital cost and may be important to overall commercial viability, since they can provide firm capacity.

Thermochemical and photochemical systems are not included explicitly because the only actual commercial systems, solar stills and solar cookers, will probably not impact overall energy usage. The direct generation of hydrogen from water by photoelectrochemical systems is in a very early research stage, and costs are not well known.

Current Markets

The status of the solar commercial markets is determined by several factors, but the most important is the relative availability and cost of the conventional options for meeting the particular energy need at each specific site. Gone are the times of the oil crises when the general belief in dramatic and continued oil price increases led to very short payback times for solar energy systems. Today, with the low current conventional energy prices and projections of moderate increases, without recognition of the external effects of energy decisions, the payback for solar systems are much longer and the potential markets much more difficult to address.

Markets continue to exist nevertheless for low-temperature solar thermal systems where the cost of conventional energy is relatively high due to taxes or where the government incentives to encourage the use of solar energy have not been removed. So far the majority of sales are in the developed countries, and limited mostly to solar water heating systems, mainly in homes and hotels. Solar space heating systems have been demonstrated but are not generally commercial due to their long payback time for the limited heating season in most climates. Markets for space cooling have also not developed for the same reason, which is aggravated due to the added cost of the absorption refrigeration equipment. Space heating and cooling, which are the major thermal energy demand, can be addressed best by solar architecture, which minimizes the energy needs by passive means and therefore reduces or even eliminates the need for active solar components. But, although it is often cost effective, it is not broadly used, due to limited awareness by design professionals and customers.

The needs for stand-alone electric power have been best addressed by photovoltaic systems. The main characteristic of these markets is that the energy requirement is small and therefore the cost of conventional electricity supply, either through grid extension or by diesel-generator sets, is very high. The photovoltaic systems, on the other hand, are less affected by economies of scale, and their low maintenance requirements are an added advantage. The different applications can be classified in three broad categories according to the end-user: professional, consumer products and rural. The first two have reached fully commercial status worldwide but their potential market is very small. The potential rural market is immense and it is being addressed mainly through co-operation programmes, of which the European Union photovoltaic programme in the African Sahel is an important example.

Solar systems have not achieved cost-effectiveness for grid-connected electric power – not even in the peaking power markets that allow a higher energy cost. The systems that have been built have required either R&D funding support or government incentives to overcome the higher capital cost. The deployment of natural gas-assisted solar parabolic trough power stations by Luz has been more successful, as mentioned above. One often stated reason for the Luz success is the modular nature of this technology, which allows an incremental construction and start-up of the plant, thereby reducing the time that the invested capital is unproductive. Nevertheless, the termination of the remaining tax credits in California has also led to financial problems for Luz. The grid-connected solar electric power market has been shifting away from the USA to Europe, and from solar thermal electric power stations to dispersed photovoltaic systems, like the current photovoltaic roof programmes in Germany (recently cancelled) and Switzerland.

It must be emphasized that the high capital cost of all solar systems is a major problem which curtails the development of expanded markets because, in many parts of the world, the concept of life-cycle cost is not applied for decision making. This practice automatically eliminates any chance for the solar system which always requires a high investment but relatively low annual operating costs, compared to the low investment and high operating costs of the conventional system. And even when life-cycle cost evaluation is used it is not applied to its full extent, which would require including all the environmental and other "external" costs of the competing energy options.

DEVELOPMENT TRENDS FOR SOLAR TECHNOLOGY

Development History

Since the oil price shock of 1973, governments of the developed countries have spent over US$ 4 billion on R&D for solar energy technologies. This is illustrated in Figure 2.5, which also shows that the level of governmental support peaked by 1980 and has been much lower since. Over the period 1979–89, the total investment in the developed countries on renewables (of which solar received about half) has been comparable to that spent on coal, and about 15% of the investment on nuclear energy technologies[17], as shown in Figure 2.6.

The largest portion of the funding for solar energy was directed at relatively high-technology approaches. Only a relatively minor fraction was directed towards technology specifically appropriate for the developing countries, which supplemented the small amount that those countries could invest on their own. Private investment from industry added to the total development progress, and the various estimates of the non-government investment in solar energy R&D are in the range of US$ 2 billion over the past 15 years. In contrast, industry has spent many times more than governments in conventional energy technology development.

This investment in solar energy was successful in establishing a very extensive body of information on the resource, technology options, and experience with the early systems. In some applications, it caused solar energy to make the transition to commercial use. However, the rapid rise

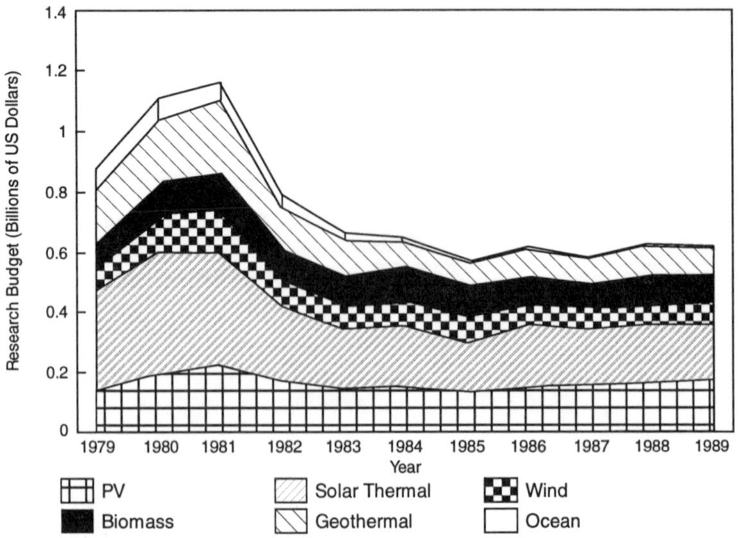

Figure 2.5 *Trends in renewable R&D funding in IEA countries*

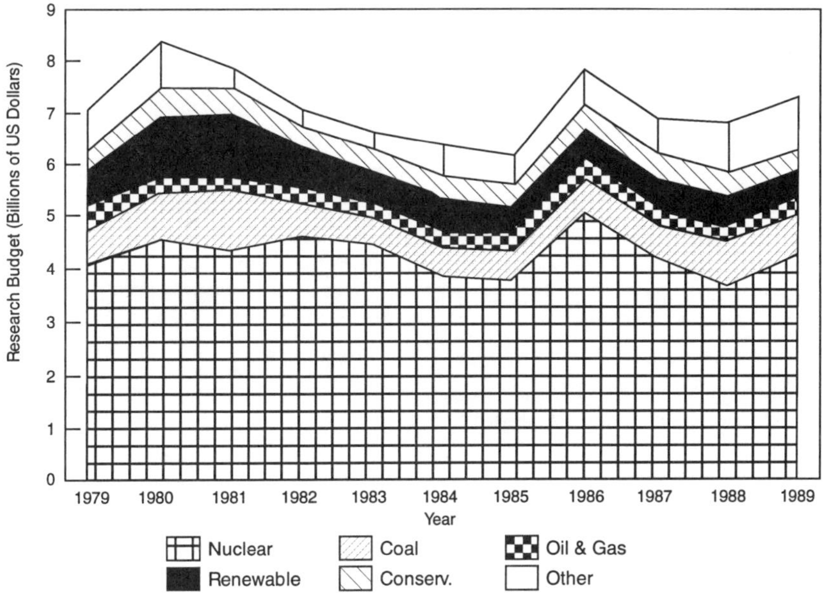

Figure 2.6 *Trends in energy funding in IEA countries*

and fall of government support in the form of R&D funding and market incentives due to the oil shocks had a major impact on the immature solar energy equipment and services industry. The initial growth was much more rapid than rational planning could support – in the order of 100% per year. During the shakeout that followed the collapse of government support, many companies left the industry and their employees went into other fields. There was also a customer backlash that resulted from the poor design of some of the systems that were installed, and poor maintenance by many of the owners, giving solar technology, especially the low-technology segment of the industry, a reputation for poor reliability. Of course, in the rapid industry growth phase there were many instances of people and companies overpromoting the benefits and economics of the technology, which raised the expectations of the general public beyond reasonable levels.

Current Trends

The core of the solar industry that survived the downturn of the 1980s has survived because it has found specialized markets for which solar is the economic choice and has taken maximum advantage of those incentives that remain in areas such as California. As the 1980s ended, there was renewed interest in developing renewable technologies in general and solar in particular, with the result that R&D support has now started to rise again. The majority is directed at photovoltaic cells and the technology is continuing to evolve rapidly. The reasons for this interest involves increased environmental concerns over the very high and rising penetration of the energy markets by fossil fuels.

The photovoltaic system market segment has been growing at 15% per year world-wide since 1985 due to viable markets for remote, small-scale power generation and power supplies for consumer electronics. The distribution of production for the past decade is shown in Figure 2.7.

Further PV growth could be spurred if electric utilities played a more active role in the stand-alone market. One electric utility has identified about 20 different remote applications ranging from water level sensors for mountain lakes above hydroelectric dams to mountain-top aircraft warning lights, and had more than 400 installations by 1989 totalling 32 kWe[18]. They have also determined that new single-family residences located more that 0.5 km from existing electric lines are more cost-effectively served in much of their territory by solar energy than extending the electric grid. Their analysis showed that a diesel-generator of equivalent capacity has a life-cycle cost almost twice that of the PV system. The economics differ significantly between California and the developing world, but the conclusions are similar. An analysis for rural areas of South Africa showed that PV systems are cost-effective for demands up to 4 kWh/day at distances more than 1 km from the grid, and had lower costs than for diesels with capacity factors less than 40%[19]. However, when only initial investment cost is considered, the diesel system is a clear choice, and the ability to finance the PV system without government assistance is very limited. Electric grid extension is economically justified for larger villages, if they are located within about 10 km from the existing system[20]. This points to an opportunity for the majority of the world's people that currently live without access to an electric grid.

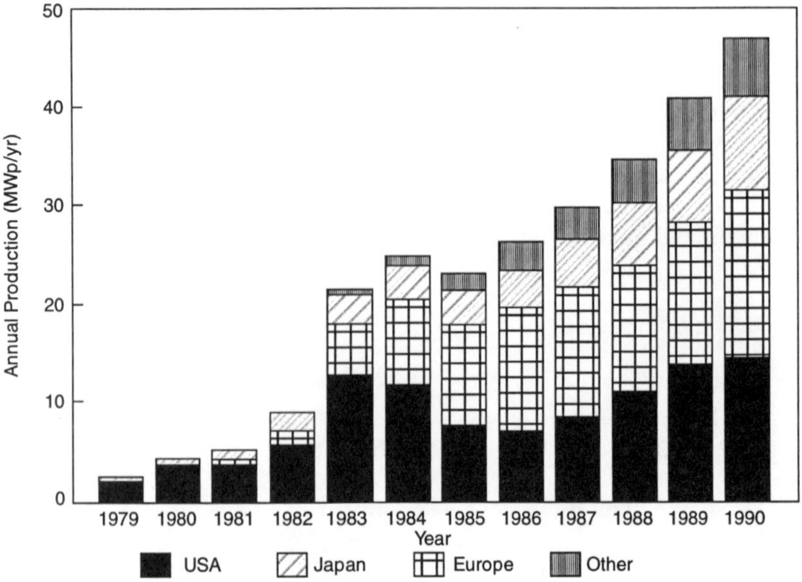

Figure 2.7 *Photovoltaic production by region*

Although it must be stressed again that some solar technologies are economically viable in selected markets and selected regions, several of the high-efficiency technologies discussed in the previous section currently represent unattractive investments to private industry due to the remaining funds needed to achieve commercial development, and the uncertainty of the timing of the potential return on the investment. Technologies that are not modular (most notable the central receiver which is not projected to be viable below about 50 MWe) appear to require government support during the establishment of the technical and market viability. They represent too large an investment risk for the private sector to carry alone, even though they show long-term economic potential for addressing the high-temperature thermal and grid-connected electric markets. Almost all types of solar technology require large investment in production facilities if they are to achieve their cost goals, which depend on mass production, and with the low current and uncertain future prices of oil and natural gas, these investments are seen as risky by the financial community. Such future price and supply uncertainty has also had a negative impact on efforts to encourage energy conservation, and this market is closely related to the market for passive solar architecture.

Potential for Technology Development

In order to understand the potential contribution of solar technology by 2020, it is required that an estimate of the potential for further technology development be prepared and the cost for such development be estimated. The current status of the technologies discussed earlier in this chapter is a good starting point for this task, and each of the types must be examined in the light of critical factors such as current research goals, extrapolated efficiency and material improvements, and volume-production cost effects. These are only estimates, because it is not possible to foresee all the technical breakthroughs, and the political and market structures of the future are quite uncertain. Table 2.5 presents the results of the estimating process, and is set out in parallel format to Table 2.4 for easy comparison.

Advanced glazing materials can result in higher performance metal and plastic flat-plate collectors. And better polymer and composite fibre materials will probably lead to plastic collectors achieving the same temperatures and efficiencies of their metal counterparts, with much lower costs possible through mass production. System efficiency improvements are expected with the general adoption of the low and variable flow operation, which is at present only sparsely used.

Developments for solar architecture may be much more dramatic. Mass produced advanced switching electrochromic, thermochromic or perhaps holographic glazings, combined with transparent insulation materials channelling solar heated air into wallboards with solid-state phase-change long-term storage, may lead to buildings that need little

Table 2.5 *Summary of projected solar technology status by 2020*

Solar technology category	*Specific technology type*	*System efficiency: sunlight to*		*Installed system capital cost (1990 US$)*	*Annual op. & maint. cost (1990 US$)*	*Cost of energy (no storage)*	
		Thermal	*AC elect*			*Thermal*[(3)] *($/kWht)*	*AC elect ($/kWhe)*
Low temperature thermal	Metal flat-plate	55–75%	N/A	200–400/sq m	[(2)]	0.01–0.02	N/A
	Plastic flat-plate	60–75%	N/A	50–200/sq m	[(2)]	0.004–0.025	N/A
	Low concentration	45–60%	N/A	500–1000/sq m	[(2)]	0.02–0.05	N/A
	Greenhouse/crop drying	30–50%	N/A	large range	[(2)]	N/A	N/A
Solar architecture	Glazing, mass storage	N/A	N/A	large range	[(2)]	N/A	N/A
Linear parabolic thermal	Solar pond	15–20%	5%	1500–2500/kWe	15/kWe	0.01–0.015	0.09–0.14
	Parabolic trough	45%	14%	1400–2000/kWe	40/kWe	0.02–0.03	0.08–0.11
	Dish/stirling engine	75&	28%	1300–2000/kWe	60/kWe	N/A	0.09–0.12
	Central receiver	50%	17%	1200–1500/kWe	35/kWe	0.015–0.02	0.07–0.09
Concentrating PV	Fresnel lens	N/A	18–25%	1700–2000/kWe	35/kWe	N/A	00.8–0.10
Non-concentrating PV	2nd generation – ingot	N/A	18–20%	1500–2100/kWe	5/kWe	N/A	0.06–0.09
	2nd generation – block	N/A	16–18%	1300–1800/kWe	5/kWe	N/A	0.05–0.08
	2nd generation – ribbon	N/A	16–18%	1200–1600/kWe	5/kWe	N/A	0.05–0.07
	3rd generation – single film	N/A	8–10%	1400–2000/kWe	10/kWe	N/A	0.06–0.08
	3rd generation – multiple film	N/A	15–20%	1000–1500/kWe	5/kWe	N/A	0.04–0.06

Notes: 1. Based on current status as presented in Tables 2.3 and 2.4. Items not repeated from Table 2.4 are the same.
2. Annual costs cover a very wide range, but are generally low.
3. Temperature varies with system type.
4. Annual system efficiencies and resulting energy costs refer to Albuquerque, NM, USA site. Refer to the "Solar Technology Status" for application to other sites.
5. Solar thermal systems delivering heat can improve their economics compared to the values above with the addition of appropriate amounts of thermal storage.

conventional energy at all for space heating and cooling. Efficient film light guides may pipe sunlight deep into indoor areas, filtering out non-desired wavelengths, and reduce the needs for electric lighting.

For the higher-temperature, concentrating thermal technologies, many parts of the systems have potential for significant improvement. Lower cost reflective surfaces using innovative structural support, such as stretched membranes, and improved materials for high efficiency receivers at high temperatures are key to their development. Improved system efficiency, better operating strategies and control, improved high-volume manufacturing and installation processes, and lower O&M costs with experience all contribute to significantly better overall performance and effectiveness. This development effort is expected to require about 10 to 20 years of consistent development, with longer development paths for the higher-temperature, higher-efficiency systems.

Mass production of Fresnel lenses can lead to low costs per unit area for the high concentration PV modules, and mass production of second generation technologies can lead to low area costs for the ingot, block, and ribbon flat-plate PV modules. If this is coupled with no further efficiency gains, but only with the maximum values achieved already by industry leaders translated into common general practice, the resulting power costs would be low and about the same for all these technologies. Mass production may also lead to low area costs for thin-film third generation technologies. But efficiency is a big issue here. Quantum physics can explain the PV effect in a single crystal but not in the amorphous state. A theoretical breakthrough is therefore possible, and could result in high and stable multijunction module efficiencies, and possibly lower power costs. Otherwise, translating only current maximum efficiency values into general practice as it has been done here for the other PV technologies would lead also to about the same power costs. The effect of lower cost, mass produced, structural panel support and better efficiency power conditioning is minor but has also been taken into account. Among the electrochemical batteries under development, polymer based and hydrogen based batteries deserve particular attention since they may lead to low storage costs (not included in Table 2.5).

It is also too early to ascertain whether photoelectrochemical research aimed at the direct generation and storage of hydrogen will be successful, and this is not included in Table 2.5. But even if it is not, hydrogen generated from solar energy may play a crucial role in shaping our future, but will probably not have a major impact by 2020.

Required Investment to Develop Solar Technology

In the light of the approximately US$6 billion that has been invested in the past 20 years to reach the level of technology development that was presented in the current technology section earlier in this chapter, the requirements to complete the development of solar technologies are substantial. The estimates presented in Table 2.6 are based on a variety

Table 2.6 *Estimated global investment to develop solar technology (US$ billions, 1990)*

Solar technology area	*R&D*	*Deployment*
Low-temperature thermal	1.0	1.0–1.5
High-temperature thermal	1.4	3.0–5.0
Photovoltaics	5.0	2.5–4.0
Various other technologies	0.6	0.4–2.0

of information from various government laboratories[10, 21], and indicate that the level of global investment is on the order of US$8 billion over the next 20 years, from both the government and private sectors. The breakdown is very approximate, with photovoltaics needing more than half because of the large amount of improvement over present day technology. The last column in the table represents the subsidies that are probably necessary to support initial deployment between the time that the technology is ready for widespread use and the time that the manufacturing economies-of-scale are achieved to compete with conventional options. This is more uncertain than the R&D estimate, and will depend on a variety of factors as seen by many different governments.

The estimated total of the investment in both categories is about US$15 to 20 billion, which is about 0.1% of the annual global GNP at the turn of the century. The amount that can be contributed by the private sector will depend on the speed of expected economic penetration, since recovery of that investment will be typically expected within the first few years of production, and will add to the capital cost of the components.

THE LONG-TERM POTENTIAL OF THE SOLAR RESOURCE TO MEET GLOBAL ENERGY DEMAND

The discussion of the solar resource at the start of this chapter included the fact that this resource is much larger than our total energy needs, while having characteristics that may limit its usefulness to directly replace our conventional energy resources that dominate current energy markets. Most studies of the long-term energy future conclude that solar energy will need to play a significant role, but *when* that role is likely to emerge is the essential question. These studies also vary in the paradigm for solar energy application, and therefore several points of view will be used to assess the upper limit of the potential contribution, even though it will not be achieved by the year 2020. Since later sections will address the constraints on achieving this potential, this analysis is based on what is expected to be technically possible, pending the further development discussed in the previous section, and the development of complementary non-solar technologies.

For the sake of illustration, three specific points of view will be discussed here, but they are by no means the only ones that could be considered. The first two envision major changes in concept, and are referred to as the centralized approach and the distributed approach. The third is based on an incremental approach to the current market structure.

Centralized Approach to a Solar Future

The centralized approach is based on a macro-scale view of the earth as an energy system. The solar energy variation at any point on the earth's surface could apparently be overcome by moving energy long distances with high efficiency. Within the next few decades, near-ambient temperature superconductors may be developed that can handle very high current. This would make it theoretically possible to link the six large solar resource areas of the earth in a massive electrical grid. Because these areas are widely spaced around the earth, as shown in Figure 2.8, it is conceivable that energy could be distributed whenever and wherever it was needed.

The combined land area of these high-solar regions – which are mostly arid desert with relatively little human, animal, plant or economic use – is about 36 million sq km, and they receive an annual average beam irradiance of 0.30 kW/m^2 on a 2-axis tracking plane. Future 200 MW central receiver thermal-electric plants are projected to have a ground coverage of 18% and therefore, assuming the 17% annual efficiency of Table 2.5, 5.4% of this desert land would be needed in order to cover the total world energy requirements in 2020, estimated to be about 13,500 Mtoe per year (18 TW) for the 1989 WEC moderate (M) scenario[22]. PV power stations, needing almost no water, can have a ground coverage of 24%. Assuming the 23% annual efficiency of Table 2.5, concentrating 2-axis or fixed flat-plate PV power stations could supply 18 TW using 3% of the desert lands, and 2-axis-tracking, non concentrating PV would need only 2.7%.

However, as it is generally believed that the thermal-electric technologies will be able to generate cost-effective electricity sooner than PV systems, a recent study for the Mediterranean region was undertaken to illustrate the near-term potential for this approach. The study examined the available land area and corresponding solar insolation, as well as comparing the technical and economic potential of solar thermal plants to address the growth in electric power demand for the region. The high-resource areas of Italy, Spain and North Africa were found to be capable of producing many times the total energy use of the region, and the economic potential was between 5 and 10% of the regional electricity demand by 2025. If the power could be economically moved to Northern Europe, a much larger contribution would be possible from this region by 2020[23].

A different variation of the centralized approach is the use of PV space power stations in geosynchronous orbit which beam microwave power to land-based rectennas. This potential solar future has received more publicity, and has the advantage that solar irradiance in space at the

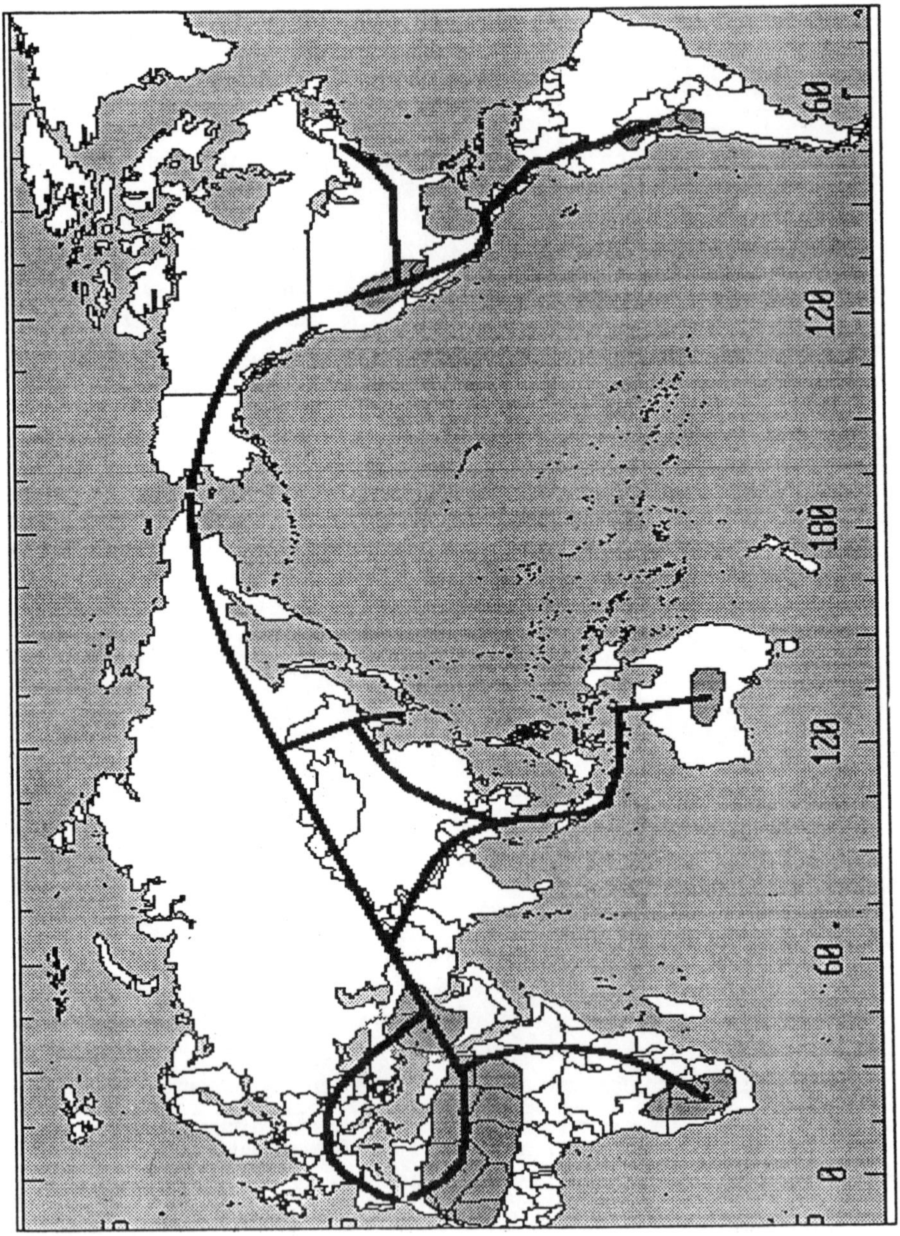

Figure 2.8 *Possible linkage of major solar resource areas*

Base map copyright 1991, Broderbund Software, Inc.

average Earth's distance from the Sun is 1.367 kW/m^2. Therefore, even allowing for losses due to conversion to microwaves and to reconversion to electricity, less global area would be needed than for desert PV deployment. But it also faces significant technical challenges related to space structures of very large size and massive energy transport through the atmosphere[1].

Another form of this concept is under investigation in Japan, which involves a network of very large floating structures in the Pacific Ocean. These would support thermal or photovoltaic collectors, and may involve ocean thermal systems as well. The energy could be transported by chemical means with tankers, or by submarine transmission lines to major Pacific Rim load centres[24].

For maximum efficiency, the centralized approach to a solar future would require that electricity was used for most energy needs. Fuels produced from solar energy, such as hydrogen, would also be required for applications where a storable and transportable energy source is required, such as air transport. It would also need large-scale energy storage technologies such as superconducting magnets to survive short-term interruptions due to weather. Of course, there are tremendous technical challenges to implementing this approach, and it would require the development of solar equipment manufacturing as one of the world's foremost industries. But the most difficult challenge may not be technical or economic but political; it would require an unprecedented level of international co-operation to be successful.

Distributed Approach to a Solar Future

The second approach evaluates the potential contribution of solar energy in the light of a restructuring of our basic development path for society. If the self-sufficient "village" is re-established as the basic modular unit of human life, the level of energy intensity and energy demand patterns can be tailored to maximize the potential contribution of solar and other renewable energy forms. This would require that the development patterns of many nations, which appear to be moving toward an urban, industrial lifestyle, be reversed.

This model can be discussed in several different types of society. In the major developed countries, urban living is giving way to the suburban lifestyle, with a more distributed work-force that is increasingly reluctant to relocate for employment, is less willing to commute long distances to urban jobs, is becoming linked via computer-based information transfer to many parts of the world, and is increasingly concerned with the quality of their life and the environment. In a striking way, this trend bears a resemblance to the rural lifestyle of the developing world, while being very different. In a sense, the path of development may appear circular, with technology making a return to an ancient pattern of living possible.

Similarly, many developing countries are interested in finding ways to reverse the movement from the rural, agricultural areas to the major cities. The development concept of dispersed, renewable-energy-driven communities in developing countries with access to the modern world through television and modern communications may be attractive. The ability to improve the health, education and welfare of the people without forcing them to abandon their traditions and values is a very powerful dream.

This "soft path", distributed approach attempts to address many of the major environmental concerns, mostly by reducing the local stresses of intensive human activity. If a heavy use of locally generated solar hydrogen was made to provide seasonal energy storage, this path could also lead to a 100% solar society. An overall collector area larger than in the desert or space centralized approach would of course be necessary. But the surface area required could be made available on rooftops and building facades is solar access became an important element in city planning. Other changes in human industrial and economic structure would need to be very significant. Neither the market-oriented nor the centrally planned nations appear to be moving in a direction to make this approach likely to succeed at a rate fast enough to prevent other development trends from dominating the next 30 years.

Incremental Approach

The third approach, and the most likely to receive attention from the important international and regional leaders, is to evaluate the potential of solar energy to penetrate the existing markets. This approach leads to the lowest of the three evaluations of the potential contribution of solar energy, being that which appears to be technically feasible in the current sectors of energy use, but still represents a valid upper limit consistent with the point of view.

It requires an extensive analysis of current and future projected energy demand data, and since a major planning simulation is beyond the scope of this book, an approximate approach is chosen to evaluate the solar energy potential. However, the accuracy of a more detailed analysis is more important for evaluating the specific costs and benefits for each national situation. For the world as a whole, the approximate solution is adequate to gain the important and critical perspectives.

For this approximate analysis, the world's nations were divided into 19 groups ranging from one to 40 countries each. The division was chosen to cover the range of different factors which are expected to impact the penetration of solar energy, from population density to per capita GDP to solar resource level. A range of data was assembled to represent these regions and is presented in Table 2.7.

The breakdown of final energy demand was developed and examined for each of the regions. A finer breakdown was desired to differentiate, for example, between agricultural use and urban residential use by type of fuel, but that type of data is not consistently available for all the countries.

Table 2.7 *Estimated long-term penetration with incremental approach*

Country/region	Area (1000 sq km)	Population (millions) 1990	Population (millions) 2000	Population (millions) 2020	Population data (1990) Growth (%/yr)	Population data (1990) Density (/sq km)	Population data (1990) Urban (%)	1990 GNP (US$) Per capita	1990 GNP (US$) Total (billions)	GNP growth (%/yr)	2020 GNP (US$) Per capita	2020 GNP (US$) Total (billions)
Canada	9,977	27	29	32	0.9	3	76	15,852	428	2.3	26,705	847
USA	9,373	250	273	295	0.9	27	74	19,724	4,931	2.3	33,064	9,753
North-western Europe	4,786	171	171	184	0	36	93	14,602	2,497	2.3	26,773	4,939
South-western Europe	4,408	190	196	211	0.3	43	73	10,132	1,925	2.3	18,028	3,808
Eastern Europe	1,277	141	148	160	0.5	110	60	5,206	734	3.1	11,448	1,834
USSR	22,401	291	315	341	0.8	13	65	9,058	2,636	3.1	19,335	6,587
Middle East	6,817	200	270	358	3.0	29	55	1,990	398	4.1	3,716	1,329
North Africa	6,017	120	157	208	2.7	20	45	1,233	148	4.1	2,379	494
Central America	2,709	145	177	234	2.0	54	64	1,338	194	4.1	2,764	648
So. America – North	4,087	87	110	146	2.4	21	70	1,736	151	4.1	3,447	504
So. America – South	5,203	60	68	91	1.3	12	83	1,917	115	4.1	4,241	384
Brazil	8,513	158	200	266	2.4	19	71	2,114	334	4.1	4,199	1,115
Australia + NZ	7,956	20	22	24	1.2	3	86	10,550	211	2.3	17,245	417
SE Asia + Oceania	12,994	457	568	753	2.2	35	24	505	231	4.1	1,024	771
Japan + Asian NICs	510	197	213	229	0.8	386	74	12,091	2,382	2.3	20,547	4,712
China + N. Korea	9,723	1138	1282	1564	1.2	117	41	301	342	5.1	976	1,528
India/Pakistan Area	4,486	1117	1380	1945	2.1	249	25	296	331	4.4	614	1,193
South Africa	1,220	37	46	50	2.2	30	56	1,784	66	2.3	2,631	131
Central Africa	23,025	497	681	960	3.2	22	20	292	145	4.4	545	523
WORLD TOTAL	145,483	5,303	6,307	8,050	1.7	36	45	3,432	18,199	2.8	5,157	41,515

Table 2.7 *Contd.*

	Energy demands by category in 1990 (%) (Note 2)							1990 energy statistics				(Note 1) total
		Industrial			Other							
Country/region	Transport	Primary comm.	Non-comm.	Elec-tricity	Primary comm.	Non-comm.	Elec-tricity	Total (MTOE)	Per cap. (TOE)	Growth (%/yr)	Density (/sq km)	in 2020 (MTOE)
Canada	26	28	0	9	24	0	13	200	7.41	0.9	0.02	280
USA	34	27	0	2	25	0	12	1,800	7.20	0.9	0.19	2,400
North-western Europe	22	28	0	8	33	0	9	600	3.51	0.9	0.13	800
South-western Europe	28	28	0	8	29	0	7	450	2.37	0.9	0.10	600
Eastern Europe	6	39	0	6	50	0	13	400	2.84	1.5	0.31	700
USSR	15	32	0	7	36	0	10	1,200	4.12	1.5	0.05	2,000
Middle East	30	18	0	2	40	0	10	150	0.75	2.7	0.02	375
North Africa	25	17	0	5	45	0	8	70	0.58	2.7	0.01	175
Central America	30	36	4	4	14	8	4	150	1.03	2.7	0.06	375
So. America – North	35	27	2	7	15	7	7	175	2.01	2.7	0.04	440
So. America – South	30	25	6	4	20	10	5	120	2.00	2.7	0.02	300
Brazil	21	12	4	8	29	20	6	125	0.79	2.7	0.01	310
Australia + NZ	37	30	1	7	14	0	11	60	3.00	0.9	0.01	90
SE Asia + Oceania	35	23	10	7	10	12	3	150	0.33	2.7	0.01	375
Japan + Asian NICs	21	38	0	11	25	0	5	450	2.28	0.9	0.88	700
China + N. Korea	6	33	0	9	19	30	3	750	0.66	2.7	0.08	1,875
India/Pakistan Area	15	30	3	5	2	40	5	400	0.36	2.7	0.09	1,000
South Africa	3	39	0	19	34	5	8	50	1.35	0.9	0.04	80
Central Africa	10	5	1		13	70	1	100	0.20	2.7	0.00	250
WORLD TOTAL	22	29	1	6	27	7	9	7,400	1.40	1.6	0.05	13,125

Table 2.7 *Contd.*

Country/region	*Solar resource level*	*Potential solar penetration by category in 2020 (%)*							*Total potential solar contribution*		*Comments*
			Industrial			*Other*					
		Transport	*Primary comm.*	*Non-comm.*	*Elec-tricity*	*Primary comm.*	*Non-comm.*	*Elec-tricity*	*(MTOE) (Note 3)*	*Regional (%)*	
Canada	Low	5	25	0	25	35	0	35	66	24%	Latitude>49 Deg.
USA	Med/high	7	35	0	40	50	0	50	747	31%	
North-western Europe	Low	5	25	0	25	35	0	30	195	24%	Latitude>48 Deg.
South-western Europe	Medium	7	35	0	40	45	0	50	189	32%	
Eastern Europe	Med/low	3	30	0	45	45	0	45	301	43%	
USSR	Med/low	3	30	0	35	50	0	50	710	36%	
Middle East	High	6	65	0	50	70	0	75	188	50%	
North Africa	High	5	65	0	50	70	0	70	91	52%	
Central America	Med/high	6	60	60	45	65	60	65	165	44%	
So. America – North	Medium	6	55	55	40	60	55	60	167	38%	
So. America – South	Med/high	7	60	60	50	65	60	65	135	45%	
Brazil	Medium	6	50	50	40	55	50	60	130	42%	
Australia + NZ	High	7	45	45	50	60	0	50	31	34%	
SE Asia + Oceania	Medium	2	50	50	40	55	50	65	125	33%	
Japan + Asia NICs	Med/low	5	25	0	25	30	0	25	154	22%	Pop/energy dense
China + N. Korea	Medium	3	50	0	40	70	50	60	945	50%	
India/Pakistan Area	Med/high	3	65	65	50	75	65	75	557	56%	
South Africa	High	5	50	0	50	65	65	65	48	60%	
Central Africa	Medium	2	60	60	40	85	60	80	144	58%	
WORLD TOTAL (MTOE)		163	1461	51	289	1738	537	552	5,086		
(% sector penetration)		6%	38%	57%	37%	50%	56%	49%			
(% of world total)		1%	11%	0%	2%	13%	4%	4%		39%	

Notes:
1. Based on overall 1989 WEC Scenario M.
2. Breakdowns from variety of sources, using WEC, World Bank, and other published data if available, and extrapolation where necessary.
3. Does not include widespread use of storable chemicals (eg hydrogen) from solar energy.

Such a breakdown would currently require too much extrapolation to be reasonable. The non-commercial use was separately identified because there are major concerns about the sustainability of these traditional biomass uses. The ability of solar to displace these fuels was therefore also assessed. Then, for each region and each sector, the technical limit for penetration of solar energy was assessed with regard to the region's solar resource and many other factors whose impact is discussed below.

The transport sector presents a major challenge for the direct use of solar energy, unless electrical energy or solar-generated fuels is a significant percentage of the transportation energy use. For dense urban areas, which may depend heavily on electric train systems, a maximum of 25% electric penetration of the transport sector was assumed, as opposed to only 10% for dispersed regions, and the ability of solar energy to impact this electric portion was estimated at up to 30%. The advent of battery-powered electric vehicles in significant numbers may occur within the next several decades, but the logical time to recharge is overnight, since most travel occurs during daylight hours. Hydrogen and related fuels can be produced from solar energy, and may be commercially produced at some point in the future, especially if global warming makes it prudent, but for the near term, hydrogen is at an economic disadvantage compared to carbon-based fuels. Therefore, the estimated penetration is small (less than 10% for all the regions) with greater penetration assumed for those regions with high solar resource, highly urban population patterns, and low indigenous resources of oil and natural gas.

The industrial sector has much different prospects for the use of solar energy, and higher penetration is possible than for the transport sector. The use of primary energy, of both commercial and non-commercial types, is mainly for heating of materials being processed, and surveys of the temperature range of this process heat and the size of the typical facilities have been carried out for developed countries with solar energy use in mind. About 65% of the total primary energy in industry is used in facilities of less than 100 MWt, and the remainder is used in very large complexes. Over 50% of the total energy is used at temperatures below 335°C (550°F), which is well within the capability of parabolic trough solar collectors, and an additional 25% is used between 335 and 700°C (550 and 1200°F), a range within the capability of central receiver and parabolic dish collectors[25–27]. Since short-term thermal energy storage in tanks is possible, the maximum penetration of solar was estimated to be 80% for a very good site to about 50% for an average solar resource, using a 12-hour day as average in the industrial sector. For high reliability, fossil fuel-based backup systems are also expected to be necessary, and could supply the remaining energy demand at the facility. The overall penetration on a regional basis was estimated to range from 65% to 25% penetration, due partly to the lack of space for solar collectors adjacent to many industrial facilities.

For developing countries, the mix of industrial thermal energy use was assumed to be different, with 65% less than 335°C (550°F) and an additional 25% between 335 and 700°C (550 and 1200°F). This industrial use is consistent with a higher concentration of food, building materials and related industries. The same 80% to 50% range for maximum penetration was assumed, depending on solar resource. Also, for all regions, greater penetration was assumed for lower population density, since land availability and cost is a factor that is more significant to the use of solar energy than to conventional energy forms. The penetration of non-commercial energy use in industry is expected to be controlled by the same factors as for commercial use, so the same approach was applied.

For the electrical energy used in the industrial sector, this energy is usually supplied from a major grid, and the ability of solar generated electricity to penetrate the electric grid supply is limited by several factors. Therefore, a maximum of 50% penetration was assumed for high resource regions, with 25% used for poorer resource regions.

The other sectors are more complex to analyze, since they include a variety of activities such as agriculture, commerce, public services, and household uses. The broad variety of cultural and developmental differences lend further texture to the problem and cause each region to be estimated separately to take advantage of their solar resource levels. Because the data to differentiate the end-use further is not widely available, these uses are lumped together for reporting, but further breakdowns were used for analysis where available.

The agricultural sector uses energy for water pumping, tractive power, fertilizer production, drying of crops and related needs. This sector is generally a small contributor to a nation's total demand, with some countries as high as 10% but many in the 1–3% range. Because the energy requirement is decentralized, solar energy can be used to fill many of the requirements where good resources exist, except those such as motorized tractors. The timing of many agricultural uses can be adjusted to match the solar timing. Therefore the penetration of this sector in developing countries is expected to be relatively high, with somewhat lower potential in the developed countries.

In the commercial and public service sectors, the energy is used primarily for buildings, and electricity can be a significant fraction of the total energy use. Solar energy can penetrate this sector to a moderate degree, with less potential in dense, urban areas. Solar architecture, as well as structural integration of photovoltaics, can make penetration of this sector more feasible.

The residential sector has significant differences between developed and developing countries, and between rural and urban dwellings. This sector is generally as large as the agricultural, commercial and public service sectors combined for the developed countries, and dominates total energy use for some developing countries. It consumes the largest fraction of the non-commercial energy, generally in the form of fuelwood and charcoal used for cooking and heating. For those climates where space heating is a significant demand, experience with demonstrations of solar use have concluded that solar can reasonably supply about 80% of the total requirement at high-resource sites and about 50% at less favourable sites.

Conclusions on Long-term Solar Potential

The upper limit of the contribution possible for solar energy varies with the framework used for the evaluation. The centralized approach concludes that solar energy can supply all of the world's needs, as long as we act like one engineering system. The distributed approach concludes that solar energy along with conservation and the sustainable use of other renewable resources can supply essentially all the energy that the world really needs. The incremental approach concludes that solar energy, if it could overcome a variety of constraints, could penetrate our existing energy markets to about the 40% level. With the extensive production, storage, and use of solar-generated fuels considered in the other two approaches, this penetration could also reach much higher levels. The sectoral results show little penetration in the transportation sector and penetration to about half of the industrial and other sectors. One important thing to note is that each of these approaches concludes that solar energy could eventually supply a very significant fraction of the total energy, much more than most people would expect, since their thinking is oriented to the conventional energy markets.

CONSTRAINTS TO WIDESPREAD USE

Although the solar energy resource has the potential to make a major contribution to the future energy supply picture for the world, there are numerous reasons why its contribution could be limited to a significantly lower value, especially in the next quarter century. The constraints are evident in every aspect of the energy supply situation, and their significance varies greatly over the range of use patterns and national priorities.

The range of constraints are displayed in summary fashion in Table 2.8. They are divided into major categories of technical, institutional, economic, sociocultural, and educational. Each category has a range of factors to be considered, and the list included in the table is only representative of the range of possibilities. In the technical category, the resource is significantly impacted by weather, as well as its cyclic variations, and in high-latitude areas, the lack of a good solar resource creates greater energy demand for human activities. Among the contraints of the institutional category, the immature state of the supply industry, the lack of standardization, and the ability of our energy infrastructure to accept a variable, distributed energy source are all major factors. The economic constraints are perhaps the most significant, including the capital-intensive cost structure, the need for storage or fossil-fueled backup for high reliability, and the relatively poor resource in some high energy use areas. Other types of constraints include limited public awareness of the real potential, and lifestyle changes necessary for maximum utilization.

Examination of the constraints points out the complexity of the markets and institutions that control the decision making for energy supply systems. Although solar is perceived to be a desirable choice in many situations, it will not be applied to all the uses, nor to the extent that is possible.

Table 2.8 *List of constraints for solar energy utilization*

Constraint	*Impacts*	*Most affected situations*	*Potential for mitigation*
TECHNICAL			
Low energy density	Low energy value per unit area of structure; concentration needed for many end-uses	High temperature, high energy density uses; areas with high wind or seismic loading	Low cost support structures, distributed uses, integration with building structures, increased efficiency
Resource available only during day	Energy delivery only during day unless storage is used at additional expense	All types of terrestrial systems using solar energy	Improved, low-cost energy storage; orbiting space stations with power transmission to earth; low-cost superconducting transmission over long distances
Random weather and clouds	Lower total energy available; interrupts concentrator operation, makes backup or storage necessary for some applications	Systems using direct insolation, high reliability uses, sites with significant cloud cover	Storage or fossil backup with oversized collector (limits use of concentrating systems)
Atmospheric attenuation	Reduced energy at low sun angles, and with airborne pollution, dust and moisture; change in frequency spectrum with air mass	High latitude sites, humid or urban environments, frequency sensitive PV cells	High altitude sites, improved frequency sensitivity
Inverse relationship of temperature and efficiency	Practical limits on amount of energy collection possible to meet high temperature uses	High temperature uses involving concentrating collectors	Improved thermal insulation and selective surfaces
Exposure to damage by freezing	Need for controls or operator attention during operating outages or cold weather operation	Thermal systems with water as the working fluid, or high temperature fluids that freeze above ambient	Careful design and control system operation
Need for space heating greatest in high latitude areas	In these areas, the resource is lower while the needs are greater and heat losses are higher	Active and passive space heating systems and structures	Solar can play a more limited role unless it becomes the dominant criteria for buildings

Table 2.8 *Contd.*

Constraint	*Impacts*	*Most affected situations*	*Potential for mitigation*
TECHNICAL *(continued)*			
Material limitations	Reduced lifetime or high cost of reflective surfaces, high flux surfaces, selective coatings, or photo-active materials. Chronic wastage and high costs in achieving requisite quality of silicon	Some types of photovoltaic cells, high temperature thermal systems, and concentrating systems	R&D on improved materials to reduce limitation
Incomplete knowledge of the site-specific resource	Reduced ability to design cost-effective installations in many areas, and performance uncertainty	Many rural and remote areas in developing countries	Increased awareness of the need for a good data base, and use of satellite data to get an overall picture
Hazardous materials used in some solar energy collections systems	Environmental impacts of released gases such as silcane and arsine and disposal of solids such as cadmium or thermal storage salts	Photovoltaic cells may use hazardous materials in manufacturing, thermal systems may use hazardous working fluids	These potential impacts are relatively minor compared to other energy systems, and can be mitigated with awareness
Limited ability of solar energy to impact transportation sector energy use	Fuels for motor vehicles are difficult and costly to produce from solar unless electricity is used for transportation	Transportable liquid fuels are critical to many economies, from highly developed to remote and rural	R&D on improved processes for production of solar fuels, and electric powered transportation
INSTITUTIONAL			
Rational rate of industry growth	Practical limits on the growth of the supplier/distributor/installer/maintainer industry due to technology transfer and personnel training	All solar technologies that have a very much larger potential market than the near-term industry can support	Develop incentives to allow industry growth prior to full market development; effective planning for industry growth
Lack of standardized components and systems	Difficult to specify and maintain systems with unique components; high cost of inventory	All types of solar systems without industry standardization	Co-operation to define standard components and system configurations

Table 2.8 *Contd.*

Constraint	*Impacts*	*Most affected situations*	*Potential for mitigation*
INSTITUTIONAL (*continued*)			
Variability of government incentive programmes	Large swings in market demand for systems and services driven by incentives, and resultant quality variations	Any system which has incentive-driven market	Careful planning, consistent policies and co-operation for rational industry growth
Distributed systems require extensive maintenance/service network	Movement of spare parts and qualified service technicians to distributed systems requires extensive infrastructure	Critical for small, remote systems with high-tech controls or parts	Careful design for appropriate levels of technology and reliability, and user training
Ability of electric grid to absorb variation in supply	Too high a penetration of intermittent supply sources into an interconnected grid leads to operational problems, lower reliability, and higher costs	Areas where production of bulk electric power not well timed to match electric utility system peaks	Energy storage, wheeling of power to other areas, or improved operating practices and controls in conventional power plants
Energy investment can absorb only a minor fraction of total investment for development	Many developing countries have great needs in all economic sectors, and solar energy investment must compete for available funds	Most developing countries, especially those with rapidly expanding populations	Developing indigenous resources such as solar can liberate funds for other uses that now finance imported fuels
ECONOMIC			
Low conventional energy prices	Market for relatively high priced solar energy and willingness of industry to invest are limited	Subsidized energy markets, and areas where conventional energy transportation costs are low	Reduced cost of solar systems with higher production and improved technology
"External" costs not considered in energy decisions	Conventional energy sources are not penalized during economic analysis for a variety of costs needed to sustain their use and clean up the consequences	All types of energy systems without serious external consequences	Revise decision-making methodology so that all important issues are taken into account in economic terms
High capital costs of solar systems	Longer payback period compared to low capital cost, high operating cost fossil-fired systems	All types of solar systems may represent greater financial risks than alternatives	Long-term financing, lower capital cost through improved technology and high volume production

Table 2.8 *Contd.*

Constraint	*Impacts*	*Most affected situations*	*Potential for mitigation*
ECONOMIC (*continued*)			
High cost of transportation of electric or thermal energy from solar	Currently prohibitive to generate solar electric or thermal energy and transport it to distant uses	Use of high temperature thermal or electric energy to displace fossil fuels in industrial facilities in urban areas	Improved conductors and thermal insulation or increased prices for conventional fuels
High cost of storage of electric or thermal energy from solar	Expensive to generate solar electric or thermal energy and store it until needed at a later time	Use of thermal or electric energy for base-load applications	Improved storage technologies or increased prices for conventional fuels
Need for backup supply during solar outages for many applications	Additional capital cost to completely replace the equivalent fossil-fired system with the same availability	Major impact on remote and small systems, less on grid connected electric generation	Improved storage technologies a partial solution
Expense of site-specific building analysis compared to value of energy	Relatively crude ability to estimate payback without careful analysis, especially for passive systems	For custom built buildings, a good system design is complicated, and needs professional judgement	Training of local professionals on computer techniques and handbooks of experience and good design practice
Non-modularity of some technologies makes development and demonstration expensive	Some systems need relatively large demonstrations to prove that the technical risks are small	Central receiver systems are difficult to build privately in economic sizes until a few plants are built with government help	Government risk sharing may be necessary to overcome the financial hurdle
Reliability over useful lifetime	Unless the system performs as designed throughout its useful life, the economic performance can be greatly reduced	All systems that are capital intensive and have no or small fuel costs during operation	Proven design practices and standardized systems for remote areas with planned maintenance
SOCIOCULTURAL			
Lifestyle changes for maximum utilization	For large penetrations of solar energy into the energy mix, energy-intensive activities must adapt to solar availability unless storage is used	Any activities that are difficult to do only in daytime, such as cooking and base-load industries	Storage and long-range transmission of electric energy or solar-derived fuels may help, but some behaviour modification necessary

Table 2.8 *Continued*

Constraint	*Impacts*	*Most affected situations*	*Potential for mitigation*
SOCIOCULTURAL (*continued*)			
Change from traditional energy supplies	Rural and household traditions will change as traditional fuels become unavailable or very expensive	Those situations which are very social; activities such as cooking need to be done differently	A significant amount of time and education required for people to change their habits
Movement of rural population to urban areas	The distributed nature of energy use in the developing countries is becoming more intensive and concentrated in urban areas	Solar systems are not as well adapted to concentrated urban uses	The information revolution in the developed world may lead to more distributed energy use
EDUCATIONAL			
Limited knowledge/experience by local professionals	Feedback over time on the long-term performance of solar systems must be factored into the design and installation of distributed systems for good success	Many developing countries do not have the trained professionals needed to implement successful solar programmes	Much effort must be given to training and technology transfer
Consumer expectations raised beyond reasonable limits	Many people have believed optimistic media reports on the ability of solar energy to replace our conventional energy use but don't understand the difficulty or cost of doing it	There are significant gaps in public understanding of the true potential role of this energy resource	Public education and factual presentations on the potential such as the WEC reports
Need for individual awareness of environmental protection	Many of the major environmental problems could have been avoided with a different attitude toward sustainable development and resource use	General education on environmental awareness will help people understand the proper role of all energy resources	Promote general information exchange on energy and natural resource conservation

IMPACT OF THE CONSTRAINTS ON THE USE OF SOLAR ENERGY

This section examines the impacts of the constraints discussed in the previous section on the expected use of the solar resources within the next 30 years, as contrasted to the solar potential discussed earlier in this chapter. Several of these constraints can be analyzed individually, in an attempt to discover the ones that will have the most impact unless effective measures are taken to overcome them. The ones selected for discussion here are:

- economic competition with fossil fuels;
- risk equivalence between capital and future operating costs;
- total capital available for investment;
- rational growth of manufacturing production;
- penetration of intermittent sources into energy systems;
- changes required in lifestyle and education.

Each of these constraints will impact on both the timing and extent of penetration of solar energy into the energy markets. These and other constraints can be addressed by incentive strategies, which are discussed in the next section.

Economic Comparison with Conventional Energy Technologies

One of the major constraints facing the increased use of solar energy is the relative economic performance of early-generation solar systems compared to mature fossil-fueled systems. The analysis of the comparative economics combines the installed capital costs of plant equipment, converted to an annualized basis (with a 10% fixed charge rate), along with the operating and maintenance cost, the fuel cost and any other relevant annual costs such as waste disposal. For this book, the market fuel prices contained in recent WEC projections[22] were used to project future fuel prices, based on 1989 Scenario M (moderate growth). The capital and O&M costs of the facilities were taken from the recent US Interlaboratory White Paper on Renewable Energy Potential[21]. That report derives much of its data from a Solar Energy Research Institute (SERI) report[28], which is a compilation of data from other well-known sources.

The analysis was done for grid-connected electric power applications, since there is a considerable amount of information available for this end-use sector. Figure 2.9 presents the projected costs for peaking, intermediate load, and base load conventional systems. The first two are assumed to be natural gas or oil-fired, with base load using coal with flue gas desulfurization. The high–low range shown for each type of plant is caused by the range in the WEC Scenario M fuel prices, combined with typical capital and operating costs as shown in Table 2.9.

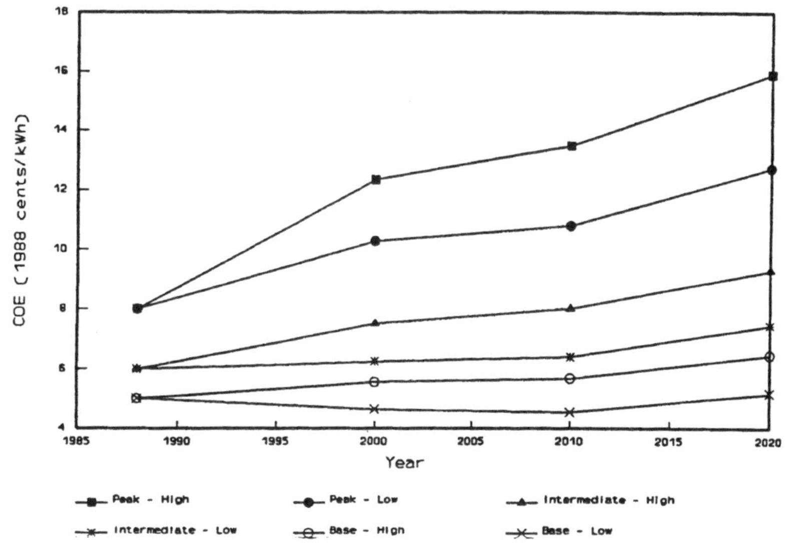

Figure 2.9 *Conventional electricity prices for comparison*

Table 2.9 *Assumptions for fossil-fueled generation plants (US$, 1988)*

Type	*Capacity factor*	*Capital cost*	*O&M cost*	*Heat rate (kJ/kWh)*
Peak load	10%	$340/kWe	1.2%/yr	10,995
Intermediate	35%	$625/kWe	2.0%/yr	8,270
Base load	65%	$1525/kWe	2.5%/yr	9,335

The cost of electricity (COE) rises with time due to fuel cost increases greater than general inflation. The assumption is made that capital and operation/maintenance (O&M) costs keep pace with general inflation, and therefore don't change in constant dollars.

The comparison of the projected cost of solar thermal electric and photovoltaic generation with the conventional fossil fuel plants is shown in Figures 2.10 and 2.11, respectively.

The solar thermal electric curves show that even the best situations in 1990 are not competitive with the large conventional power systems. The success of the Luz plants in California was partially due to the ability to burn low cost natural gas for a significant fraction of their generation at maximum economic advantage. However, the low end of the solar thermal spectrum starts to compete with peak load generation in the mid-1990s, and the mean should start to penetrate the peaking market around 2000. Since with thermal storage, solar thermal can achieve capacity factors of at least 50% at good sites, it starts to compete with intermediate load plants about 2010 for the mean, and is competitive over a large range of conditions by 2020. There appears to be some potential to displace base-load plants about 2020, although these solar thermal plants will need additional capital cost for a fired heater to achieve the equivalent reliability at 65% capacity factor.

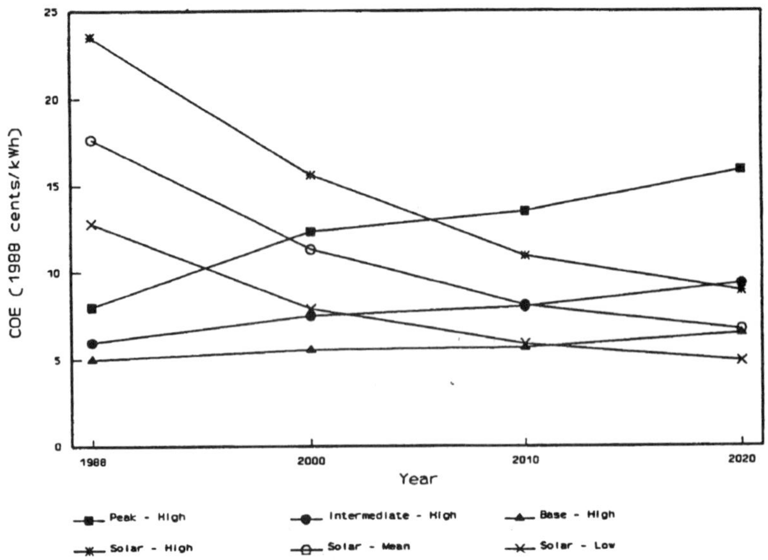

Figure 2.10 *Solar thermal electric cost comparison*

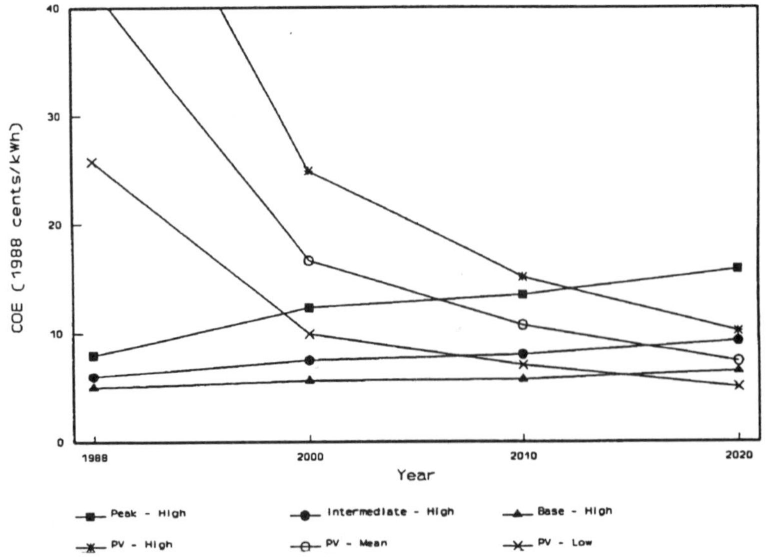

Figure 2.11 *Photovoltaic electric cost comparison*

The situation with photovoltaics, as shown in Figure 2.11, starts with more expensive solar electricity, and the best situations are not competitive for large-scale applications until about 2000. The mean curve starts to compete with peaking generation before 2010 and PV is projected to be widely competitive for peaking by 2020. However, for higher capacity factor needs, the extra costs of electric storage for PV plants means that a competitive situation for intermediate load will not develop until at least 2020.

Table 2.10 gives the cost and capacity factor values used in the above comparisons. These are based on the best currently available information

Table 2.10 *Assumptions for solar generation plants (US$, 1988)*

System type	*Resource*	*Year*	*Capacity factor*	*Capital cost*	*O&M cost*
Thermal	High	1988	50%	$4500/kWe	2.5%/yr
		2000	50%	$2500/kWe	2.2%/yr
		2010	50%	$1700/kWe	2.0%/yr
		2020	50%	$1500/kWe	1.5%/yr
	Low	1988	15%	$2800/kWe	2.0%/yr
		2000	15%	$1800/kWe	1.8%/yr
		2010	15%	$1200/kWe	1.7%/yr
		2020	15%	$1200/kWe	1.2%/yr
PV	High	1988	25%	$8000/kWe	1.0%/yr
		2000	28%	$3500/kWe	0.5%/yr
		2010	28%	$2100/kWe	0.5%/yr
		2020	28%	$2100/kWe	0.4%/yr
	Low	1988	15%	$5500/kWe	0.5%/yr
		2000	17%	$2325/kWe	0.2%/yr
		2010	17%	$1625/kWe	0.2%/yr
		2020	17%	$1150/kWe	0.2%/yr

of large-scale solar technology. The decrease in cost over time is the result of both improvements in technology characteristics with experience, as well as improved production costs with higher volume.

The simplistic nature of these projections includes assumptions which might be handled differently in a more rigorous study. For solar thermal systems, plants in high insolation areas have significant thermal energy storage to increase the capacity factor, while in low resource areas, no thermal storage is included, resulting in lower capital cost and lower capacity factor. Due to thermal inertia effects, performance and energy cost for systems in low resource areas is expected to be lower than indicated in this section, and the differences will depend on local weather patterns. For PV systems, fixed tilt arrays are assumed, although a similar set of numbers could be presented for 2-axis tracking (concentrating or non-concentrating) arrays. These alternate system configurations would have higher capital costs due to the tracking mechanisms and supports (also lenses for concentrators), but also higher capacity factor. Within the gross accuracy in this illustrative discussion, the projected costs of these alternatives for PV are judged to be equivalent.

As solar electric costs become competitive with conventional generation, it is also useful to compare the cost of bulk solar energy to displace the energy-only costs for gas and coal. This is very important for intermittent resources such as solar thermal without storage and photovoltaics. Figures 2.12 and 2.13 show these comparisons, and demonstrate that the mean solar thermal costs will start to displace gas energy about 2020, while PV will not achieve this until after 2020.

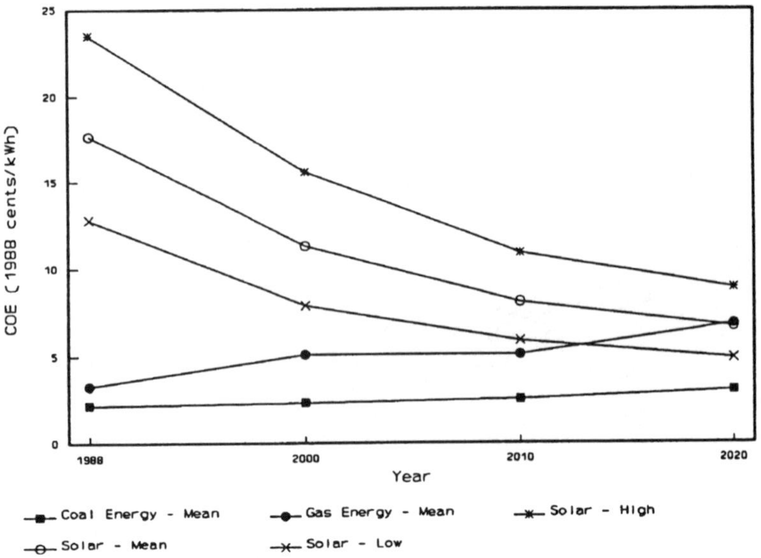

Figure 2.12 *Solar thermal energy displacement potential*

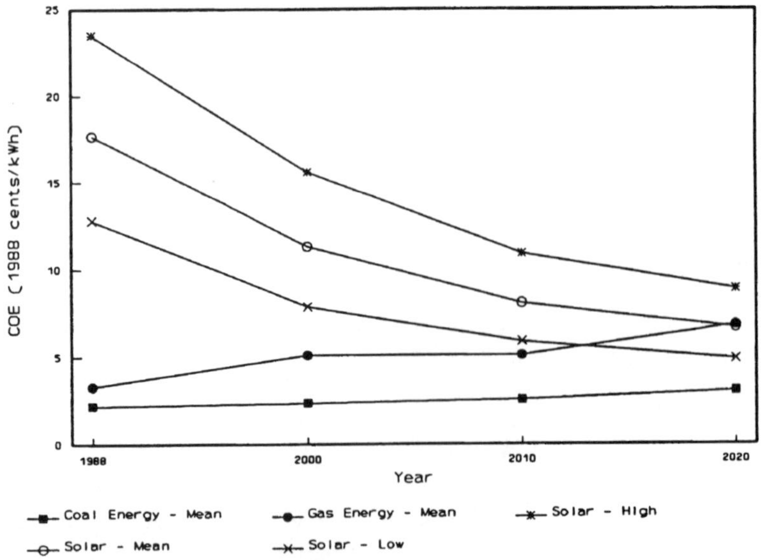

Figure 2.13 *Photovoltaic energy displacement potential*

Although the numbers have not been specifically calculated here, some preliminary conclusions can also be drawn about the ability of solar thermal to penetrate the large-scale industrial and commercial markets for bulk energy by displacing gas and coal. The cost of solar heat compared to the cost of gas burned in a boiler is roughly equivalent to the comparison shown in Figure 2.12, and that results in the conclusion that penetration of the large-scale, direct thermal use sectors will start before 2020, but be limited in significance by that time.

There is a high uncertainty in the preceding curves, because they represent projections of future events with many assumptions about the escalation of conventional energy facilities, as well as the development progress of the solar technologies. The cost reductions for solar are impressive, but within possibility when they are compared to other situations such as microelectronics. The pace of technological change is increasingly rapid in high technology areas, for which solar thermal reflective materials and photovoltaic cells are valid examples. It is also true that these technologies are now beginning to compete economically, but they are not yet mature and well understood in terms of the technical and financial risks.

For smaller and more remote applications of electric power generation, solar is far more competitive than for larger installations at this time, especially photovoltaics. And this trend will accelerate as conventional fuel prices rise and the costs for solar equipment decrease with greater volume production. For these applications, even with the penalties for energy storage or the cost of conventional backup, small-scale solar energy use will increase rapidly unless limited by other factors.

Risk Equivalence between Capital and Future Operating Costs

Although not solely a factor for solar energy comparison with conventional fossil-fueled plants, there is a disadvantage due to the risk comparison between options that differ in the timing of costs, using classical microeconomic theory such as discount rate. For a decision maker who must choose between a fossil-fueled facility and a solar system, for which he must pay for the equivalent of the fuel cost in advance, the solar system represents a greater apparent risk. Unless payback of the additional capital cost is achieved within the first several years from savings in operating costs, a conservative investor will be reluctant to invest in the solar plant. The result of this constraint is that solar systems that reach equivalence of life-cycle cost with systems with escalating fuel costs will be delayed in their penetration of the market due to discounting of future fuel cost risks and short-term economic decision making. Sufficient demonstration in real-life situations over significant operating periods will be necessary to reduce these perceived risks.

Total Capital Available for Investment

There is already significant concern over the ability of many countries to find the capital resources to finance their total development needs in the coming decades. This problem is especially acute in the developing countries, where the projections for capital requirements to finance electric power development identify a need for US$100 billion per year for at least the next 10 years[29]. Electric power represents approximately 70% of the world's total capital investment for development of the energy sectors at the turn of the next century, and half of the requirement for the electric power sector is for transmission and distribution facilities[30]. The developing countries already spend about 25% of their public sector budgets on power development[31], and their ability to expand this percentage is limited. The World Bank, other multilateral agencies and bilateral funding agencies are estimated to be able to provide only 20% of this need for power development. The private sector must play a much larger role in financing energy development, and the indigenous private capital resources must be developed to the maximum extent possible.

This represents a significant challenge for widespread solar energy utilization, as well as for other capital-intensive technologies including nuclear and other renewables. Many countries have incurred significant foreign debt loads in the past two decades, due to importing conventional fuels. The same level of energy usage with higher capital cost technologies would mean higher foreign debts unless a large fraction of the equipment was produced domestically. The investments for manufacturing facilities also represents a significant drain on cash flow.

The overall impact of capital limitations on the penetration of solar energy is difficult to predict. On the positive side, there is a real potential for solar PV to supply off-grid power in rural and remote areas at lower cost than delivering imported fuels or extending the transmission systems. For grid-connected power, and urban/industrial situations, solar represents higher costs in the next decade or two, except for the best regions. So widespread use of solar may cause lower total growth in energy usage, and investment in increasing energy utilization efficiency still must be made. The total cost of development and demonstration will also subtract from the investment available for expansion of the penetration of the total market, but this development cost is likely to be borne by the developed countries.

To put these considerations in a quantitative framework, the total world GNP from Table 2.7 was about US$18 × 10^{12} in 1990 and about $42 × 10^{12} in 2020. The estimated investment in total energy development for 2000 is equivalent to be about 5% of the total GNP. The fraction of total GNP for the South varies from about 20% in 2000 to 30% by 2020. For typical capital investments of $2500/kWe, $1800/kWe and $1200/kWe in 2000, 2010, and 2020, respectively, and using a maximum of 10% of the total energy development investment allocated for solar, the quantities of energy supplied can be estimated. For 2000, the annual installation of solar facilities (assuming PV technology) is 40 GWe, which equates to an energy contribution of 20 Mtoe/yr. This increases by 2020 to 150 GWe/yr additions, with energy contribution rising by 75 Mtoe/yr. The cumulative effect over the next 30 years is approximately 2000 GWe of solar capacity, and an energy production equivalent to 900 Mtoe in 2020, which is about 6.5% of the world total. The corresponding percentage contributions for 2000 and 2010 are 1% and 3.3%, respectively.

If the 10% limit of total energy development funding allocated for solar assumed above was removed, using these simplified calculations, to reach the technical potential estimated in Section 2.4 over the next 30 years would require in the order of 4% of total world GNP, equivalent to the projected expenditures for all energy development.

Rational Growth of Manufacturing Production

There is very little existing manufacturing capacity to produce solar hardware. Each segment of the technology has different requirements, from the very basic processes to produce flat-plate collectors and solar cookers, to the very sophisticated processes to produce concentrating photovoltaic cells. But historical attempts to build massive manufacturing capacity very quickly have generally been very inefficient unless there was an overriding social goal such as national defence. There appear to be practical limits to the rational growth that can be sustained in the private sector.

The experience in the US flat-plate collector industry, which reached 100% per year growth during the height of the tax incentives in the early 1980s, showed that excessive growth had many undesirable consequences. The photovoltaics industry has been growing at about a 15% rate for several years, and that level of growth has not led to the same excesses. This may be due in part to the nature of the PV market, which was not created overnight by government action. Experience in many markets also shows that high growth rates are possible in the early years, when the base is very small, but cannot usually be sustained when the base becomes much larger. This is due to a number of factors, including the diversification from a few firms to many firms, and production in a few countries to many countries.

For the purposes of this section, the possible limitations of growth rates starting at 50% per year and declining to 25% per year by 2020 were examined. PV, solar thermal electric and solar water heating were examined. For PV, the world production in 1990 is about 50 MWe, and growth at 50% per year for the next 10 years would result in a capacity of about 1900 MWe/yr by 2000. By 2010 the production would be 33 GWe/yr if a 33% per year growth were sustained, and by 2020 the annual production would be 390 GWe/yr if the growth slowed to 25% per year. The cumulative production by 2020 would be 1900 GWe, equivalent to 750 Mtoe annually at a 20% capacity factor. This scenario of growth would result in PV being able to supply about 5.5% of the total energy supply by 2020.

For solar thermal electric, the current capacity is about 100 MWe/yr, and following the same scenario as for PV, the contribution by 2020 would be as much as 19% of the total energy supply, and an annual production of 775 GWe. For flat-plate solar collectors, the maximum US capacity in the early 1980s was about 2 million m^2/yr, which can produce about 0.5 Mtoe increase per year. Assuming that capacity could be regained by 1995 world-wide, with growth of 50% per year for the next 5 years, and growth as discussed above between 2000 and 2020, the contribution potential by 2020 is about 6700 Mtoe, or about 50% of the total energy needs. Overall, growth in manufacturing capacity is not likely to be the limiting factor by 2020.

Penetration of Intermittent Sources into Energy Systems

Solar energy is an intermittent resource, which has both predictable day/night variation, and random interruptions due to weather. The impact of the former, along with the annual average weather impact, has been included in the discussion of technical potential earlier in this chapter, since it cannot be overcome without storage or long-distance transmission of solar-derived energy. But the random unavailability causes upsets when large penetrations of conventional energy systems are attempted. For example, a typical electric utility grid has many generators that are dispatched to meet a varying load. The impact of an unplanned outage on an individual generator is accommodated by having a reserve margin (installed spare capacity) in the system. For a fuel-supply system such as a pipeline, the equivalent backup is provided by spare pumps or storage tanks to plan for an equipment failure or supply interruption.

For small penetrations of an intermittent resource into an energy supply system, there is no practical economic disadvantage because of the margin that usually exists. For higher than about 10 to 15% penetration, the situation must be examined carefully. The correlation between the intermittent interruptions and the energy demand cycles can be determined statistically for a given region. For example, in sunny regions of the south-western US, one of the primary driving forces for energy use is air conditioning, which is caused by solar energy input. The correlation in this case is very high. However, random thunderstorms or weather fronts may have little correlation with energy demand changes, and the fact that solar energy plants are by nature distributed over a wide area is the only reason that a sudden complete outage of the solar plant output is unlikely.

For the range of typical situations in the electric power generation sector, the correlation between the availability of the intermittent resource and the need for power can vary from about 20% to 70%. This also represents the capacity value assigned to the intermittent plant. For very high penetrations by the intermittent source, the system must add reliable capacity to maintain sufficient operating margin. When capacity credit is low, the amount of additional capacity is high, significantly increasing the effective cost of the energy supplied by the intermittent resource. When capacity credit is high, little backup capacity or storage is needed, with a smaller impact on the equivalent cost of the intermittent energy. To use a quantitative example, a PV plant (with a capacity factor of 25%) that has a capacity credit of 50%, can only displace 50% of the equivalent conventional plant, and if half of the cost of a peaking plant capital cost ($170/kWe) was added to the capital cost of the PV plant ($1250/kWe in 2020), the effective cost of the energy produced by the PV plant would be increased by 14%. For the same example, if the capacity credit was only 20%, the effective energy cost from the PV plant would increase by about 22%.

The direct impact of this issue would be to limit the economic penetration in those areas with below average or marginal solar resources, since the cost improvement due to the technology would be offset by the lack of capacity credit given to additional solar capacity, above the level of 10 to 15% penetration which can be accommodated by operating margin in the conventional systems. The situation is somewhat better for those utility systems which have significant control of demand, because they can influence positively the correlation of the intermittent resource with the energy demand.

Changes in Lifestyle and Education

The above constraints deal primarily with the technical and economic aspects of the situation, but to contribute in a significant way to the overall energy supply of the world, many people must be aware and active. For each, solar energy must fit well into the cultural context of their lives, and must be important to them as individuals. There are wide ranges in these issues, from relative indifference if solar displaces conventional electricity generation on an established grid, to very significant if everyday cooking must change from personally gathered firewood to a solar cooker. For those changes that require more flexibility and commitment from the end-user, with loss of convenience or the need for new skills, significant penetration will take several years or even generations. For those who do not currently use commercial energy, the changes are beyond the impact of solar, and this is important for isolated village power. Education of the people towards understanding energy issues in general and the advantages of solar energy is a major global task, which will take significant investment by governments and the private sector.

The other part of the social equation for solar energy is the need to develop the technical skills among the designers, installers, and maintainers of the large amount of equipment that would be widely distributed throughout the world. One significant advantage for solar is that it has the potential to generate significant employment opportunities in rural areas, offsetting some of the pressure to move to urban areas. However, from the experience in India, where 86% of the rural energy use is for cooking and agriculture, the lack of trained manpower to implement the New Energy Sources Strategy is considered the weakest link[32]. Advances in training techniques with video, expert systems and improved communications to remote areas will help to improve this situation, but the penetration of these technologies will itself take time.

The net impact of the social and educational constraints will be delays in the penetration of new technologies, especially if the technology requires some significant behaviour modification, or large numbers of trained people.

Probable Penetration of Solar Energy by 2020

The constraints to rapid penetration of solar energy into the total global energy supply are significant and complex. Many of these constraints will tend to delay the market development and the technical capability to meet the potential market. A business-as-usual, free market approach to allowing solar energy to penetrate the total energy market on its own momentum will result in a minor contribution by 2020.

The trends discussed above lead to the conclusion that total capital investment available will limit solar to less than 10% in the next 30 years. The comparison with conventional fossil fuels leads to the conclusion that the penetration of large-scale solar into many markets will be delayed until after 2000, and likely until 2010 in many market segments. Solar will not be likely to displace direct energy from fossil fuels until at least 2020. Growth of manufacturing capacity is not likely to be a more significant constraint on the total contribution by 2020. Limits on penetration in regional energy systems will come into play in the most promising areas, placing upper limits on penetration in the near term, until the price of conventional fuels rise significantly[33–36]. Risk of capital investment and social issues will only add to potential delays in the penetration of the global market.

Although quantitative evaluation of these factors borders on the impossible, an estimate for the current policies scenario with regional breakdowns is presented in Table 2.11. The total penetration by 2020 is almost 1% of total global energy, which is about 30% of the total contribution included in the 1989 WEC Scenario M for all new sources of energy. There is clear indication that penetration will increase after that 2020, due to lower costs achieved through high-volume production and increasing constraints on other options.

The highest regional penetration (more than 1.5%) may occur in areas with low per-capita consumption, good solar resource, and significant growth in total energy consumption such as North and South Africa, and the Middle East. Penetration of more than 1% may occur in Australia/NZ and the USA, and also in some developing regions with rural population patterns like the India/Pakistan area, Central America and Central Africa. Penetration for the remaining regions is projected to be lower than 1%.

The 2020 contribution by solar energy will be distributed over the entire range of solar technologies. The relative contribution is expected to change over time, and become much different than the more than 90% now contributed by low-temperature solar heating. The shares of the solar market may evolve toward the following estimated distribution:

- solar thermal electric 30 to 35%
- photovoltaic 30 to 35%
- solar heating/cooling 30 to 35%

Table 2.11 *Estimated impacts of constraints on solar energy use*

Country/region	Projected total energy use (MTOE) [1]				Solar resource	Solar Penetration Estimate (2)					
						2000		2010		2020	
	1985	2000	2010	2020	level	(MTOE)	(%)	(MTOE)	(%)	(MTOE)	(%)
Canada	200	230	250	280	Low	0.3	0.1%	0.7	0.3%	1.5	0.5%
USA	1,800	2,060	2,250	2,400	Med/high	4.7	0.2%	10.3	0.5%	26.5	1.1%
North-western Europe	600	685	750	800	Low	0.6	0.1%	1.7	0.2%	3.9	0.5%
South-western Europe	450	515	565	600	Medium	0.8	0.2%	1.7	0.3%	5.4	0.9%
Eastern Europe	400	500	580	700	Med/low	0.7	0.1%	1.6	0.3%	3.2	0.5%
USSR	1,200	1,500	1,740	2,000	Med/low	1.5	0.1%	3.8	0.2%	7.9	0.4%
Middle East	150	225	295	375	High	1.1	0.5%	2.3	0.8%	5.8	1.5%
North Africa	70	105	135	175	High	0.7	0.7%	1.7	1.3%	3.3	1.9%
Central America	150	225	295	375	Med/high	0.6	0.3%	1.5	0.5%	4.4	1.2%
So. America – North	175	260	340	440	Medium	0.4	0.2%	1.1	0.3%	3.4	0.8%
So. America – South	120	180	235	300	Med/high	0.5	0.3%	1.3	0.6%	3.1	1.0%
Brazil	125	185	240	310	Medium	0.3	0.2%	1.0	0.4%	2.4	0.8%
Australia + NZ	60	70	75	90	High	0.3	0.4%	0.6	0.8%	1.3	1.4%
SE Asia + Oceania	150	225	295	375	Medium	0.3	0.1%	1.1	0.4%	3.2	0.9%
Japan + Asian NICs	450	515	565	700	Med/low	0.4	0.1%	1.2	0.2%	3.0	0.4%
China + N. Korea	750	1,115	1,450	1,875	Medium	1.5	0.1%	3.9	0.3%	14.1	0.8%
India/Pakistan Area	400	595	775	1,000	Med/high	1.8	0.3%	4.5	0.6%	12.5	1.3%
South Africa	50	55	60	80	High	0.3	0.5%	0.6	1.0%	1.4	1.8%
Central Africa	100	150	195	250	Medium	0.8	0.5%	1.4	0.7%	3.0	1.2%
WORLD TOTAL	7,400	9,395	11,090	13,125		18	0.2%	42	0.4%	109	0.8%

Table 2.11 *Contd.*

WEC region	*Solar penetration estimate* [3]							
	1990		*2000*		*2010*		*2020*	
	(MTOE)	*(%)*	*(MTOE)*	*(%)*	*(MTOE)*	*(%)*	*(MTOE)*	*(%)*
North America	3	(25%)	5.0	(28%)	11.0	(26%)	28.0	(26%)
Latin America	1	(8%)	1.8	(10%)	4.9	(12%)	13.3	(12%)
Western Europe	1	(8%)	1.4	(8%)	3.4	(8%)	9.3	(9%)
USSR & East Europe	2	(17%)	2.2	(13%)	5.4	(13%)	11.1	(10%)
Mid East/No. Africa	1	(8%)	1.8	(10%)	4.0	(10%)	9.1	(8%)
Sub-Saharan Africa	1	(8%)	1.1	(6%)	2.0	(5%)	4.4	(4%)
Pacific/China	1	(8%)	2.5	(14%)	6.8	(16%)	21.6	(20%)
Central/South Asia	2	(17%)	1.8	(10%)	4.5	(11%)	12.5	(11%)
WORLD TOTAL	12	(100%)	18	(100%)	42	(100%)	109	(100%)

Notes:
1. Based on overall 1989 WEC Scenario M.
2. Percent contribution refers here to total global primary energy use.
3. Percent contribution refers here to total global solar energy use.

These estimates are very rough, and are based on projections of the significant manufacturing growth needed and regional penetrations of the technologies.

EXPERIENCE WITH INCENTIVE PROGRAMMES

This section discusses part of the world-wide experience with financial and regulatory incentives for promoting solar energy installations, identifies some of the lessons learned, and outlines important characteristics for future programmes.

Incentive Approaches

To accelerate the development, demonstration, and deployment of solar energy systems, local and national governments and international development organizations can undertake various types of incentive programmes. These programmes can include:

- **Economic assistance**. These programmes remove some or all of the initial investment risks of solar energy systems relative to conventional energy systems. The assistance may be in the form of investment tax credits, accelerated depreciation schedules, low interest rate loans, project construction grants, equipment capital cost sharing, or partial government financing of manufacturing facilities.
- **Research, development and demonstration**. These programmes use government funds to share the cost of R&D programmes to develop concepts, components, systems, manufacturing techniques, and demonstration projects prior to the demands of the market. Thus, domestic manufacturers can be ready to meet the needs of a rapidly expanding market, with a minimum of financial risk, schedule delays, and development costs.

- **Government regulation of energy markets**. To ensure a customer for the output of a solar electric facility, utility companies can be obligated to purchase energy at preferred rates such as avoided cost, which can include time of day cost structures. In addition, the existence of multi-year power purchase agreements helps to secure project equity and debt financing. Another option is to place requirements on new residential or commercial building construction, or government facilities, such as mandating the use of passive solar design and energy conservation features, or stipulating that water heating in buildings or swimming pools be done primarily by solar energy.
- **Demand control for conventional energy**. The demand for solar energy systems can be increased by discouraging, through taxes or allocations, the use of conventional energy supplies. This is one way to account for the external costs of conventional energy use that are not normally included in private economic decision making. Unfortunately, many nations are doing the opposite, subsidizing the costs of fossil fuels.
- **Technology transfer**. Other incentive programmes include technology transfer from government and international research institutions to domestic industry, solar resource assessments, funding of project feasibility studies, publishing of information on technical assessment and application experience, and training in equipment operation and maintenance.

These and other types of incentive programmes can be developed to target the overcoming of many of the constraints identified earlier. It is also expected that all incentive programmes will require effective public information/education programmes to overcome inertia toward new ideas and changes in habits. In this way, solar energy utilization and energy conservation are very similar.

Specific Examples

There have been numerous attempts to institute incentive programmes which promote the use of solar energy instead of conventional energy. A few specific examples are given below:

- Czechoslovakia: concessionary loans for 20 years, with a 2.7% interest rate; 0% equipment sales tax; and investment subsidies of 30% for solar water heating systems in agriculture[37].
- Federal Republic of Germany: investment subsidies of 7.5% in the industrial sector; and a depreciation allowance of 10% for 10 years in the private sector[37].
- Europe: the European Union has supported up to 40% of the cost of PV demonstrations within its member countries[38].
- USA: state and federal investment tax credits (many of which expired in 1985); accelerated depreciation allowances; the 1978

Public Utility Regulatory Policies Act (PURPA), which obligates utilities to purchase energy from qualifying facilities at avoided cost; federal R&D programmes, with technology transfer to industry; federal cost sharing of programmes to improve current manufacturing methods; and industry R&D programmes subsidized by various states, including California, New Mexico, Florida and New York[8].

- Republic of Korea: loan incentives from Energy Conservation Promotion Fund with 10–11.5% interest and payback of 5 to 8 years; 6–10% reduction in income tax from total investment; 50% special depreciation allowed over initial investment[9].
- Thailand: government co-investment in a solar cell factory; 50% reduction in import duty on oil-saving equipment; and financial support for PV lighting in remote communities[9].
- India: flexible/concessionary loans to manufacturers of solar equipment; up to 33% subsidy for purchase of solar cookers and 50% for solar domestic water heaters[39].
- Israel: 65% penetration achieved of domestic hot water market because use of solar required for all new residences up to 4 storeys high[40].
- UNIDO: the Division of Industrial Operations Technology implemented three medium-sized demonstrations of the use of solar energy for low-temperature water heating systems in Sierra Leone, based on the successful example of a hospital in Tanzania, and is identifying local institutions and industries to carry out similar applications[41].

A significant part of the well-documented experience comes from the extensive USA incentive programmes initiated in the late 1970s. Here, the PURPA and solar investment tax credit programmes fostered the installation of 1,000,000 residential hot water systems, more than 350 MWe of solar thermal power plants, 30 MWe of photovoltaic facilities, and numerous industrial process heat projects.

Of these installations, one of the most successful examples of the effective use of incentives is the series of Luz International Solar Electric Generating Station (SEGS) plants. Projects completed to date include:

- SEGS I (13.8 MWe) and SEGS II (30 MWe), using negotiated power purchase agreements with the local utility.
- SEGS III through VII (each 30 MWe), using Standard Offer 4 power purchase agreements with guaranteed capacity and energy payments for 10 years.
- SEGS VIII and IX (each 80 MWe), using a Standard Offer 2 power purchase agreement with guaranteed capacity payments for 10 years.

Future projects are under development, although the Standard Offer 2 and 4 contracts are no longer available, and Luz was trying to work out other arrangements for new plants. It should be recognized that they have succeeded in dramatically reducing their cost of energy over the eight plants. Unfortunately, their high-risk, highly leveraged strategy was finally derailed when the government did not further extend the critical solar tax credits.

Lessons Learned

In market economies with taxation of income, investment tax credits and accelerated depreciation allowances are effective tools for accelerating the use of solar energy systems by (a) reducing the effective investment cost to values more comparable to those for conventional energy systems, and (b) attracting private venture and investment capital to finance the early, high-risk demonstration projects, and establish the commercial manufacturing capabilities.

Large tax incentives, instituted quickly, can also lead to negative influences in the use of solar energy systems. A minor percentage of the residential hot water systems installed in the US between 1978 and 1985 suffered from poor workmanship, exaggerated performance claims, and excessive pricing. Also, tax benefits to the consumer were based only on the initial cost of the system rather than the amount of conventional energy saved by the solar equipment. One way to minimize abuse of tax shelter incentives which are aimed at effectively reducing initial capital cost is to monitor performance over useful lifetime and require repayment if low energy delivery is due to poor quality of design or maintenance.

It is also interesting to note that many of the private sector companies that have been successful in producing solar energy equipment, such as Luz, have been relatively small, entrepreneurial firms focused on the renewable energy market, willing to "bet the company" on the success of critical early projects, and able to take a relatively long-term view of the market. Without consistent government leadership and stable incentive policies, this approach is too risky for most private sector firms.

International development organizations and their bilateral equivalents such as the US Agency for International Development have financed or subsidized many renewable energy projects in developing countries. The general experience shows that few projects resulted in large-scale use of renewable systems or significant energy production. This was partially due to the following:

- Many projects were oriented towards research and development, and direct dissemination of commercial technologies was not a goal. The early-generation equipment often performed below expectations in a real-world setting.
- Many projects were hampered by inadequate attention to the local capabilities/needs for operation, maintenance, repair, spare parts and technology transfer.
- Adoption of renewables is severely constrained when fossil fuel prices are heavily subsidized, renewable energy equipment receives unfavourable import tariff treatment, and the convenience of conventional energy systems exceeds that of renewable systems.

All nations need to learn from the variety of experience that has been gained to date within the solar energy industry and the funding organizations. This information is available but needs wide dissemination and future experience needs good summary documentation and individual contact identification.

Recommendations for Future Programmes

Incentive programmes can and should be developed to address a large variety of end-use applications and attempt to overcome a number of important constraints which will tend to limit the contribution of solar energy. It is recommended that these programmes, where appropriate, should have some or all of the following characteristics:

Technical

1. Develop regional and local databases to support planning.
2. Invest in RD&D to speed achievement of technology potential, but the private sector must be effectively involved.
3. Use/share the results of RD&D programmes in developed countries.
4. Emphasize good quality of design and manufacture for equipment and systems; don't widely distribute developmental hardware and software.
5. Work towards continuous improvement based on valid experience.

Economic

1. Target incentives for promising market niches, especially to establish new energy supplies.
2. Leverage private sector funds to the extent possible for the application.
3. Consider the external costs of conventional energy when setting and quantifying subsidies.
4. Tie incentives to energy delivered as well as capital cost, if practical.
5. Don't target fuel substitution unless there are valid national reasons, such as foreign exchange, that make it a priority.

Institutional

1. Balance foreign equipment purchase/sales with technology transfer.
2. Develop domestic manufacturing industry at a rational rate in advance of market growth.
3. Incentives need to be applied to applications that are an appropriate part of an overall development strategy, not derived from another cultural context.
4. Work with domestic institutions, such as utilities, as agents for subsidies but involve the private sector in ownership.
5. Keep programmes constant long enough to be effective.
6. Use leverage of national funds with international and bilateral organizations.

Educational/ Sociocultural

1. Link programmes with energy conservation initiatives.
2. User/public information and education is critical to success.
3. Develop the needed human resources in parallel with the equipment.
4. Offer free assessment services to end-users by trained professionals.
5. Don't oversell the potential energy contributions and other benefits.

Experience has shown that the use of incentives to influence market forces is much more likely to succeed than massive government distribution of free equipment. It is important that the end-user be an active participant, and be committed to success of the program. This also applies to the manufacturing industry that produces the technology. The eventual success of solar energy utilization depends heavily on long-term integration with the economy.

THE POSSIBLE IMPACT OF INCENTIVES ON ACHIEVING WIDESPREAD USE

In the previous sections, rough boundaries have been established for the maximum technical role of solar energy and the nominal role it will achieve without special attention and promotion. The next and most crucial step in the analysis is the estimation of the possible role that is achievable within the next 30 years. There are a great variety of needs and ambitions among the countries of the world, and solar energy cannot meet them all. However, it is too easy to dismiss its possible role as insignificant, without examining the critical issues that hold the key to success.

The key factors which control the bright potential of solar energy are:

- access to financing of high capital cost technologies;
- growing environmental concerns related to fossil fuel use;
- gesire for economic and social development distributed over a wide geographic area;
- rapidly increasing production to make a significant impact within three decades;
- near-term advantages to use of solar in a hybrid mode with conventional energy;
- diversification of the energy supply picture; and
- expansion of R&D into the private sector.

The wise use of these factors in the overall development programme of individual nations, along with a large emphasis on international cooperation, can lead to increased use of renewable energy sources such as solar.

The developed nations are heavily dependent on imported fossil fuels, and need to look ahead towards the time when those fuels will become increasingly scarce and expensive, probably being reserved for feedstock and transportation markets. The developing countries have energy development as a priority, but only one of many. The integration of solar energy development into overall planning concepts can help slow the migration to the urban areas, and retain a strong agricultural sector in the economy of rural areas. The development of indigenous production facilities will help foreign exchange restrictions. The examination of several of these factors will aid the estimation of the possible expanded role of solar energy by 2020.

Financing of High Capital Cost Technologies

Compared to the fossil-fueled alternatives, solar energy systems typically have a higher capital cost, even though a life-cycle cost analysis may show that applications such as low-temperature thermal and remote PV are the economic choice. As other types of applications become economic, they will still face this investment barrier unless government action is used to facilitate financing. Providing access to long-term financing represents an additional cost to governments, but this cost may be justified by factors such as reduced pollution or regional development. It is likely that unless such support is provided, the penetration of even the currently economic applications will be significantly delayed.

In the developed countries, there are numerous examples of pilot programmes to encourage renewable energy use through improved ability to finance the capital cost. Regulated utility companies have been allowed to offer low-cost financing for promising applications of both renewable energy and energy conservation where initial capital cost was a barrier. Governments have emphasized the use of economic systems in their own buildings and facilities. Elimination of subsidies or increased taxes on imported fossil fuels, and tax credits on solar energy systems have been used to reduce the payback time to the user. Measures such as these also help to develop a *market pull* for renewable systems, to supplement the *technology push* efforts of the manufacturers and advocates.

For developing countries, where investment is a major barrier for most developmental thrusts, the sources of financial support and the justification may be quite different. International financing institution's support and bilateral aid programmes or developmental loans are often needed to overcome the capital cost barrier. Rationale for providing assistance for renewable energy can range from improved economic development in rural areas, to coincident educational and health-care benefits.

Measures to Mitigate Carbon Dioxide Production

The nations of the world are becoming worried about the possible effects of increasing CO_2 concentrations in the atmosphere due to fossil energy use. At this time, we can observe the symptoms but do not fully understand the consequences. Theories range from insignificant to dire, and the experts are sharply divided. However, if it becomes generally accepted that it is necessary to reduce the total man-made CO_2 emissions, the cost will be significant, and will change the balance of energy usage in favour of technologies such as nuclear and renewables. At this time, only very crude estimates of this change in the balance can be made.

Some recent investigations of the mitigation measures have shown that the impact of offsetting or capturing the CO_2 production from a modern coal-fired power plant ranges from roughly equivalent to one-third of the cost of electricity produced, or about equal to the fuel cost today, to more than equal to the current cost. Technically, the lower range can be achieved through capturing the CO_2 in the fuel gas from a coal-gasification combined cycle plant (the disposal method significantly affects costs) or by planting enough new trees to absorb the entire CO_2 produced by the plant[42]. For illustration here, the penalty for mitigation of CO_2 production is set at 50% of present-day fossil fuel-based cost of electricity, and does not escalate over time. The resulting costs for conventional plants are shown in Figure 2.14. This is a graphic example of one way to economically deal with the "external" costs of fossil energy use. The impact on the attractiveness and penetration of renewable energy is very large if it is found necessary to adopt this strategy in the next few years.

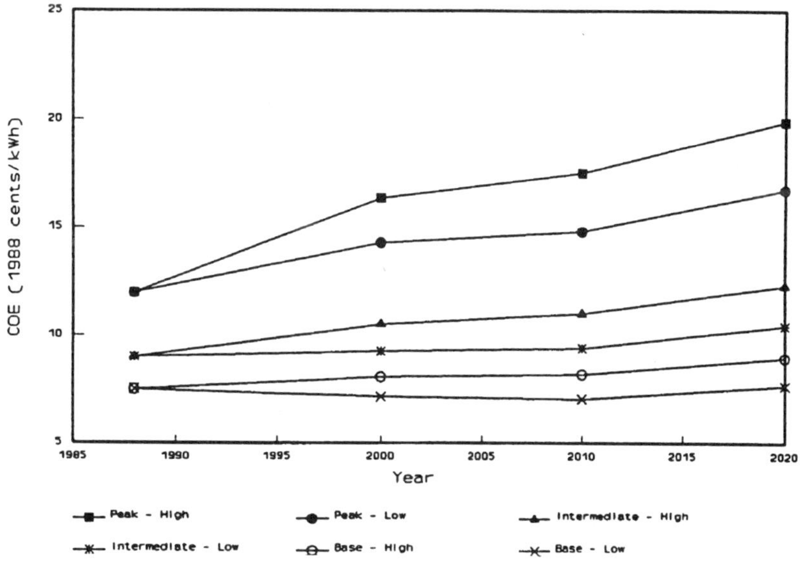

Figure 2.14 *Conventional electricity price with penalties*

The comparisons between electricity from solar energy and conventional power plants given in the "Impact of the Constraints" section above, can be modified to incorporate the example CO_2 penalty. The curves for solar thermal electric and photovoltaics are shown in Figures 2.15 and 2.16, respectively. As can be seen by comparison with the earlier section, the projected cost-effective penetration is brought forward in time, and therefore the overall penetration by 2020 should increase significantly.

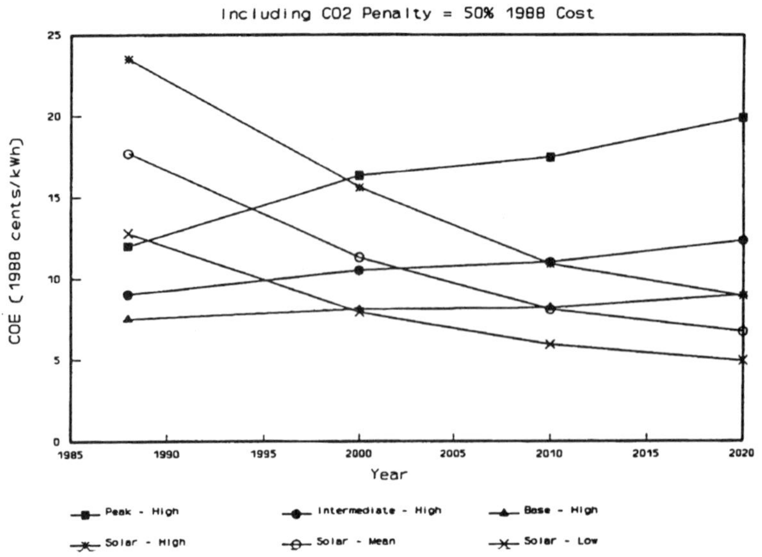

Figure 2.15 *Solar thermal electricity incentive cost comparison*

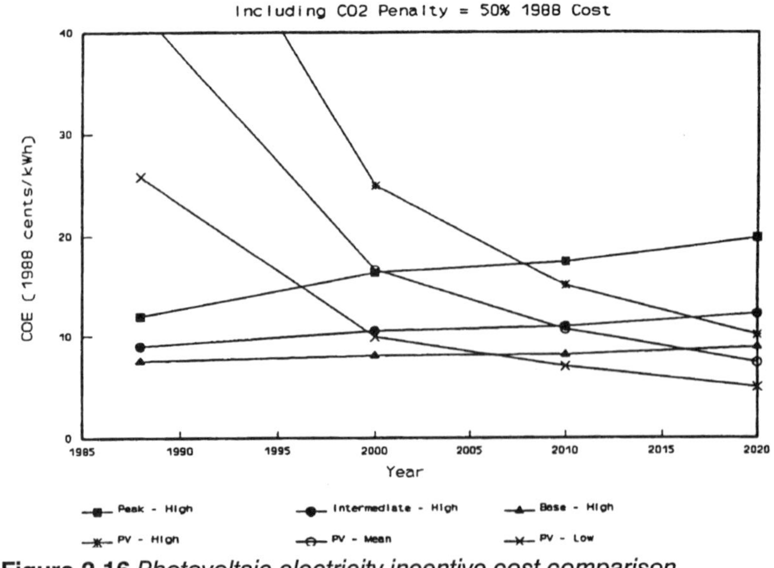

Figure 2.16 *Photovoltaic electricity incentive cost comparison*

Under this scenario, the best solar thermal electric situations appear to be competitive today, with the average case competitive before the year 2005. After 2010, there is potential for widespread displacement of conventional plants, except for base load. For PV, competitive economics start after 1995, with good penetration of the peaking market about 2005. In fact, when the energy displacement potential of gas and coal is examined, as shown in Figures 2.17 and 2.18, the best solar thermal electric plant starts to displace gas energy about 2005, and the average plant would begin to displace gas energy about 2015. The best PV plant would begin to displace gas about 2010.

The discussion presented here in quantitative terms for the electric power sector is impressive, but it understates the potential of solar

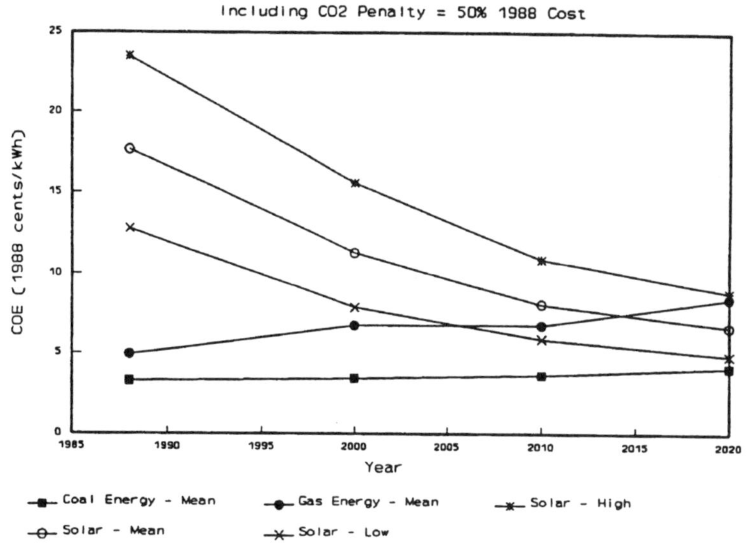

Figure 2.17 *Solar thermal increased energy displacement potential*

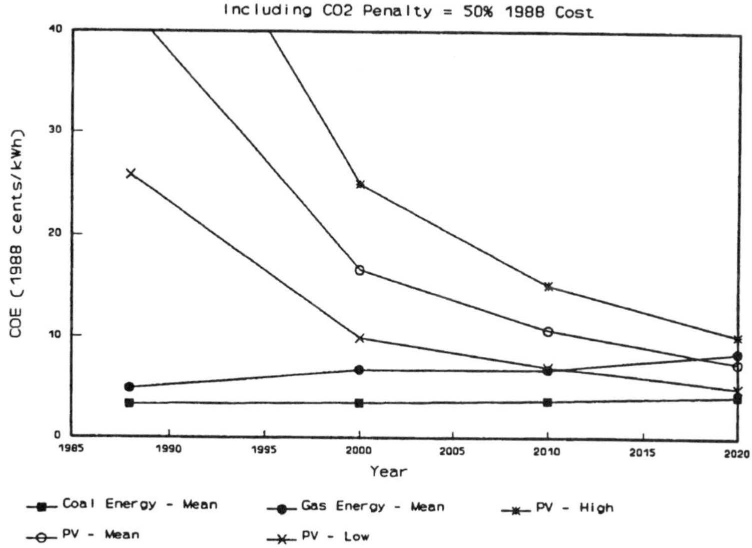

Figure 2.18 *Photovoltaic increased energy displacement potential*

energy to contribute to the displacement of other, less efficient uses of fossil energy. The smaller, distributed uses in the residential, agricultural and transportation sectors have a potential for larger improvement compared to current practice, especially when coal is displaced. This type of penalty on fossil fuels would be likely to structurally change the energy markets in ways that are difficult to predict. However, such changes are expected to be relatively slow, and consensus on the need for near-term reductions in CO_2 emissions appears to be a low probability. One promising sign of progress is that momentum appears to be building toward collective action on the ozone-depletion issue.

An analysis of the effect of accounting for all external costs, not just the impact of CO_2, has been published by Olav Hohmeyer. The results, using the market in Germany for the basis of calculation, show that the effect of external costs, calculated to be 2.4 to 5.5 US cents/kWh for coal-fired units, can accelerate the penetration of wind and PV systems by about 7 to 10 years compared to conventional economic decision making[43].

The concern for creating a better future for the planet is starting to be taken seriously. One prominent example is the debate over a plan by the European Union to institute a carbon tax. But looking at the magnitude of the potential problem, and its dimensions in the future as the South expands its use of energy, this is a small step in what could become a global action plan. A preview of the larger role is the conclusion of a recent study in Sweden that a more cost-effective way to reduce emissions affecting that country was to invest in emission reductions in Eastern Europe.

Planned Growth of Manufacturing Industry

There is a constraint that limits the market development for technologies that depend on high-volume production for their cost competitiveness. This has been called the "commercialization hump", since it represents a hurdle that must be crossed. A financial institution, either public or private, must be willing to fund the non-economic portion of the investment in early manufacturing facilities until the market is established. Solar technologies face this dilemma since they generally have high capital costs for limited production volume, and the market expands relatively slowly under normal conditions.

Incentives to aid in reducing the risks or sharing the costs of the early manufacturing facilities are an appropriate type of programme for maximizing solar energy contribution in the next few decades. Recently, for example, the US Department of Energy announced a new programme to provide US$55 million of support to domestic PV manufacturing firms[44]. To be eligible, the companies must have established production or prototype manufacturing lines, commercial products, and corporate commitment. They can submit proposals for cost-sharing of R&D programmes to improve yields or process rates, reduce material costs, increase automation, or improve any other process element critical to reducing costs and risks prior to scaling up production.

Another area where manufacturing development is necessary before commercialization can be completely successful is in the non-modular technologies such as the central receiver. For deployment of the first few plants at sizes up to the commercial threshold of about 100 MWe, the cost of heliostats in limited quantities is too high for competitive cost of electricity. Subsidies are expected to be necessary for these early plants, gradually reducing to zero as the track record for the technology is established and the market for heliostats is self-sustaining.

Given the high wastage rates (current order of magnitude 90%) and costs in producing silicons of the requisite quality for PV applications, there are both severe economic and environmental challenges which technological innovation needs to overcome.

Remote Village Power Systems

It is fortunate for the development of PV technology that it has competed well in the remote small power supply market for the past 10 years. There is significant interest in expanding this role for PV and other solar and renewable technologies into the remote village market, also known as mini-utilities. A typical system consists of several power generation devices, each supplying part of the load, connected by an energy storage device such as batteries, and a control system for operational and demand control. It appears likely that such an integrated system, if built in quantity with standardized components, is competitive with diesel engines or multi-kilometre extensions of the electric grid. The United Nations has estimated that over two million villages world-wide are without electric power for water supply, refrigeration, lighting, communication equipment, and other basic needs[40]. A typical village of several dozen families might need several kilowatts of peak capacity to meet the normal requirements. This type of application is cost effective if the capital cost can be financed.

One specific example of progress to date is in the Indonesian province of East Kalimantan on Borneo, which includes more than 1000 villages with about 2 million inhabitants. Less than 10% of the villages now have access to electricity. While only 46 villages were supplied with PV solar systems between 1984 and 1988, an additional 41 were supplied in 1989[45]. At this rate, all the villages would have access to electricity by 2020. There are also significant examples of bilateral co-operation in the application of PV, such as the Philippine–German Solar Energy Project, the French programme for rural refrigerators and lighting in Zaire, several large installations supported by Japan in Indonesia and Thailand, and numerous programmes supported by USAID and similar organizations[46].

There have been courageous proposals to implement these village power systems on a large scale. The Global Energy Society, for example, has proposed the "Africa 1000" Project, which has a goal of supplying renewables-based power to 1000 African villages in the next 5 years and another 5000 villages by the turn of the century. It is attempting to arrange financial and political support from organizations in many countries, and has assembled a significant advisory panel of African leaders. The total cost of the first 1000 villages is estimated to be about US$100 million. Such ambitious projects could make a significant difference in the lives of thousands of people, but there are literally billions in need of help.

Near-term Use of Hybrid Systems

For the current decade, the cost of conventional energy is likely to remain less expensive than solar-derived energy. However, the benefits of economics and environmental acceptability can be shared in systems that use both sources to meet part of the requirements. For example, the Luz plants use gas as a supplementary fuel to maximize the revenue of their plant while using primarily solar energy. This works well with the particular solar resource in Southern California and the time-of-day pricing structure for the electricity sold. Other examples are systems that use conventional fuels as a backup when solar energy is insufficient or not available. Many thermal systems are readily adaptable to this type of design, and the emissions per quantity of total energy delivered can be much lower than conventional systems. This is especially important is some areas such as Los Angeles, California, where concern over urban smog is stimulating discussion of many radical approaches to reducing use of fossil fuels in the air basin. Other major population centres such as Mexico City are experiencing frequent curtailment of fossil fuel burning due to pollution concerns.

Conclusions on the Possible Potential of Solar Energy by 2020

The consideration of the above factors in the context of the overall global energy supply can lead to a broad range of conclusions. Starting from the projections of the current policies scenario presented in Table 2.11, the improvement can be small or very large, leading up to a 10% solar penetration into global energy supply, according to some published projections. Extensive concurrent energy conservation programmes can also tend to increase the relative solar penetration by reducing the overall energy use.

Table 2.12 presents penetration estimates that are considered possible within the next 30 years through policies that actively promote the expanded use of solar energy. The penetration estimates, referred to as the ecologically driven scenario, are in general from 2 to 4 times that estimated in the current policies analysis presented in the earlier "Impacts of the Constraints" section. This is in agreement with the recent US National Energy Plan analysis that examined the penetration of renewables with incentives related to accelerated RD&D as well as a premium of 2 cents/kWh on conventional fossil fuels[21]. To put this penetration into perspective, it is about the same in percentage of total energy as nuclear energy in 1990, and would be achieved in about the same number of years of sustained development as has been necessary to get nuclear energy to today's level of penetration. This amount of solar contribution, about 3 times the amount under the current policies scenario, would displace an additional 250 to 500 million tonnes of CO_2 annually that would otherwise be generated by fossil energy use. This is approximately 4 to 7% of the total emissions projected in 2020 under the 1989 WEC Scenario M. This would supplement the larger reductions that would occur since one of the main elements of a complete ecologically driven scenario would be a major improvement in efficiency of energy use.

The expanded penetration of solar will change the technology mix somewhat from that estimated in the "Impacts of the Constraints" section. The major effect is expected to be a faster development of the solar ther-

Table 2.12 *Estimated impact of incentves on solar energy use*

Country/region	*Projected total energy use (MTOE)* [1]				*Solar resource level*	*Solar penetration estimate* [2]					
						2000		*2010*		*2020*	
	1985	*2000*	*2010*	*2020*		*(MTOE)*	*(%)*	*(MTOE)*	*(%)*	*(MTOE)*	*(%)*
Canada	200	230	250	280	Low	0.3	0.1%	0.9	0.4%	3.2	1.1%
USA	1,800	2,060	2,250	2,400	Med/high	5.7	0.3%	23.1	1.0%	82.3	3.4%
North-western Europe	600	685	750	800	Low	0.8	0.1%	3.4	0.5%	10.5	1.3%
South-western Europe	450	515	565	600	Medium	1.6	0.3%	5.5	1.0%	15.4	2.6%
Eastern Europe	400	500	580	700	Med/low	0.7	0.1%	3.0	0.5%	9.7	1.4%
USSR	1,200	1,500	1,740	2,000	Med/low	2.0	0.1%	5.8	0.3%	19.8	1.0%
Middle East	150	225	295	375	High	1.2	0.5%	2.6	0.9%	8.1	2.2%
North Africa	70	105	135	175	High	0.7	0.7%	2.9	2.1%	9.6	5.5%
Central America	150	225	295	375	Med/high	0.6	0.3%	2.4	0.8%	9.7	2.6%
So. America – North	175	260	340	440	Medium	0.4	0.2%	2.0	0.6%	8.6	2.0%
So. America – South	120	180	235	300	Med/high	0.3	0.2%	1.8	0.8%	7.1	2.4%
Brazil	125	185	240	310	Medium	0.3	0.2%	1.6	0.7%	7.5	2.4%
Australia + NZ	60	70	75	90	High	0.3	0.4%	1.2	1.6%	5.2	5.8%
SE Asia + Oceania	150	225	295	375	Medium	0.5	0.2%	2.7	0.9%	9.5	2.5%
Japan + Asian NICs	450	515	565	700	Med/low	0.4	0.1%	2.1	0.4%	6.4	0.9%
China + N. Korea	750	1,115	1,450	1,875	Medium	1.7	0.2%	11.3	0.8%	55.7	3.0%
India/Pakistan Area	400	595	775	1,000	Med/high	3.1	0.5%	14.2	1.8%	63.9	6.4%
South Africa	50	55	60	80	High	0.4	0.7%	1.4	2.3%	4.6	5.8%
Central Africa	100	150	195	250	Medium	1.0	0.7%	4.1	2.1%	18.6	7.4%
WORLD TOTAL	7,400	9,395	11,090	13,125		22	0.2%	92	0.8%	355	2.7%

Table 2.12 *Contd.*

WEC region	*Solar penetration estimate* [3]							
	1990		*2000*		*2010*		*2020*	
	(MTOE)	*(%)*	*(MTOE)*	*(%)*	*(MTOE)*	*(%)*	*(MTOE)*	*(%)*
North America	3	(25%)	6	(27%)	24	(26%)	86	(24%)
Latin America	1	(8%)	2	(7%)	8	(8%)	33	(9%)
Western Europe	1	(8%)	2	(11%)	9	(10%)	26	(7%)
USSR & East Europe	2	(17%)	3	(12%)	9	(10%)	30	(8%)
Mid East/No. Africa	1	(8%)	2	(9%)	6	(6%)	18	(5%)
Sub-Saharan Africa	1	(8%)	1	(6%)	6	(6%)	23	(7%)
Pacific/China	1	(8%)	3	(13%)	17	(19%)	77	(22%)
Central/South Asia	2	(17%)	3	(14%)	14	(15%)	64	(18%)
WORLD TOTAL	12	(100%)	22	(100%)	92	(100%)	355	(100%)

Notes:
1. Based on overall 1989 WEC Scenario M.
2. Percent contribution refers here to total global primary energy use.
3. Percent contribution refers here to total global solar energy use.

mal electric and photovoltaic technologies, and the greater penetration of PV into a variety of markets. PV is expected to account for about 40% of the solar contribution by 2020, with solar thermal electric and solar heating/cooling each supplying about 30% of the total solar share.

The regional breakdown of solar contribution will show a larger contribution in the developing world than anticipated in the current policies scenario. Penetrations above 6% of total energy use are expected in the Sub-Saharan Africa and Central/South Asia regions, and almost 6% in North Africa and Australia. The projected contribution for the US is about 3.5% of total energy use.

As with any of the estimates presented in this book, although a single number is reported, it actually should be thought of as a range of values around the mean. It must be stressed that these projections cannot be achieved by each nation acting alone with its limited resources, but require extensive international co-operation to be achieved.

OPPORTUNITIES FOR INTERNATIONAL CO-OPERATION

The economies of the world are being intertwined into a new concept – the global economy. Each nation and each private company must learn to see itself in the context of global resources and competition. However, people live in a locality and although they can communicate with the rest of the world efficiently, they are dependent for their lifestyle on their local environment. That environment is being threatened by local and global trends, from ground water pollution to stratospheric ozone depletion. As the global economy matures, there will need to be more consideration given to the global commons and the balance between environmental and short-term economic priorities. The emerging market sectors in the expanding population of the developing world need to be supplied with solutions as well as products. There is a real possibility of major social unrest and conflicts unless the developed countries co-operate to prevent the causes from becoming overwhelming.

Technology

Solar technology has made important advances over the past two decades, as a result of significant government investment in R&D, as well as entrepreneurial efforts by many private companies. Much of this technology is in the public domain, and most of the remainder is available under commercial arrangements. Although significant work remains to be done to reach competitive levels in a number of markets, some types of solar technology are now generally competitive in specialized markets.

The potential for improving and transferring solar technology to widen its market opportunities is great. The lower temperature thermal systems can be manufactured in most countries, and the higher technology systems can be implemented over time as capability is developed. There is considerable data published on the issues of application assessment, system design, and lessons learned from early operating experience.

Much of this information and technical know-how is readily available from the developed/industrialized countries. Expertise exists in many universities and research laboratories. International development organizations, bilateral aid organizations, and private industry can benefit from this massive database. Development of new production facilities, assessment of solar potential, and sharing of practical know-how are excellent ways to promote technology co-operation among developed and developing countries.

Institutional

There are limits to the capital available for investment by governments of most countries. One way to expand these limits is to attract private investment capital and leverage government funds allocated for solar energy investment. Luz was successful in doing this in the US, as have been several PV companies. The environment is very risky for private investment in numerous countries, especially when government sector funds and approvals are involved. Measures to increase the attractiveness to private investors will enhance the atmosphere for development of maximum solar potential.

Economic

It has been suggested by the World Bank that some types of grants, such as for electrification of rural villages with a variety of technologies including solar, are more effective in poor developing countries than loans or other assistance. A large number of people can be positively affected by this type of programme, but it needs to be complete co-operation, not just a crate of equipment delivered with an instruction book in an unfamiliar language.

Incentives are effective in a number of ways. There should be special attention paid to those that help accelerate the growth in demand for solar energy systems, to help build volume that will decrease costs. One example of an approach is to require use of renewable technologies where possible in new or refurbished government facilities. Also, it may be very effective to develop domestic manufacturing capability in advance of the market development, through incentives directed at manufacturing companies.

Educational/Sociocultural

There is a tremendous amount of information available on solar energy application to all energy sectors. Like many other subjects, the ease of sharing this information and the quantity available in electronic format is expanding rapidly. In fact, the amount of GNP allocated to information industries is rapidly expanding beyond that devoted to energy industries. There are many opportunities for communication and co-operation due to these expanded media resources.

The overall benefits of solar energy use fit well with other development objectives, since solar is by its nature decentralized and widely available. However, many of the early plants have been technically sophisticated and fragile, not well suited to widely distributed operation and maintenance. The appropriate technology mix needs to be applied so that the cultural context and perceived needs of the end-users are met. This is both an educational challenge and a great opportunity. Some of the necessary changes will take decades or a generation, and must be balanced with the retention of rich cultural traditions. The developed world has much to learn in this area, but has much to offer.

Conclusion on International Co-operation

The issues discussed above are sophisticated and complex. Many regional situations can be found around the world, some very unique. There needs to be a matching of resources and needs, technology and users, funds and benefits. International co-operation is needed as a minimum, and as shown in the earlier "Long-term Potential" section, perhaps to an unprecedented degree if solar energy is to meet its long-range potential.

CONCLUSIONS AND RECOMMENDATIONS

The review of the potential of solar energy presented in this chapter gives a feeling of the diversity and complexity of this resource. Solar energy systems will probably not be a simple replacement for our existing energy supply systems based on fossil fuels. They have unique characteristics that will impact their use, both positively and negatively. Their development and use is relatively immature at this stage, and therefore it is difficult to make comparisons to more mature technologies with clarity. However, a few powerful conclusions can be drawn from this review:

- Solar energy is a large enough resource to be of serious interest for world energy supply.
- Solar is a preferred resource from the environmental perspective, and has broad public support.
- Solar energy technology has undergone significant development effort and made significant progress in the past 20 years. Many people may be surprised by the role it plays today. In some specialized markets, solar energy is the resource of economic choice. Several solar technologies, including photovoltaics and solar architecture, still have the potential for significant further technical progress.
- There are numerous constraints to the widespread utilization of solar energy. Unless these constraints are addressed by positive and well-planned incentive programmes, solar energy penetration will proceed at a rather slow pace, reaching only an estimated 0.8% of total energy demand by the year 2020.
- With help from incentive programmes, solar can probably increase its contribution to an estimated 2.7% in the next 30 years. This is a very significant amount when viewed from today's perspective, and it can positively impact the lives of many millions of people. But solar is not likely to be able, in that time frame, to substitute for our heavy dependence on fossil fuel, even if that dependence is eventually proven to be a major environmental liability.
- Solar energy can and probably will play a very significant role in the long-term energy supply for the planet, but it will take many years to achieve that status.

The experience of the past 20 years of solar energy development, especially in the light of the trend towards open markets for energy supply, has taught us several interesting lessons. There has been a trend toward more diversity in the energy supply picture, with open markets allowing all resources and technologies to find their best situations. This has been especially noticeable in the US since the passage of PURPA. Solar will be one of the new types of energy systems to take advantage of this.

To be successful, incentive programmes to stimulate the increased use of solar energy must include a well-planned integration of technology push and market pull. There are several major market penetration issues to be overcome, and solar energy must compete for priority with many other development priorities throughout the world. In addition, it will be necessary to tailor incentive policies for each technology and market segment. The market can work to foster the further use of solar energy, but only if the external costs of energy use are included in the decision making.

For the long-term benefit of the human race, we should all take a more proactive attitude toward the promising future of solar energy, since it has a great contribution to make if we let it, and it offers many benefits if we are open to accept them.

It is encouraging to note that the manager of electric supply for the largest investor-owned electric utility in the US, serving an area that would rank 13th among the world's economies, recently stated to the US

Congress that the key elements for meeting growth of demand in the next decades were energy conservation, renewable energy supplies and competition in the energy markets[47]. It is true that California is one of the best areas for solar energy use, but the fact that a conservative company would find this policy reasonable is a possible precedent that others may follow in the coming years.

ACKNOWLEDGEMENTS

This chapter was prepared by the Solar Subcommittee of the WEC Study Committee on Renewable Energy Resources. The principal authors are J.R. Darnell of the USA and A. Oistrach of Spain. A great many people reviewed, commented and provided input for the work, and specific recognition is appropriate for W.C. Turkenburg of the Netherlands, D.L.P. Strange of Canada, T. Horigome of Japan, C.J. Winter of Germany, B. Devin of France, M.M. Koltun and G.V. Tsykauri of Russia, and G.W. Braun, R.L. San Martin, T.D. Bath, W.J. Stolte, B.D. Kelly and R.L. Lessley of the USA. The contributions of all were greatly appreciated.

REFERENCES

1. H. Z. Tabor (editor). *Solar Power: Report of the WEC Solar Power Committee*. Montreal, Canada: 14th Congress of the World Energy Council, Sept. 1989, p. 6.
2. A. B. Meinel and M. P. Meinel. *Applied Solar Energy*. Reading, MA: Addison-Wesley Publishing Company, 1976, pp. 39–115.
3. A. Oistrach. The SURYA Solar Model, to be published in *Solar Energy*, ISES, Pergamon Press, New York, USA
4. J. F. Kreider and F. Kreith (editors). *Solar Energy Handbook*. New York: McGraw-Hill Book Company, 1981, pp. 2-1 to 2-78.
5. World Meteorological Organization. *Meteorological Effects on the Use of Solar Radiation as an Energy Source*. Geneva, Switzerland: WMO No. 557, 1981.
6. C. J. Winter, R. L. Sizmann and L. L. Vant-Hull (editors). *Solar Power Plants: Fundamentals, Technology, Systems, Economics*. Berlin, Germany: Springer-Verlag, 1991.
7. J. R. Darnell, et al. *Solar Distillation and Water Pumping*. San Francisco, CA: Bechtel National, Inc, Research and Engineering, December, 1978, pp. 5-1 to 6-12.
8. S. Williams and K. Porter. *Power Plays: Profiles of America's Independent Renewable Electricity Developers*. Washington, DC: Investor Responsibility Research Center, 1989, pp. 341–360 and 387–401.
9. Economic and Social Commission for Asia and the Pacific. *New and Renewable Sources of Energy for Development*. Bangkok, Thailand: United Nations, Energy Resources Development Series No. 30, 1988, pp. 10–13.

10. R. J. Holl. *Status of Solar-Thermal Electric Technology*. Palo Alto, CA: Electric Power Research Institute, EPRI GS-6573, December, 1989.
11. D. A. Andrejko (editor). *Assessment of Solar Energy Technologies*. Boulder, CO: American Solar Energy Society, May, 1989, pp. 1–38.
12. S. C. Carpenter and T. Caffell. Active Solar Heating in Canada to the Year 2010. Toronto, Canada: *Energy, Mines and Resources Canada*, November, 1989, p.30.
13. B. Givoni. Characteristics, Design Implications, and Applicability of Passive Solar Heating Systems for Buildings. *Solar Energy,* ISES, Vol 47, No. 6, June 1991, pp. 425–435.
14. T. Beardsley. Bright Future for a New Photovoltaic Cell. *Scientific American,* January, 1992, pp. 138–139.
15. J. P. Rollefson. *Canadian Solar Energy Review – March 1986*. Ottawa, Canada: National Research Council, May 1986, p. 25–26.
16. U. Sprengel and W. Hoyer. *Solar Hydrogen: Energy Carrier for the Future*. Stuttgart, Germany: German Aerospace Research Establishment (DLR), 1991.
17. International Energy Agency. *Energy Policies and Programmes: Annual Review*. Paris, France: 1980–1990.
18. C. Jennings. *PG&E's Cost Effective Photovoltaic Installations*. Pacific Gas and Electric Company, San Ramon, CA, 1989, pp. 1–6.
19. A.A. Eberhard and M.L. Borchers. *Financial Costs of Stand-Alone Photovoltaic Systems, Diesel Generators, and Electricity Grid Extension*. Cape Town, South Africa: Energy Research Institute, University of Cape Town, Report No. REP-025, September, 1990, p. 60.
20. B. McNelis, A. Derrick and M. Starr. *Solar-powered Electricity: A Survey of Photovoltaic Power in Developing Countries*. London: Intermediate Technology Publications, Ltd., 1988, pp. 66–69.
21. Solar Energy Research Institute, et al. *The Potential of Renewable Energy: An Interlaboratory White Paper*. Golden, Colorado, USA: SERI/TP-260-3674, DE90000322, March, 1990, p. 20 and Appendices E, G, and H.
22. WEC Conservation and Studies Committee. *Global Energy Perspectives 2000–2020*. London, UK: 14th Congress of the World Energy Conference, Montreal, Canada, September, 1989, p.11.
23. H. Klaib, F. Staib and C. J. Winter. *Systems Comparison and Potential for Solar Thermal Installations in the Mediterranean Area*. Presented at the Workshop entitled "Prospects for Solar Thermal Power Plants in the Mediterranean Region", Sophia-Antipolis, France, 25–26 September, 1991.
24. Personal Communication with T. Horigome (Tokyo University of Agriculture and Technology) on the *Solar Energy Conversion Plant on Ocean (SEPO) Project*, 18 October, 1991.
25. J. J. Iannucci. *Survey of US Industrial Process Heat Usage Distributions*. Sandia National Laboratories, Livermore, CA, USA, SAND80-8234, January 1981, pp. 12–14.

26. K. C. Brown, et al. *End-use Matching for Solar Industrial Process Heat.* Solar Energy Research Institute, Golden, CO, USA, SERI/TR-34-091, January 1980, pp. 1–11.
27. *Survey of the Applications of Solar Thermal Energy Systems to Industrial Process Heat.* Battelle Columbus Laboratories, Ohio, USA, TID-27348/1 January 1977, pp. 8–25.
28. Lynn R. Coles. *Comparative Characteristics of Electric Generating Technologies, Volumes I and II.* Golden, Colorado, USA: SERI/MR-160-3444, December, 1988, p. C-1, C-17, and C-22.
29. A.A. Churchill, and R.J. Saunders. *Financing of the Energy Sector in Developing Countries.* Montreal, Canada: 14th Congress of the World Energy Conference, 17–22 September 1989, pp. 2–3.
30. H.K. Schneider, and W. Schulz. *Investment Requirements of the World Energy Industries 1980–2000.* Cannes, France: 13th Congress of the World Energy Conference, 5–11 October 1986, p. 24.
31. *Power Shortages in Developing Countries: Magnitude, Impacts, Solutions, and the Role of the Private Sector.* Washington, DC, USA: US Agency for International Development, March 1988, p. vi and 24–27.
32. V.K. Bhansali. *Human Resources Development in the Non-conventional Energy Sector of India.* Montreal, Canada: 14th Congress of the World Energy Conference, 17–22 September 1989, p.15.
33. S.C. Carpenter, and T. Caffell. *Active Solar Heating in Canada to the Year 2010.* Ottawa, Canada: Energy, Mines and Resources Canada, November 1989, pp. 37–39.
34. Scanada Consultants Ltd. Passive Solar Potential in Canada 1990–2010. Ottawa, Canada: *Energy, Mines and Resources Canada,* 11 September 1989, (Draft) p. 46.
35. J.R. Frisch, et al. *World Energy Horizons 2000–2020.* Paris, France: Editions Technip, 1989, p. 50.
36. The Prospects of Renewable Energy Sources. *United Nations Economic Bulletin for Europe*, Vol. 40, No. 1, 1988, p. 69, 100, 139.
37. United Nations Economic Commission for Europe. *Economic Bulletin for Europe: The Prospects of Renewable Energy Sources.* Pergamon Press, Oxford, UK: Vol 40, No. 1, 1988, p. 11.
38. Commission of the European Community, Directorate-General for Energy. *Community Demonstration Programmes in the Sector of Photovoltaic Energy.* Brussels, Belgium: 1989, p. 3.
39. M. Dayal and G.D. Sootha. *Renewable Energy Technologies – Indian Experience.* Department of Non-conventional Energy Sources, Ministry of Energy, New Delhi, India: 1989, p. 12.
40. M.C. Brower. *Cool Energy: The Renewable Solution to Global Warming*. Union of Concerned Scientists, 1990, p. 31.
41. A. Tcheknavorian-Asenbauer. *Energy Imperatives and Priorities in Developing Countries: Point of View of UNIDO.* UNIDO, Vienna, Austria: presented at the 14th Congress of the World Energy Council, Montreal, Canada, September 1989, p. 9.

42. Submittal by Luz Development and Finance Corporation to the California Energy Commission on SEGS 9 and 10, covering the value of emissions displaced in the South Coast Air Quality Management District, November, 1989.
43. O. H. Hohmeyer. *Social Costs of Energy Consumption: External Effects of Electricity Generation in the Federal Republic of Germany*. Berlin, Germany: Springer-Verlag, Commission of the European Communities No. EUR 11519, 1988.
44. Solar Energy Research Institute. *Science and Technology in Review*. Golden, Colorado, USA: Vol. XII, No. 2, Spring 1990, p. 6.
45. Solar Power in Southeast Asia. *International Power Generation*. February, 1992.
46. A.F.L. Slob et al. (CEA). *Solar Photovoltaics for Developing Countries: Final Report*. Netherlands Ministry of Foreign Affairs, Directorate General for Development Co-operation, October, 1989.
47. G. Rueger. Testimony before the US House of Representatives Subcommittee on Energy and Power (on *The Development of a National Energy Strategy*). San Francisco, CA, 7 March 1991.

CHAPTER

3

Wind Energy

GENERAL INTRODUCTION

Wind energy, in common with other renewable resources, is broadly available, but diffuse. It was widely used as a source of power before the industrial revolution. In the course of this revolution wind energy was displaced by fossil fuels, because the use of these fuels was cheaper and more reliable. The first oil crisis, however, triggered renewed interest in wind energy technology for grid-connected electricity production, water pumping and for power supply in remote areas. In recent years this interest has been stimulated by environmental problems and the threat of a global climate change related to the use of conventional energy sources. Recently there has also come a greater realization that wind turbines, especially where they are large and numerous and placed in sensitive landscapes, can cause severe visual intrusion. In some instances, noise continues to pose a significant problem.

Since 1975 enormous progress has been made in the development of wind turbines for electricity production. Around 1980 the first modern grid-connected wind turbines were installed. Also the first multi-megawatt market was created in California. At the end of 1990 about 2000 MWe wind power for grid-connected electricity production was in operation world-wide, producing over 3200 GWh of electricity per year. Nearly all this production took place in the USA (California) and Denmark. Various other countries such as the Netherlands, Germany, the UK, Italy and India, have initiated national programmes for the development and commercial introduction of wind energy.

During the last decade the cost of energy production by wind turbines has decreased considerably. The wind turbines are now more efficient and reliable than 15 years ago. However, the widespread introduction of Wind Energy Conversion Systems (WECS) has not yet been started. Many constraints will have to be removed before wind energy can be implemented on a large scale for electricity production.

A special application of wind energy is water pumping. After the widespread use of wind pumps in the last century and the first half of this century, a rather sudden decrease occurred in the 1950s and 1960s. Wind pumps lost the market to engine pumps. However, wind pumps are still sold, mainly in China, South Africa, Argentina and the United States. It is estimated that over 1 million wind pumps were still in operation in 1991. The wind pumps that are mainly used are powered by the classical multi-blade wind turbines. Technology in this field has also developed progressively during the past 15 years. Any further large-scale introduction of wind pumps will need new collaborative action.

The contribution to the global energy supply made by small-scale systems for battery charging and water pumping will always be negligibly small. However, such systems should not be judged in terms of megawatts but in terms of the number of people who benefit from these systems for their basic needs. For example, a mere 10 MWe installed capacity (being 100,000 wind turbines of approximately 100 W) is supplying 100,000 families, or roughly half a million people, with their basic electricity needs.

In this chapter we discuss the status of wind energy, the constraints and the opportunities for the application of wind energy on a world-wide scale. The emphasis is on wind turbines for grid-connected electricity production, because this application can make an important contribution to the world electricity supply. It is estimated that, based on a "current policies" scenario for the future development of the world economy, the contribution of wind energy to the energy supply in the year 2020 could be about 375 TWh/year, generated by 180 GWe wind capacity. In an "ecologically driven" scenario this contribution could be about 970 TWh/year in the year 2020, generated by 470 GWe wind capacity, if a policy committed to a strong and intensified development of renewable energy sources is followed. The ultimate potential of wind energy as a long-term power source is estimated to be about twice the present global electricity consumption.

WIND ENERGY RESOURCE

Before the possible contribution of wind energy to the energy supply can be analyzed, the characteristics of the wind energy resource need to be investigated. In this section we discuss the origin of the wind, the global distribution of the wind, the availability and the variability of the wind energy resource and our knowledge about wind regimes. Moreover, the potential of wind power production is discussed for various countries. Targets that have been set to realize wind power are given. Finally the global potential of wind power is analyzed.

The Origin of Wind

Winds develop when solar radiation reaches the earth's highly varied surface unevenly, creating temperature, density and pressure differences. In the tropical regions there is a net gain of heat due to solar radiation, whereas in the polar regions there is a net loss. This means that the earth's atmosphere has to circulate to transport heat from the tropics towards the poles. Ocean currents act similarly, and are responsible for about 30% of this global heat transfer. On a global scale, these atmospheric currents work as an immense energy transfer medium. Rotation of the earth further contributes to the establishment of semi-permanent, planetary-scale circulation patterns in the atmosphere.

Besides these major forcing agents, other factors such as topographical features and local temperature gradients alter wind energy distribution. For example, the difference in heat capacity of land and water along a coastline can create sea breezes. In valleys and mountains similar processes occur, creating local wind.

Global Distribution of Wind

Between 30°N and 30°S, air heated at the Equator rises and is replaced by cooler air coming from the south and the north. This is the so-called Hadley circulation. At the earth's surface this means that "cool" winds blow towards the Equator. The air that comes down at 30°N and 30°S is very dry and moves eastward, due to the fact that the earths' rotational speed at these latitudes is much less than at the Equator. At these latitudes desert areas, like the Sahara, are found. Between 30°N(S) and 70°N(S) predominantly western winds are found. These winds form a wavelike circulation, transferring cold air southward and warm air northward. This pattern is called Rossby circulation, as illustrated in Figure 3.1[1].

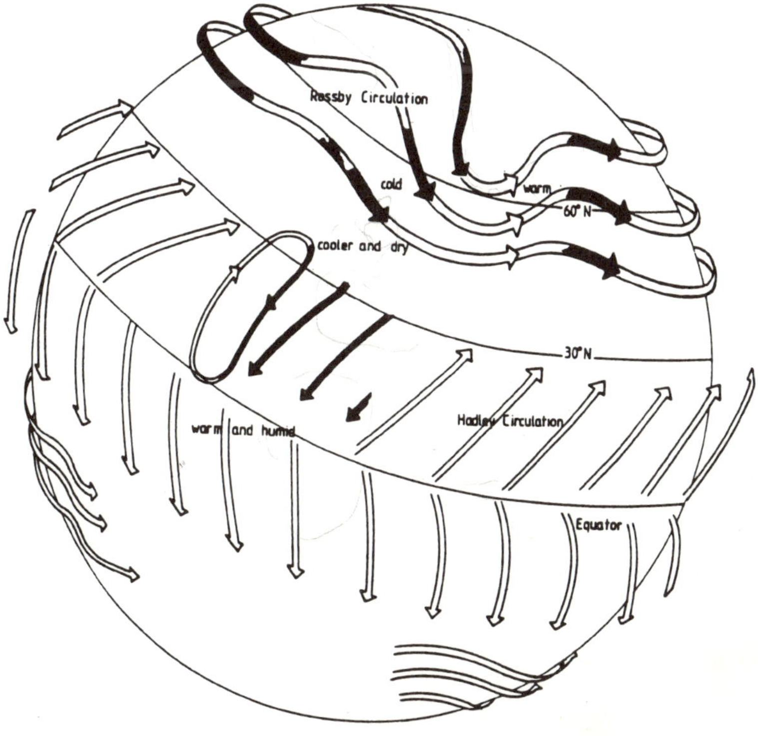

Figure 3.1 *Global circulation of wind over the earth*

Resource Availability and Variability

In a study for the World Meteorological Organization (WMO) in 1981, the Pacific Northwest Laboratory (PNL) developed a world-wide wind resource map, showing the mean wind speed and mean wind power density for different regions. The analysis was based on wind measurements taken by meteorological institutes[1]. It should be noted that the mean wind speed may differ by as much as 25% from year to year. In most areas there are also significant seasonal differences. Generally, wind speeds are higher in winter, although there are exceptions. In California (USA), for example, summer winds are stronger due to local topography and sea breeze effects.

Due to seasonal variations, the potential of wind energy for power production can be significantly higher than the annual mean wind speed would indicate. Therefore not only the mean wind speed but also the wind speed frequency distribution, commonly described by a Weibull distribution, have to be taken into account in order to calculate the amount of electricity that can be produced by wind turbines in a certain region.

The wind speed varies with height, depending on surface roughness and atmospheric conditions. The increase in wind speed with height is usually described in terms of a power law or by a logarithmic expression.

There are also daily and hourly variations in the wind speed. These variations are important for electric utilities making use of wind energy, since they have to adjust the output of the conventional power stations to meet these variations while matching electric demand. On time scales of minutes and seconds the variations in wind speed (turbulence) are important for wind turbine manufacturers as they influence the optimum design of a wind turbine.

Detailed Knowledge of Wind Regimes

In the past decade, studies have been carried out in many countries in order to estimate the regionally available wind energy resource. Some of these studies have developed into wind atlases – eg the *Wind Energy Resource Atlas of the USA*[2], the *European Wind Atlas* (for EU countries) and the *Atlas Eólico Preliminar de America Latina y el Caribe* (South and Central America). Wind maps have been published for China, Spain, Peru, Egypt, Jordan, Somalia, the Sahel countries, Ethiopia, parts of the C.I.S., etc. A wind map has also been made for the world as a whole[1].

Measurement of the Potential of Wind Power

On the basis of current knowledge about the regionally available wind resources, the potential of wind energy as a power source has been studied for different regions. In comparing the outcome of these studies, different types of potential need to be distinguished. Here we divide these types into the following five categories[3]:

- **Meteorological potential**. In essence this figure is equivalent to the available wind resource.
- **Site potential**. This figure is based on the meteorological potential, but is restricted to those sites that are geographically available for power production.
- **Technical potential**. The technical potential is calculated from the site potential, taking into account the available technology (efficiency, turbine size etc).
- **Economic potential**. The economic potential is the technical potential that can be realized economically.
- **Implementation potential**. In this figure constraints and incentives have been taken into account to assess the wind turbine capacity that can be implemented within a certain time frame.

The Potentials and Targets of Wind Power in Various Countries

Several countries have investigated their wind energy potential with regard to large-scale power production, and have formulated targets for the use of wind energy in the near future. Some figures relating to the wind energy potential and targets are indicated in Table 3.1. In most studies, no clear definition of "potential" is given, so the results presented in this table are a mixture of different types of wind energy potential analyses.

In 1988 a survey was published, indicating which developing countries might be the most promising ones for the application of wind energy[4]. In this survey not only climatological and economic factors were assessed, but also institutional and technological factors. It was concluded that the most favourable developing countries for the deployment of wind energy are Jordan, India, Pakistan and China, followed by Mauritania, Morocco and Chile.

The Global Potential of Wind Power

The global potential of power production using wind, which can be defined as the ultimate technical potential, has been analyzed in several studies. In 1981 the International Institute of Applied Systems Analysis (IIASA) estimated the ultimate technical wind potential world-wide to be 26,000 TWh/year (3 TWe), with a limitation to continental areas within 1,000 km off the coast line, and between 50 degrees Northern and Southern latitude. Because of economic, aesthetic and physical planning limitations it was assumed that roughly one third of this potential could be realized, or nearly 9,000 TWh/year (1 TWe)[5].

Table 3.1 *Estimated potential and targets for wind energy power production*

Country	*Estimated potential*	*Target (installed capacity)*
China[6]	1,600 GWe	100–200 MWe in 2000
Denmark[7]		1,000 MWe in 2000, 2,000 MWe in 2010
Finland[8]	11–16 TWh/yr	20–35 MWe in 2000 800 MWe in 2010
Germany[9]	2.7 GWe(economic potential)	250 MWe in 1995
Greece[10]	6.4 TWh/yr	150 MWe in 2000
India[11]	20 GWe	
Italy[12]		300 MWe in 2000
Jordan[13]		50 MWe in 2010
The Netherlands[14]		1,000 MWe in 2000, 2,000 MWe in 2010
Norway[15]	14 TWh/yr	
Spain[16]		100 MWe in 1993
Sweden[17]	30 TWh/yr	100 MWe in 1996
United Kingdom[18]	45 TWh/yr onshore; 230 TWh/yr/offshore	
USA[19]	2,500 GWe	4,000–8,000 MWe in 2000
CIS[18]	2,000 TWh/yr	

From the world-wide resources map[1] we estimate that about 27% of the earth's land surface (107×10^6 km^2) is exposed to an annual mean wind speed higher than 5.1 m/s (11.5 mph) at 10 metres above the surface. Table 3.2 shows the land surface exposed to different mean wind speeds above 5.1 m/s, totalling about 3×10^7 km^2.

If it would be possible to use this area for the installation of wind farms having a generating capacity of 8 MWe/km^2, 240,000 GWe installed wind turbine capacity could be realized. Of course, this is a fictitious number because land is in use for other purposes. In reality, probably just 4% of the area which is exposed to wind speeds higher than 5.1 m/s can be used for wind farms. This figure of 4% can be derived from detailed studies about the potential of wind power in the Netherlands and the USA[20, 21, 22, 23]. This means that on average about 0.33 MWe wind turbine capacity per km^2 can be placed. If we assume that the wind turbines have an average energy production of 2000 MWh/MWe per year (ie a capacity factor of 23%), the global potential of onshore wind power production is estimated to be 2.3 TWe (ie 2.3 TWyr/yr ≈

20,000 TWh/yr). For comparison, the total world energy consumption in 1987 was about 12.5 TWe (400 EJ). The total world electricity consumption was about 1 TWe.

It should be noted that the figure for the global potential of wind power depends strongly upon the assumed mean wind power production per square kilometre. For this value we chose 0.33 MWe. We derived this value from some regional potential studies. Further studies are needed to arrive at a more precise figure for different regions. Therefore, it is only a first rough estimate. It is possible that the ultimate potential can be some factors higher or lower than the figure presented. For example, the potential estimated by Grubb and Meyer[25], in which a reduction factor based on the population density is incorporated, is about 53,000 TWh/year. This is about 2.5 times the global potential estimated in this study.

Furthermore, it should be pointed out that this global potential is an estimate for large-scale, grid-connected wind turbines. Wind pumps and battery chargers can already be economical in areas where the mean wind speed is above 3 m/s. The land surface that is exposed to an annual mean wind speed between 4.4 m/s and 5.1 m/s (class 2), is about 50% of the earth's surface, which means that the application of small-scale wind turbines can be practical in many parts of the world.

Table 3.2 *Estimated world resource of wind energy*

WEC region	*Total land surface 10^3 km^2*	*Land surface with class 3–7* %	*Land surface with class 3–7* 10^3 km^2
North America	19,339	41	7,876
Latin America and Caribbean	18,482	18	3,310
Western Europe	4,742	42	1,968
Eastern Europe and CIS	23,047	29	6,783
Middle East and North Africa	8,142	32	2,566
Sub-Saharan Africa	7,255	30	2,209
Pacific	21,354	20	4,188
(China)	(9,597)	(11)	(1,056)
Central and South Asia	4,299	6	243
Total	106,660	27	29,143

Note:
Based on References 1, 2 and 24. For different wind classes the total exposed surface of the eight WEC regions (in thousands km^2 and as a percentage of total land surface) is indicated; class 3 represents an annual mean wind speed at 10 m height between 5.1 and 5.6 m/s; class 4 between 5.6 and 6.0 m/s; and class 5–7 between 6.0 and 8.8 m/s.

WIND TURBINE APPLICATIONS

As already mentioned, most of the wind turbine capacity installed over the past few decades has been coupled to the electric grid. Sometimes wind turbines are used in off-grid applications, like remote power production and battery charging. An important application is also the production of mechanical power for pumping water. In this section these applications are discussed in more detail.

Wind Pumps

Applications of water-pumping wind turbines are: water supply for livestock in remote regions; small-scale irrigation; and low-head pumping for aquatic breeding. Water supply for livestock is the main application of wind pumps. Over a million of these pumps are in use today – eg in Argentina, the USA, South Africa and Australia. About 300,000 wind pumps have been installed in South Africa, Botswana, Namibia and Zimbabwe. The energy production of these wind pumps is equivalent to about 50–75 MWh per year[26]. In South America more than 600,000 wind pumps have been installed, especially in Argentina[27].

A global evaluation of wind pump programmes has been conducted in a study commissioned by the World Bank and the United Nations Development Programme[28]. Figure 3.2 illustrates the situation of wind pumps for the year 1970 in the evaluated countries[5]. Since 1975, so-called modern wind pumps, intended for local production, have been developed in several countries, especially for low-lift and medium-lift

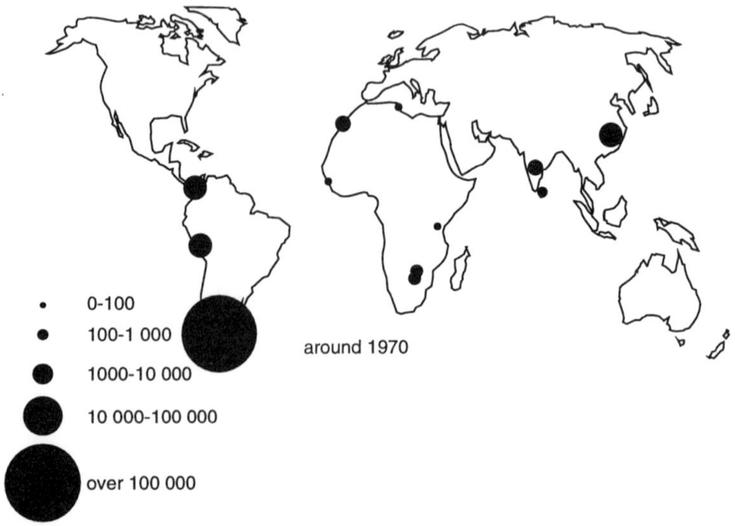

Figure 3.2 *Estimated number of wind pumps in selected countries*

applications (up to 50 m). About 10,000 of these pumps are now in operation. The pumps are being used for supply of domestic water and for small-scale irrigation (eg Sri Lanka, Colombia, Brazil, Mozambique, Cape Verde Islands). In China, there is a large demand for wind pumps that can be used for high-volume, low-lift applications for irrigation, prawn breeding and land reclamation. Optimum designs for such wind pumps are not available at the moment.

The number of wind pumps installed decreased sharply in the 1950s and 1960s, since in this period cheap fuel became available world-wide and the investment costs of small internal combustion engine-driven pumps went down considerably. The present rate of installation is estimated to be just over 10,000 wind pumps per year. A conservative estimate of the total world-wide potential of wind pumps would be 100,000 installations per year.

Off-grid Applications

A wide variety of wind turbines are used in off-grid applications. These applications can be divided into the following groups:

- **Battery charging.** In most instances this application involves small wind turbines for individual residential use. Most of the low and medium-priced wind turbines with a rotor diameter of 3 m (40–1000 W) belong to this category.
- **Remote power production with high reliability.** This involves small, unattended applications of wind turbines for highly reliable remote power. The wind turbine is usually associated with battery storage and might be used in conjunction with other power sources such as photovoltaics or diesel generators. Typical uses include powering marine navigation aids and telecommunication.
- **Water heating.** These systems are used for private residences. Typically the wind turbine is connected directly to an immersion heater or to an electric radiator.
- **Other remote uses.** These consist of village power production, powering mini-grid systems, commercial refrigeration and desalination.

The application that dominates the off-grid wind turbine market is the use of wind turbines for battery charging. Usually the rotor diameters of these wind turbines are less than 5 metres and their normal capacity is less than 1kWe. Battery chargers represent a growing market, with a trend towards smaller, less expensive systems and increased manufacturing in developing countries. Especially Chinese wind turbine manufacturing has grown rapidly to serve a huge domestic market. The market increased from 7,500 units shipped in 1986 to about 14,000 in 1987[29] and about 30,000 in 1990[6]. By the end of 1989 over 100,000 battery chargers had been installed in China. New markets have developed in regions like Argentina, Brazil, New Zealand and Morocco. The market in North America and Europe decreased sharply in 1987.

The world wind turbine market for high-reliability remote power systems has been stable over the period 1981–87. This market represents about 100 wind turbines per year, with a total annual capacity of 0.3 MWe[29, 30]. It seems that photovoltaic power systems have replaced wind systems for a number of high reliability applications, particularly marine navigational aids.

There is also a small stable market for water heating applications, especially in Europe. A market for 100 wind turbines per year with a total capacity of 1 MWe existed in the period 1981–86. In 1987, however, the market decreased to 48 units per year with a total capacity of only 322 kWe.

Other remote applications are stand alone wind turbines, wind-diesel systems or mini-grid applications. In 1985 about 20 wind turbines were shipped for this market. In 1986 this number increased to 70, with a total capacity of nearly 2 MWe. Nearly 1 MWe was shipped to Africa, in particular to Egypt. The growing awareness of wind technology options on the part of development funding organizations was the major reason for this increase. In 1987, however, the market decreased to 20 units, which was the level in 1985[24, 25].

Grid-connected Applications

There are generally two applications of grid-connected wind turbines:

- **Solitary wind turbines.** These provide power for residential, commercial, industrial or agricultural on-site electric loads. The electric load is near the wind turbine, and the load is also grid-connected. In most cases, this means that a wind turbine is placed near a farm or a group of houses. Typical sizes of these wind turbines are between 10 and 100 kWe.
- **Wind farms.** This application averages a centrally operated multiple wind turbine installation, designed to provide power primarily for distribution through the power grid, rather than to serve a specific local electric load. Typical sizes of these wind turbines are between 50 and 500 kWe.

The dispersed, grid-connected market is determined by national power generation policies. In the United States, Denmark, the Netherlands and Germany individuals have been able to interconnect privately owned wind turbines to the power grid and sell their excess generation to the local utility. Now, there is a movement towards larger units. The total shipments dropped from more than 900 in 1981 to about 500 in 1986, whereas the total rated power increased from 11 MWe to 38 MWe in the same period. In 1987 this trend persisted: the total shipment was 423 wind turbines, with a total rated power of 41 MWe[29, 30].

In the period 1981–86, the wind farm world market was dominated by the California market. The rapid growth of this market between 1981–85 was made possible by several economic and financial factors. Since 1986

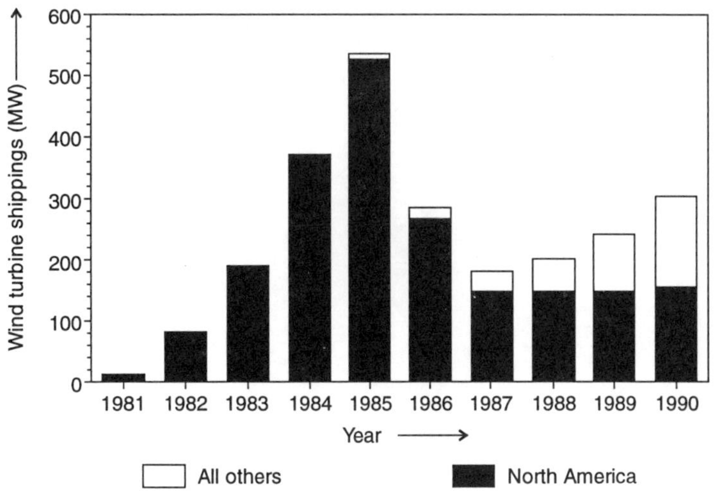

Figure 3.3 *Wind farm world market by region*

the sales of wind turbines in California have declined, because of the expiration of the tax credits. In this sector there is also a trend towards larger units; however, most of the wind turbines sold are in the range of 10–20 m rotor diameter, or about 25–100 kWe. In 1987, the market for wind turbines used in wind farms decreased to 1,495 units, with a total rated power of 178 MWe. Over 300 MWe was installed in 1990, equally dispersed over Europe and North America. Figure 3.3 shows that the market nowadays is on the same level as in 1986[23, 29, 30, 31]. However, the market has shifted somewhat to the European countries – particularly to Denmark, the Netherlands and Germany. The International Energy Agency (IEA) *Wind Energy Agreement Annual Report* estimated the world-wide capacity at the end of 1991 to be 2,200 MWe, amounting to a total of 21,000 wind turbines[32].

State of the Art and Installed Capacity

An overview of the state-of-the-art of wind turbines for different applications is given in Table 3.3. The total installed grid-connected wind turbine capacity in 1990 is shown in Table 3.4, as well as the electricity production and the mean capacity factor. Various sources have been used to compile these tables.

WIND TURBINE MANUFACTURERS

Expressed in units produced, China is the largest manufacturer of wind turbines. In 1987 more than 20,000 wind turbines were produced world-wide, 13,600 of these in China. In 1990 the Chinese production was

Table 3.3 *State-of-the-art for different applications*

	Rotor size (diameter)	*Number of turbines installed*	*Number of manufacturers*	*Investment costs (US$, 1989)*	*Energy costs (US cents/ kWh, 1989)*	*Technological phase*
Battery chargers	<5m	>100,000	50	300–1,000 per m^2 no tower/batteries		Commercially available
Water pumps	<8m multi-bladed >8m	>100,000 >10,000	50 (local (production)	400 per m^2 classical type 200 per m^2 modern type	1–3 per kWh^h* 0.25–0.75 per kWh^h*	Classical: commercially available. Modern: commercially available and in further development
0–100 kW	<20m	>10,000	50	600–1,000 per kW	0.05–0.10	Commercially available
100–300 kW	20–30m	>2,000	20	800–1,200 per kW	0.05–0.10	Commercially available
300–750 kW	30–40m	>100	20	1,000–1,500 per kW	0.05–0.10	Ready for mass production
>750kW	>40m	24	10	Commercially not yet available	–	Only prototypes available

* 1 kWh^h is the amount of energy needed to pump up 367/z m^3 over a height of z metres.

Table 3.4 *Estimated grid-connected wind turbine data*

Country/region	*Installed capacity (MWe) in 1990*	*Electricity produced (GWh) in 1990*	*Mean capacity factor (%)*
USA[19]	1,500	2,520	19.2
Denmark[31]	343	604	20.1
The Netherlands[33]	47	44	10.7
Germany[33]	23	26	12.9
Rest of Europe	20	n.a.	n.a.
India[11]	(1988) 31	n.a.	n.a.
China[34]	(1987) 7	n.a.	n.a.
Rest of the world[18]	(1988) 15	n.a.	n.a.
Total	about 2,000		

about 30,000[6]. Nearly all Chinese wind turbines are battery chargers to serve the domestic market. Outside China, 3,100 battery chargers were produced, mainly in Argentina and the United Kingdom[29,30].

The classical type wind pump is produced mainly in Australia, Argentina and the United States. Over the past two decades new simple wind pumps have been designed which can be produced in developing countries, where the main market is. However, this approach has not yet been a great success, mainly due to three factors. First, the design of more modern wind pumps still has some flaws, which are amplified when manufactured locally. Secondly, lack of funds and know-how in developing countries inhibit the setting up of large-scale production and technical maintenance. Also the market has not yet developed[28].

Grid-connected wind turbines are produced mainly in the USA, West European countries (eg Denmark, Germany, the Netherlands, Belgium, Italy, United Kingdom) and Japan. Due to an unstable market and ongoing research and development (R&D), the number of manufacturers in the USA and Western Europe has varied considerably. However, there is a trend towards the clustering of manufacturers in order to develop larger and better wind turbines (involving high R&D costs), which can compete with conventional electricity generating sources. At the end of 1990 the total manufacturing capacity was about 250 MWe per year. Table 3.5 presents an overview of the main manufacturers and their production up till the end of 1989[35].

Table 3.5 *Main manufacturers of grid-connected wind turbines*

Manufacturer	*Country*	*Number of wind turbines produced (until 1989)*
US Windpower	USA	3,500
Mitsubishi	Japan	500
Vestas/DWT	Denmark	2,800
Micon		1,600
Bonus		1,250
Nordtank		1,100
Danwin		300
Windworld		102
HMZ/Windmaster	Belgium/NL	269
Nedwind-Bouma	The Netherlands	58
Nedwind-Newinco		68
Lagerwey		125
Holec		19
MAN*	Germany	321
Enercon		35
MBB		29
Elektromat		15
HSW		9
WEG	United Kingdom	27
WEST	Italy	35 **
Riva Calzoni		50 **
Ecotècnia	Spain	?
Voest	Austria	?

* MAN has recently stopped its activities in this field.
** Figure for end of 1991

WIND TURBINE TECHNOLOGY

Exploiting wind energy by wind turbines is a very old concept. The earliest wind energy systems were used in ancient China and the Near East. When economic activities increased in Western Europe from the 15th century, wind turbines became important to provide mechanical power for pumping water and grinding grain. In the 19th century wind turbines contributed greatly to the economic development of countries like the Netherlands, Denmark and the USA. For example, by the early 19th century about 10,000 large wind turbines with rotor blades measuring up to 28 metres, were in operation in the Netherlands. In Denmark, wind energy prevailed until the latter half of the 19th century. At that time, about 3,000 wind turbines with a capacity of about 150–200 MWe were in operation and Danish industry relied on wind power for one-quarter of its energy demand.

After the industrial revolution, and especially in the 20th century, the use of wind energy declined everywhere to a negligible level, due to the exploitation of coal, oil and gas resources and rural electrification. The old wind turbines were no longer economically competitive. Thereafter, very little research was done to develop new and more efficient wind turbines. The development of wind turbines to generate electricity was resumed in the mid-1970s in response to the energy crisis of 1973. As a result, wind turbine technology improved considerably, and the cost of electricity produced by wind turbines decreased dramatically.

In this section an overview will be presented of some technical aspects of modern wind turbines (see also 36 and 37). The focus will be on wind turbines that produce electricity and that are connected to the grid. The state-of-the-art and the future development of these wind turbines are discussed. Thereafter, some attention is given to wind turbines that are used for water pumping and for charging batteries.

Design Components for Grid-connected Wind Turbines

Modern wind turbines can be divided into two basic configurations: horizontal axis and vertical axis (or Darrieus type) wind turbines (HAWTs and VAWTs, respectively). HAWT are the most common units manufactured. Both types use aerodynamic lift to extract power from the wind. They also have the same sub-systems, illustrated in Figure 3.4[38].

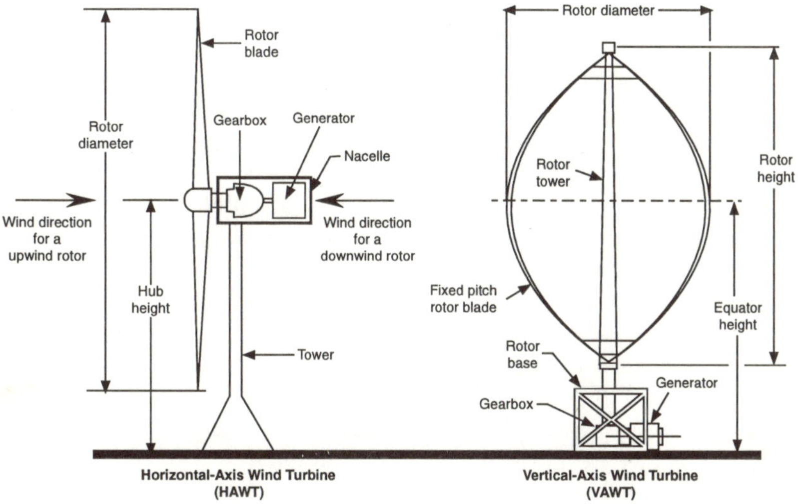

Figure 3.4 *Schematic overview of wind turbine components*

- rotor, which is the assembly of blades, hub and shaft;
- drive train, which usually includes a gearbox, a braking mechanism and an electric generator;
- tower, to support the rotor system;
- control and safety systems;
- others (electric interconnection, service facilities, supporting structures).

Rotor

Most HAWTs use two or three blades in an upwind design. When the scale of HAWTs is increased, two blades are preferred, due to a substantial weight penalty. There are also some counter-weighted single-blade HAWTs, which were developed in Germany and Italy. VAWTs are mostly two-bladed.

Blades can be manufactured from fibreglass-reinforced polyester, wood laminates, aluminium or steel. Fibreglass-reinforced polyester blades are used by most wind turbine manufacturers. These comparatively lighter blades exert less stress on bearings and rotor hubs. Other manufacturers use steel blades, because of the ease of fabrication, greater strength and lower cost. Sometimes wood laminate blades are used because they have excellent fatigue resistance properties. Most VAWT manufacturers use extruded aluminium blades.

Modern three-bladed HAWTs usually have cantilever blades, attached at their root end to a cast steel or cast iron hub (see Figure 3.4). One of the new design options for larger HAWTs, is to use a two-bladed teetered rotor. Teetering is the name given to the rocking motion that results when a two-bladed rotor is attached to the low speed shaft via pins and bearings, it allows the plane of rotation to tilt backwards and forwards by a few degrees away from the vertical. Teetering significantly reduces the loads on the blades.

Power control

One of the most crucial decisions to take when designing a wind turbine, is how to limit the power output at high wind speeds. The two main options for constant speed machines are “stall regulation” and “pitch control”.

The simplest option, stall regulation, uses fixed blades, and has been used both for HAWTs and VAWTs. If the wind speed increases, the angle of attack of the airflow over the blades increases, until flow separation (ie stall) occurs. This results in a loss of lift, an increase in drag and a levelling off of the output of the wind turbine. The effect of this process can be influenced by appropriate choice of the blade profile, the thickness and chord distribution and the blade twist. The great advantage of stall regulation is its simplicity and relatively low costs.

If provisions are made so that the blade pitch can be altered, either along the whole blades or at the outer portions, then at high wind speeds the angle of attack can be progressively reduced by changing the pitch. As a result, the power output can be held constant at the rated level of the turbine power. Pitch feathering has the advantage of good aerodynamic control, but the disadvantage of extra costs and complexity associated with the mechanism to change the pitch. Pitch feathering is used for HAWTs only.

In the past decade experience has shown that small wind turbines, up to about 25 metres diameter, mostly use stall regulation. Larger wind turbines now being sold mostly feature variable full span blade pitch control. Recently, large variable-speed machines have begun to enter the market, featuring lower stresses on the blades and hub as well as less energy loss due to power regulation.

Generators

The two main options for the generator used in constant-speed wind turbines are asynchronous (induction) or synchronous generators. Most grid-connected wind turbines installed so far use induction generators. These turbines have to be connected to the electricity grid before they can generate electricity. The generator is sometimes used as a motor to run the turbine up to synchronous speed, a feature that is utilized by stall-regulated wind turbines. Induction generators are simple and inexpensive. Their major disadvantage is that they draw reactive power from the grid system. Synchronous generators do not require reactive power and so they are favoured by utilities. However, these generators are more expensive.

Virtually all wind turbines installed in California and Europe over the past decade, have used induction generators and this trend seems to be continuing. For the variable-speed designs, ac–dc–ac power (electronic) conversion systems are used for grid interconnection and power control.

Gearbox

Wind turbines operate at relatively low speed. Blade tip speeds are typically in the range of 55 to 90 m/s, regardless of rotor size. Typically, the blades of a turbine of about 30 metres diameter rotate at about 35 to 50 rev/min. A speed-increasing gearbox is therefore required to provide an output shaft speed at the generator's synchronous speed (typically 1,500 rev/min). The two main gearbox types are with a planetary shaft or a parallel shaft. Parallel shaft designs are simple but relatively heavy, and the output shaft is usually offset. Planetary gearboxes are lighter and more compact and have an output shaft in line with the input shaft. For larger wind turbines (diameter >25 m), the cost and weight advantages of the planetary gearbox become increasingly significant.

Tower

"Lattice" or "tubular" towers are the two main types. In California, the installed wind turbines are about evenly equipped with these types of towers. The lattice tower is cheaper. The tubular tower is generally regarded as visually more pleasing. Access for maintenance to the nacelle during bad weather conditions is much more practical from within the shelter of a tubular tower. Most future European wind farms can be expected to feature turbines with tubular towers.

Characterization of wind turbines

To specify the characteristics of a wind turbine, it is common practice to specify the power output (P) as a function of the wind speed (v), which results in the so-called P-v curve. This curve gives a good indication of the performance of a wind turbine. It can be used to predict the maximum power output with a certain wind regime. Standard procedures have been developed to measure these P-v curves[39]. Figure 3.5 gives some typical examples[40]. To characterize a wind turbine, information is needed about the P-v curve, the rotor diameter, the hub height and typical elements such as control systems and generator type.

Status of Technology and Potential Developments of Grid-connected Wind Turbines

The majority of wind turbines currently installed in the California wind farms represent designs developed in the mid-1980s. Most of the design practices employed around 1980 were focused on steady-state operation and the predicted design loads did not take into account the random, or turbulence-induced, wind fluctuations under which wind turbines operate. Fluctuating aerodynamic forces can lead to fatigue failures in the blades, rotors and other components. Even today, these aerodynamic forces and their interaction with wind turbines are not well understood. However, further knowledge is being gained through specially designed testing procedures.

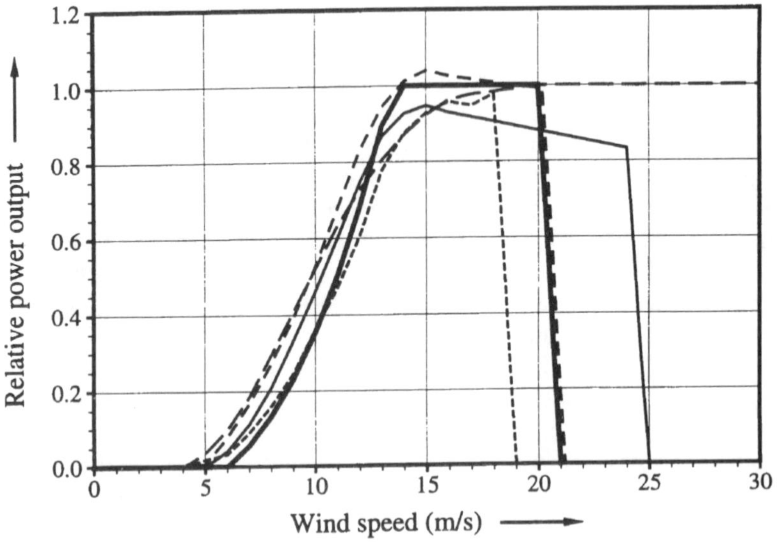

Figure 3.5 *The relative power output versus wind speed for modern wind turbines*

For all installed wind turbines in California the energy output has increased from an average of 400 kWh/m^2 (swept area) per year in 1982 to about 600 kWh/m^2 per year in 1989. Though some improvement has come from better design of the turbine and better siting, the main gains were achieved as a result of improved reliability. Due to design improvements, wind turbines with a rating of 250 to 500 kWe (25 to 35 m diameter) are commercially available. The energy output of these wind turbines is about 100 kWh/m^2 per year.

It is important to investigate the possibilities of reducing costs and improving reliability and efficiency. In a study carried out by the Solar Energy Research Institute (SERI, but name recently changed to National Renewable Energy Laboratory, NREL)[41], two general configurations for advanced turbines for the near term are presented. Fatigue life and reliability have been identified as key problems with existing designs, so both advanced systems address these issues. Concept 1 incorporates a variable-speed rotor, using power electronics to supply constant frequency power to the utility. It also uses advanced control systems to take full advantage of different wind characteristics and advanced airfoils for increased power output and reduced loads. Concept 2 uses a stall-controlled rotor, using passive aerodynamics to limit maximum power output. This concept also takes advantage of the improvements previously mentioned. Table 3.6 lists the expected impact of the improvements for the two concepts[41].

Table 3.6 *Performance and costs estimate for two advanced concepts*

Technical advancements	*Improvement in system costs %*	*Improvement in energy capture %*	*Improvement in annual O&M costs (US cents/kWh)*
Analysis codes			
Structural	5%	–	–
Fatigue	5%	–	–
Concept 1 – variable speed			
Power electronics	-10%	10%	0.00
Control systems	-1%	5%	0.20
Advanced airfoils	0%	10%	0.10
Drive train	4%	–	0.10
Tower (tall)	-8%	25%	0.01
Rotor hub	5%	–	0.10
Total	0%	56%	0.51
Concept 2 – stall controlled			
Aerodynamic controls	2%	3%	0.10
Control systems	-1%	5%	0.15
Advanced airfoils (rotor design)	2%	–	0.10
Drive train	2%	–	0.10
Tower (tall)	-8%	25%	0.01
Rotor hub	5%	–	0.10
Total	12%	49%	0.61

Possible innovations in wind turbine design with a high potential in the next century include:

- advanced airfoil families designed specifically for wind turbines to increase performance and allowing active or passive rotor control;
- variable speed drives or generators to allow the rotor to operate at optimum speeds over a wide range of wind speeds;
- adaptive or smart controls that adjust system operating parameters on the basis of the wind characteristics;
- hub configurations that allow for greater flexibility, thus reducing loads and increasing lifetimes;
- incorporation of advanced materials to allow the manufacture of lighter and stronger components;
- development of a damage resistant rotor, adapting aerospace techniques in the use of composite structures; and
- increased performance by a better understanding of micro-siting effects on wind characteristics, such as turbulence and wind shear.

Most of the improvements identified for the short-term design, focus on improving energy capture with minimal cost impact. The advances for the longer-term designs will probably have a more significant cost impact. SERI[41] expects that the effects of the near-term improvements can reduce the wind power generation costs from about 7–13 US cents/kWh to about 5 US cents/kWh for a typical Great Plains (central USA) site (wind speed 5.8 m/s). The longer-term improvements can reduce the wind power generation costs further, to about 3–4 US cents/kWh for a typical Great Plains site.

Development Towards Larger Wind Turbines

Wind turbine technology is still developing. Part of this development concerns the size of wind turbines. There is a definite trend towards larger wind turbines. By the end of 1989, in total 43 wind turbines with a rated power equal to or larger than 500 kWe had been installed in the IEA countries, amounting to a total capacity of 45 MWe, detailed in Table 3.7. Nine units, totalling 13 MWe, were under construction. These wind turbines were all installed for research purposes only. As shown in Table 3.7, the installed capacity is expected to increase substantially after 1989[42].

One of the key questions addressed frequently over the past ten years, concerns the optimum size of wind turbines to achieve the lowest energy production costs. System analyses made in the 1970s, indicated optimum sizes in the 60 m to 100 m diameter range (1 to 3 MWe). These results gave extra impetus to several national R&D programmes.

In California, a dramatic market for wind turbines was created in the early 1980s by legislation and tax credits. This led to a rapid deployment of small wind turbines (<50 kWe). After this stage, the mean rating

Table 3.7 *Cumulative history of large wind turbines (larger than 500 kWe)*

	Model/location (capacity (kW) in year of first rotation)	*1978–79*	*1980*	*1981*	*1982*	*1983*	*1984–85*	*1986*	*1987*	*1988*	*1989*	*1990*	*1991*
Can	Aquilo/AWTS						500						
	EOLEC/Cap Chat								4,000				
DK	Tvind	2,000											
	Nibe	630	630					5×750					
	Windane 40/ Masnedø								2,000				
	Tjæreborg												
D	GROWIAN/ K. Wilhelm – Koog					3,000			–3,000				
	Monopteros/ Jade									640	2×640		
	WKA–60/ Heligoland											1,200	
	WKA–60/ K. Wihelm–Koog												1,200
	HSW 750/ Husum												750
	AEOLUS II/Jade												3,000
ESP	AWEC-60/ Cabo Villano										1,200		
	Endesa/Cabo Villano												550
IT	GAMMA 60/ Alta Nurra												1,500
NL	NEWECS–45/ Medemblik						1,000						
	NEWINCO/ Maasvlakte										500		
	HMZ/Bergen op Zoom										500	–500	
	Holec/near Urk											500	
S	WTS–3/Maglarp				3,000								
	WTS–75/ Näsudden					2,000						–2,000	
	Howden/ Risholmen									750			
	Näsudden II												3,000
UK	WEG LS 1/ Burgar Hill								3,000				
	Howden/ Susetter Hill									750			
	Howden/ Richborough										1,000		
	VAWT 850/ Carmarthen Bay											500	
US	Mod–1/Boone	2,000			–2000								
A	Mod–2/ Goldendale		2,500	5,000				–7,500					
	Mod–2/Medicine Bow		2,500					–2,500					
	Mod–2/Solano				2,500				–2,500				
	WTS–4/Medicine Bow			4,000									
	WWG 0600/ Kahuku Pt.								14 ×600				
	VAWT Test Bed Bushland								500				
	Mod–5B/Kahuku Pt.									3,200			
Cumulative (kW)		4,630	10,260	19,260	22,760	27,760	29,260	23,010	35,410	40,750	45,230	44,930	54,930

increased from 50 kWe for wind turbines installed in 1981, to 110 kWe in 1987 and 160 kWe in 1990[19]. The trend towards larger turbine sizes is steadily continuing. Now, wind turbines with a rating of 250 kWe to 500 kWe (25 m to 35 m diameter) are commercially available.

It is very difficult to establish in advance what the optimum size of wind turbines will be. The optimum size depends on many aspects like the wind speed, the wind turbine costs, the construction and erection costs, which might be different for different terrain types, the operation and maintenance (O&M) costs, the costs of infrastructure, the costs of land use, the environmental impact and the social costs. Therefore, the optimum size might be very location-specific. However, some judgements can be made by reasoning that, in first approximation, the wind turbine costs will be largely determined by the weight of the material that is utilized. The energy captured by a wind turbine is proportional to the square of the rotor diameter, while the volume, and therefore the weight, is proportional to the cube of the diameter. Therefore, there is a weight and cost penalty as wind turbine sizes increase and this is exacerbated by the additional facilities that are required to handle larger components during erection and maintenance. The cost penalty will, at least partially, be compensated by reduced costs for infrastructure, O&M and land use.

In some countries, the need for larger wind turbines has been argued based on the fact that for many land-based wind farms it will be difficult to obtain planning consent. Therefore, large wind turbines should be built on the scarcely available locations, to obtain a high energy production per area. Sometimes placing wind turbines offshore in shallow waters is seen as an alternative. In such a situation large wind turbines will be required to justify the extra costs for foundation offshore, erection and O&M.

Figure 3.6, modified from data presented by Hau[43], shows the relationship between tower head mass per square metre of rotor disc area and the rotor diameter. The figure demonstrates that the mass per unit area of rotor disc steadily increases. However, the interpretation of costs per kWh is not as simple as suggested above. The increase in costs is not linear with weight and wind speed. Therefore the yield increases with height. So the conclusion that large wind turbines have high kWh costs must be interpreted carefully. New design options, such as two bladed teetered rotors, flexible blade materials and pitch control, can provide a route to lower costs even for large wind turbines[36].

Design of Wind Turbines for Water Pumping

Besides grid-connected wind turbines, other types of wind turbines have been developed for small-scale electricity production (battery chargers) and for water pumping. A variety of wind turbines are used for water pumping. These are all HAWTs, although in the past (and also recently) efforts have been made to develop VAWTs for pumping, especially Savonius rotors[26]. A convenient classification of water pumping wind

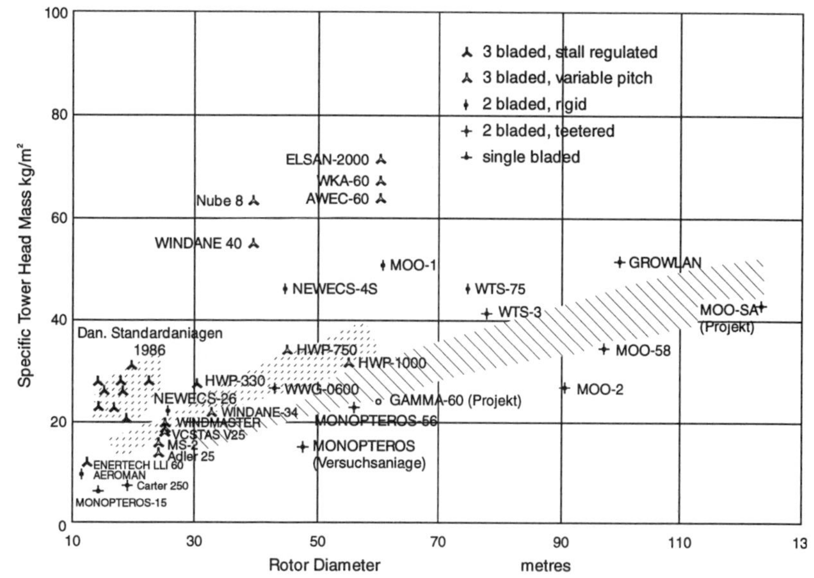

Figure 3.6 *Specific tower-head mass versus rotor diameter of horizontal axis wind turbines*

turbines can be made on the basis of the type of transmission between the rotor and the pumping device.

- **Wind turbines driving piston pumps.** The rotor is coupled mechanically to the piston pump. This is by far the most common type of wind pump. Two different types of applied technology can be distinguished. The first is known as the classical multi-bladed wind turbine. This type has 15 to 36 blades, mostly supported by a structure of spokes and rims, operating at low tip speeds (about the wind speed). The transmission normally contains a gearbox and it uses a safety system. The second consists of more recently developed wind pumps, using fewer blades (4 to 12), operating at higher tip speeds and incorporating some new and simple safety systems. This type usually does not use a gearbox but the rotor is coupled directly to the pump.
- **Wind turbines with rotating transmission.** Through a rotating transmission, the energy of the rotor is transferred to a rotating pump, for example a centrifugal pump or a screw pump. Both are used especially for low-head and high-volume applications.
- **Wind turbines with pneumatic transmission.** Wind turbines that can drive air compressors are produced by a few manufacturers.
- **Wind-driven electric pumping systems.** Wind-driven electric generators are sometimes used to drive electric water pumps directly.
- **Wind turbines with hydraulic transmission.** Several experiments have been performed on wind turbines driving water pumps by means of hydraulic transmission.

Design of Wind-driven Electric Battery Chargers

Wind-driven electric battery chargers also produce electricity but are not connected to the grid. Wind turbines are used to charge batteries from which electricity is supplied. These wind turbines are small, with rotor diameters of 1–5 metres and a rated power from 75 We to about 4 kWe. The rotor consists of 2–6 blades, made of carbon-reinforced epoxy, extruded aluminium alloy or wood. These wind turbines do not have a gearbox, and the rotor is coupled directly to the generator. Most generators used for battery chargers are permanent magnet, brushless generators. The output of the generator is controlled by an electronic device. As a storage medium, usually a bank of stationary lead-acid batteries is used[44].

ENVIRONMENTAL ASPECTS

In this section the environmental aspects connected with the deployment of wind energy on a (very) large scale will be considered (see also reference 45). First some figures are given for the energy payback time of wind turbines. Thereafter some environmental problems connected with the use of wind turbines, like safety, noise and visual impact, are described.

Indirect Energy Use and Emissions

Energy is required to produce the materials used to construct a wind turbine (steel, concrete, other materials), as well as to produce and install the wind turbine. This energy investment has to be paid back during the lifetime of the turbine. Energy analyses indicate that the so-called "energy payback time" of wind turbines varies from a few months to a couple of years at the most[44,45].

A study carried out by the University of Groningen in the Netherlands indicates that larger wind turbines have shorter energy payback times[45]. The energy payback time of a 10 kWe wind turbine varies between 0.5 and 1.5 years, whereas for a 500 kWe wind turbine it is about 2–3 months. In a study done by the Fraunhofer Institute for Solar Energy Systems, it is estimated that the energy payback time for wind turbines is about 1 year, as shown in Figure 3.7[47].

Some pollution is caused by the construction and operation of wind turbines. However, little is known about these so-called indirect emissions. San Martin analysed the CO_2 emissions during fuel extraction, construction and operation of different energy systems. Conversion of these emissions into tonnes of CO_2 per GWh electricity production gives the results presented in Table 3.8[46]. From this table it can be seen that the CO_2 emission per GWh electricity production during construction is higher in the case of wind energy than with most other electricity supply systems, but that in total the CO_2 emission over the total operating period is very low (eg 1% of the system using coal).

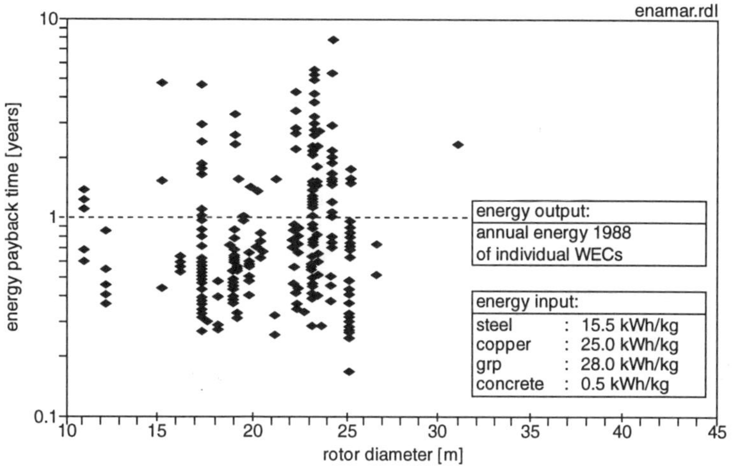

Figure 3.7 *Energy payback time for different wind turbines*

Table 3.8 *The CO_2 emissions from different electricity production technologies*

Technologies	*CO_2 emissions at various energy production stages (tonnes per GWh)*			
	Fuel extraction	*Construction*	*Operation*	*Total*
Conventional coal-fired plant	1	1	962	964
AFBC plant	1	1	961	963
IGCC electric plant	1	1	748	751
Oil-fired plant	–	–	726	726
Gas-fired plant	–	–	484	484
Ocean thermal energy conversion	na	4	300	304
Geothermal steam plant	<1	1	56	57
Small hydropower*	na	10	na	10
Boiling water reactor	−2	1	5	8
Wind energy	na	7	na	7
Photovoltaics	na	5	na	5
Large hydropower	na	4	na	4
Solar thermal	na	3	na	3
Wood (sustainable harvest)	−1,509	3	1,346	−160

– Missing or inadequate data for analysis, estimated to contribute $\leq$1%.
na Not applicable.
This analysis considered construction of new dams. According to a recent US Federal Energy Regulatory Commission report there is 8,000 MWe of small hydropower under construction or projected, much of it involving refurbishing or refitting existing dams, which would substantially reduce the CO_2 impact of small hydropower plants.

Birds

The operation of wind turbines may damage birds. This damage can be split up into bird kills as a result of collisions with tower or blades and the disturbance of breeding or resting birds in the vicinity of turbines. Research has been carried out at various institutes, including the Research Institute of Nature Management (RIN) in the Netherlands and the Game Biology Station in Denmark[40]. In a study by RIN on the effects of a 7.5 MWe wind farm (a row of 25 turbines of 300 kWe) in Urk (the Netherlands), it was found that on average 0.1 to 1.2 bird kills per day can be expected[48]. It was concluded that the number of bird kills per km wind farm was up to ten times smaller than the number of bird kills per km of high voltage transmission line, and comparable with that of one km motorway. With moves towards larger wind farms encouraged by a combination of economics and the structure of government incentives in some countries, it is feasible to envisage 80 MW wind-farms causing over 10 bird fatalities a day, or over 3,500 per year. This would widely be regarded as unacceptable. A report by Ornis Consult in Denmark states that, on the basis of observations of nine medium and small-scale wind power plants, it might be concluded that birds are not really disturbed by wind turbines[49]. Birds seem to get used to the wind turbines and learn to fly around them. This is the experience at many smaller wind farms. However, in the USA the birds are causing problems since some protected species have been reported killed by wind turbines[50].

The worst example is near Tarifa, Spain where, at the end of 1993, 269 wind turbines had been erected and a total of 2,000 turbines have been mooted. Sited on Western Europe's main bird migration route to Africa, close by the Straits of Gibraltar, this location is a monumental misjudgement. Many birds of numerous protected species have already been killed by the turbines. The Director of Spain's renewable energy agency, IDAE, has said: "What I think happened was that there was a very unfortunate lapse of memory. Nobody thought about the migratory birds." This is a rather extraordinary admission[51].

In conclusion, it seems that this problem has been small to date, although it is clear that possible damage to bird life must be analyzed carefully and might limit the application of wind energy in certain regions, such as migratory flyways.

Noise

The disturbance caused by the noise produced by wind turbines is probably the most important drawback to siting wind turbines close to inhabited areas. The acoustic emission is composed of a mechanical and an aerodynamic part. The aerodynamic part in the acoustical emission is a function of the wind speed. Analysis shows that for turbines with rotor diameters up to 20 m the mechanical component dominates, whereas for larger rotor diameters the aerodynamic component is predominant.

Several standards have been recommended for the measurement of the emission of noise from wind turbines. One example is the IEA report on wind turbine noise measurements[52].

In determining the nuisance level, the acoustic source level of a wind turbine must be calculated from acoustic emission measurements. An example of the acoustic source levels as a function of rotor diameter is given in Figure 3.8[53]. With this source level emission, noise contour lines can be determined. Using the acceptable emission levels, it can be determined whether or not a specific wind turbine site is acceptable. This noise problem can cause limitations to potential wind turbine areas.

The noise problem emerges especially in densely populated areas – for example the Netherlands. It is for this reason that the Department of the Environment of the Dutch government gives a grant for the installation of silent wind turbines.

Local topography can also have a profound effect. Near Llandinam in mid-Wales (United Kingdom) noise has affected a significant number of local residents because of the relative silence found in the surrounding valleys, refraction by meteorological gradients, wind direction and alignment of turbines, the particular turbine model, and even how stepping a few metres one way or another seems to affect the volume.

Visual Impact

The visual impact of wind turbines, although of a rather qualitative nature, can be a realistic planning restriction, particularly for areas of outstanding natural beauty, wetlands, landscapes with cultural value and in densely populated areas or countries. Several studies have been

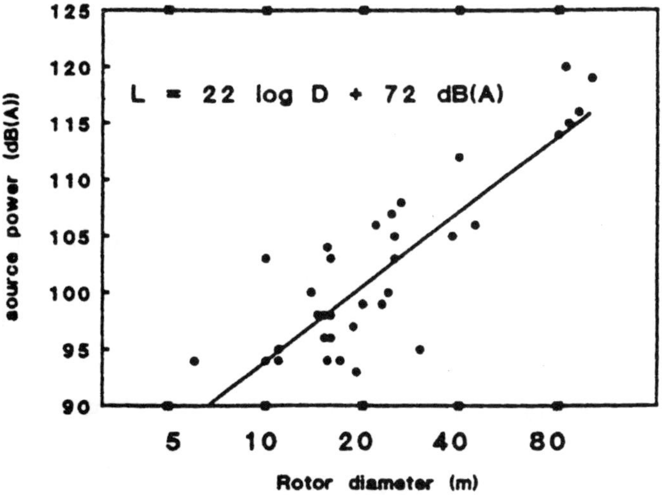

Figure 3.8 *The measured acoustic source power of wind turbines versus rotor diameter*

carried out to investigate the attitude of people to landscapes with wind turbines[54, 55]. In a recent study done by the University of Leiden (Netherlands), it was found that the public appreciation of a landscape decreases as more and more wind turbines are installed, but that the size of the wind turbines has much less effect on the appreciation. There was also a preference for wind turbines placed in line instead of in rectangular grids. This preference might be related to the linear structure of the Dutch landscape[56].

A special case of visual impact is the effect of the shadow of the wind turbine, particularly of the rotor blades. This can cause hindrance in situations where wind turbines are installed close to the workplace or residences. However, this effect can easily be predicted by calculating the hourly position of the shadow for each month. The rotation frequency plays an important role in determining the disturbance level in this case[45].

In a number of instances initial objections to wind farms on grounds of visual intrusion have been replaced by rising acceptability based upon experience (eg Delabole, Cornwall, UK). In other cases, profound concern and opposition remains. This hostility is particularly linked to upland areas of great natural beauty or decisions to place wind farms in, or within sight of, areas of great historic/heritage interest (eg St Breoch Downs, Cornwall, UK amidst renowned Bronze Age remains; Humble Hill, Kielder Forest/Tynedale, Northumberland, UK within sight of the Roman Wall named after the Emperor Hadrian). Such inappropriate siting of wind farms could severely limit the speed and scale of wind-power development.

Visual intrusion is also a function of the location of wind farms and the size and number of turbines. Where only sites yielding over 7.5 m/s mean wind speed are seriously considered, such sites are likely to be exposed and visually intrusive on approach. They are likely to be particularly sensitive to size and number of turbines. The capacity for wind farms to be economically sited where annual mean wind speeds are lower (see final paragraph of the "Global potential of wind power" section above), particularly flatter landscapes and where smaller wind turbines can be used, would greatly alleviate this problem.

Visual acceptability can be enhanced not merely by the alignment of turbines but also by use of a common size and design. Developments in California, USA have generally not conformed to this principle, with unfortunate visual consequences.

Offshore wind farms (such as at Vindeby, off the Island of Lolland, Denmark) do not in general create the same severity of visual intrusion. Care is required to avoid major bird migration routes (a proposal in the 1980s to locate a windfarm off Wells-Next-The-Sea, Norfolk, UK would have directly conflicted with this criterion) and prevent interference with shipping communications. Their additional construction, maintenance and depreciation costs would be less of an issue if internalisation of all energy provision and use costs was adopted.

Telecommunication Interference

Wind turbines present an obstacle for incident electromagnetic waves. These waves can be reflected, scattered and diffracted. This means that wind turbines may interfere with telecommunication links. The IEA has provided preparatory information on this subject[57], identifying the relevant wind turbine parameters and the relevant parameters of the potentially vulnerable radio services. Clearly, wind turbines should not interfere with telecommunication links, nor with domestic radio and TV reception[45].

Safety

Accidents with wind turbines are rare but they do happen, as in other industrial activities. Fatal accidents have occurred in the USA as well as in Europe. In most cases these accidents happened when technicians tried to stop wind turbines. From an operational point of view nearly all accidents should not have happened. For example, an accident occurred when someone climbed up the tower of an operating wind turbine in a violent storm; even experienced technicians should never do this.

Also blade throw occurs. A detailed analysis is given by Turner[58]. He concludes that, with a probability of failure of 10^{-5} per year, the probability of a person being hit near the wind turbine is 10^{-7} per year and at 600 m distance 10^{-10} per year. However, an analysis for the Dutch situation, carried out by the Dutch consultancy CEA[59], indicates that fractures of blades have occurred 2.7×10^{-2} per year on average, which is much high than indicated by Turner.

In 1987, a project was initiated by the European Community (EC) to set up a conceptual European safety standard for wind turbines. This document addresses aspects like safety philosophy, structural integrity and personnel safety. The aim of the document is to give guidance to designers so that the probability of an accident with the wind turbines they produce is at an acceptable level. The International Electrical Committee (IEC) has taken the initiative to produce an international standard on safety[60]. IEA has also produced a document on safety standards[61].

ECONOMIC ASPECTS

The economic aspects of wind energy are quite different for different applications. The largest contribution to the world energy supply will be made by the grid-connected wind turbines. In this section the economics of grid-connected wind turbines is discussed. Wind pumps are also an important application, especially for developing countries. Therefore, a brief description of the economics of wind pumps is also included.

Grid-connected Wind Turbines

In discussing the economic aspects of wind energy, the generating costs and the value of the energy produced clearly have to be treated as different subjects. In this section the generating costs of wind energy are discussed. Cost figures are given for the category of grid-connected wind turbines with a rated power between 100 and 300 kWe. At this moment, these wind turbines are the most cost effective. Smaller and larger wind turbines involve higher costs. The total generating costs for a wind turbine system are determined by:

- total investment costs;
- lifetime of the system;
- O&M costs;
- wind regime;
- energy efficiency of the wind turbine; and
- technical availability of the system.

These aspects are to be discussed in detail below. The production cost per kWh is also evaluated as a function of the wind regime.

Investment Costs

The total investment costs of wind turbines can be split up into different cost components such as manufacturing costs (ex factory), costs for infrastructure and installation, such as foundation, civil works and engineering, and grid connection costs. The relation between these cost factors varies markedly among countries, and it depends on aspects like the condition of the soil and the distance from the electricity grid. In Figure 3.9 a breakdown of costs for wind turbines placed in the USA and the Netherlands is given[45].

From this figure it can be concluded that the wind turbine component represents roughly 75% of the total investment costs. This percentage is generally valid for industrialized countries. In a country such as India,

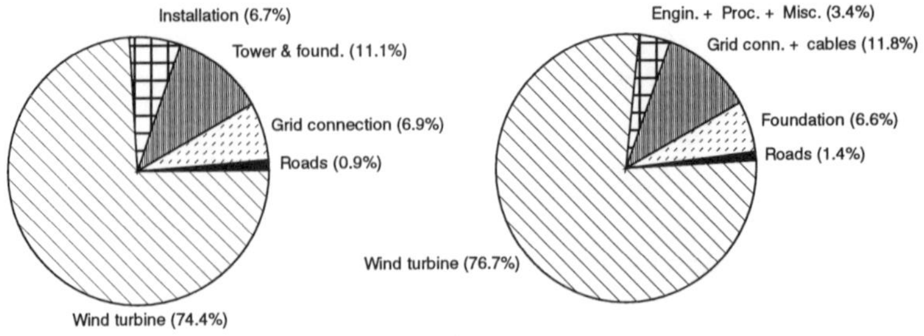

Figure 3.9 *Comparison of cost breakdown of a Danish wind turbine in a USA wind farm (left) and a Belgian wind turbine in a Dutch wind farm (right)*

the relative cost breakdown is different: about 30% to 50% of the total costs are for infrastructure, installation and grid connection[11].

It is difficult to give one figure for the total investment costs. In a study by the Department of Energy of the USA, the total investment costs for a 200 kWe wind turbine are estimated to be US$400–500 m^2 or US$1,000–1,200/kWe (1989 dollars)[62]. An EC study estimates the total investment costs to be ECU400–600/m^2 or ECU900–1,100/kWe (ie between US$1,000–1300/kWe) for wind turbines in the range between 100 and 400 kWe[63]. A Danish report asserts that the total investment costs for a 250 kWe wind turbine typically are US$800/kWe[64]. From these figures it can be concluded that in 1990 the total investment costs were in the range US$800–1,300/kWe. This is confirmed by a recent call for tenders in the Netherlands, yielding values between US$900 and US$1,800/kWe[65].

The experience in the USA has shown that a considerable reduction of turnkey costs can be realized in a short time, as illustrated in Figure 3.10[45]. It is highly probable that this reduction has been achieved by a reduction of the wind turbine costs rather than by a reduction of the infrastructure costs. In the near future it should be possible to reduce the total investment costs further by a cheaper production of wind turbines, by careful siting and by planning experience. On the other hand, the costs could rise because of higher grid connection costs, as indicated in a Dutch study[66].

Lifetime of the system

For economic analyses of wind turbines, an economic lifetime of 20 years is often used. This is equal to the figure commonly used for the

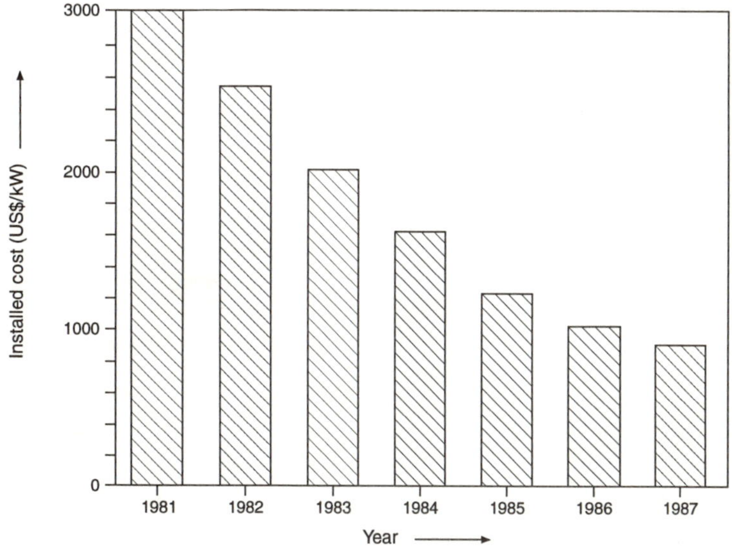

Figure 3.10 *Development of turnkey costs of USA wind plants (normalized for currency changes)*

design lifetime. However, it should be noted that the proven technical lifetime for the best wind turbines at present is about 10–15 years.

Operation and Maintenance Costs

The annual O&M costs are often discussed as a percentage of the total investment costs or of the generation costs per kWh. Quoted O&M costs based on practical experience vary from 0.5 UScents/kWh for mature European wind turbines to very high figures for projects that failed. In an EC study[63] the O&M costs are estimated to be 2% of the investment costs (about 1 US cent/kWh). A study by the US Department of Energy (USDOE) and SERI[62] also estimates the O&M costs to be 1 US cent /kWh. The Danes estimate the O&M costs for their turbines to be 0.6 US cents/kWh in the first two years, 0.8 US cents/kWh in the next three years, and 1 US cent/kWh in the following years[64].

Wind Regime

As the wind energy potential is proportional to v^3 (v is the wind speed) it is extremely important to choose sites with an annual mean wind speed as high as possible and where the wind is persistent and frequently blows during times of peak electricity demand. From the world-wide resources map[1] a rough indication of site-specific mean wind speeds can be obtained. For siting purposes this indication is not accurate enough. Information about the nature and the roughness of the surrounding landscape (land, sea or hilly terrain) and prevailing wind climate figures (such as maximum gusts, turbulence intensity, boundary layer parameters monthly variation, diurnal variation and frequency distribution of the wind speed) are required. Sometimes measurements and/or modelling techniques are required to obtain relevant figures.

An example of the wind speed distribution function is given in Figure 3.11 for three different locations[64]. It illustrates the great differences in wind climate throughout the world, and the need to analyze the wind climate for potential sites carefully. To be able to calculate the value of using wind in a utility system (see "The value of wind energy" section below), the time dependence of the wind regime should also be known. The steadiness, diurnal variation, and turbulence of the wind resource can be just as important as the mean annual wind speed.

Energy Efficiency of the System

In general, the energy output of wind turbine systems are given per year for easy comparison. The mean energy output per square metre swept rotor area over a year can be expressed by:

$$E = b\,(v^3),\ (kWh/m^2 \text{ per year})$$

in which b is the performance factor and v is the annual mean wind speed. The factor b can be seen as a measure of the performance quality of a wind turbine. However, b depends on the annual mean wind speed and the wind speed distribution. Therefore b is not constant throughout the world.

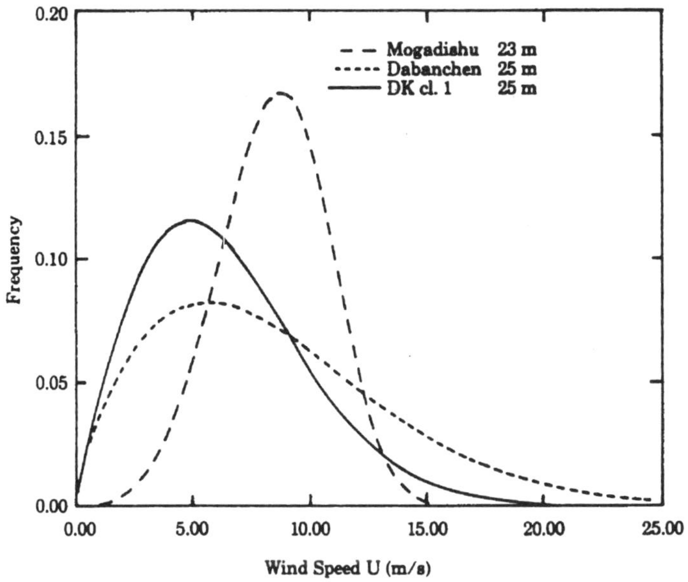

Figure 3.11 *The wind speed frequency distribution function at three locations*

The ECN test station in the Netherlands has determined *b* for several wind turbines during the period 1981–9[67]. In Figure 3.12, *b* is plotted against the year in which the wind turbine was manufactured[67]. Figure 3.12 indicates that technology improvements have enhanced the performance quite strongly. For the West European wind climate the maximum performance factor that can be achieved is about 3.5.

Two other measures are commonly used to see whether or not the wind turbine performance has improved. The first is the specific yield in kWh/m^2 (swept area) per year as a function of the year of installation of the wind turbine. An analysis based on the specific yield has been made for Danish turbines[69], and is shown in Figure 3.13. The second measure is the capacity factor; the yield in kWh per year divided by the installed capacity and the total number of hours in a year. In the USA, at very good wind sites, capacity factors of more than 50% have been achieved, although 20–25% is more typical of wind areas developed to date. It must be noted that the two measures discussed here are determined by a combination of the achievement of the wind turbine and the site-specific wind regime.

Technical Availability

The (technical) availability is the fraction of time in a year that the wind turbine is potentially able to generate energy. A wind turbine can be unavailable because of planned maintenance activities or because of unforeseen failures, incidents or accidents. Reliable figures for the availability of different wind turbine types can be presented only if they are based on experiences with a large number of wind turbines and on an operation period of many years. At this moment only the USA and Denmark have enough experience to provide these data.

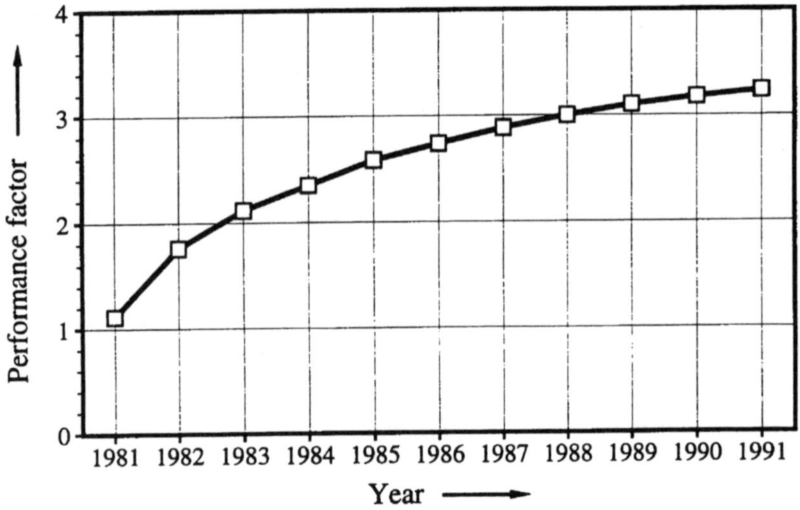

Figure 3.12 *The development of performance factor for wind turbines versus year of manufacture*

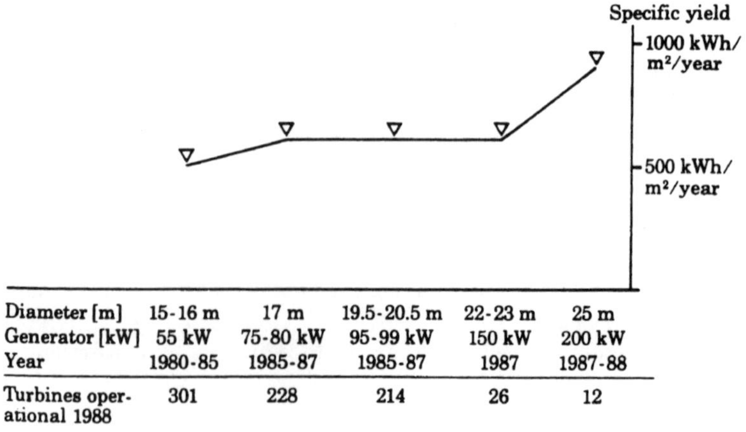
Specific yield
1000 kWh/ m²/year
500 kWh/ m²/year
Diameter [m] 15-16 m 17 m 19.5-20.5 m 22-23 m 25 m
Generator [kW] 55 kW 75-80 kW 95-99 kW 150 kW 200 kW
Year 1980-85 1985-87 1985-87 1987 1987-88
Turbines operational 1988 301 228 214 26 12

Figure 3.13 *Specific yield in 1988 for different wind turbines, corrected to "normal wind year"*

Also the development stage of the wind turbine should be considered in assessing availability. Data on availability are known from other fields of technology, such as the aircraft industry. Wind turbine manufacturers can learn from these data. For example, aircraft data indicate that the availability of new engine designs is rather low in the first years of operation. Availability figures for derivative engine designs are, however, much better. Wind turbines are still in the first development stage, but significant improvements have already been made. Experiences from the USA, such as illustrated in Figure 3.14, show that medium-sized wind turbines (up to 250 kWe) have probably reached maximum availability. The larger wind turbines (>300 kWe) are still in the initial phase. The best wind turbines in the USA have reached availability levels of 95% after 5 years of operation[45].

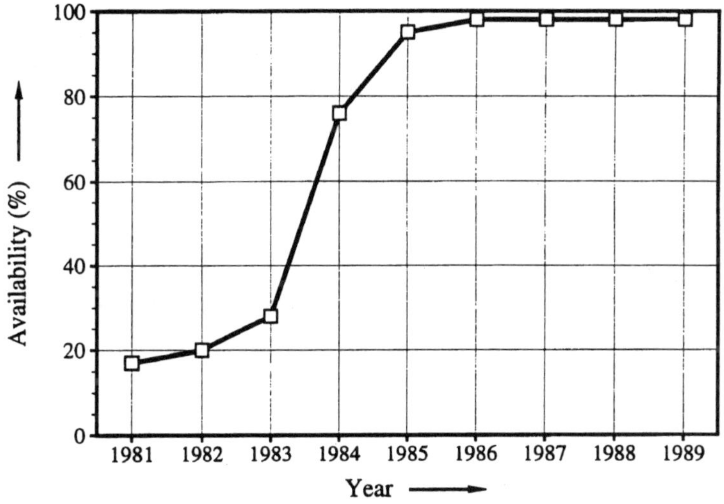

Figure 3.14 *Availability of the best wind turbines in California*

Total Generation Costs

Several studies have been carried out to get a general indication of the total generation costs. The results of three of these studies are summarized here. They give a good insight into the different approaches and the different assumptions that can be made. The important assumptions and results are summarized in Table 3.9. A study by SERI/USDOE was carried out for wind turbines in the Great Plains in the USA and uses economic assumptions derived from the local utilities[61]. EC figures are based on recent information about wind turbines placed in the European Community[62]. Danish figures are based on a typical Danish 250 kWe wind turbine[63].

Many comments can be made on these studies. For example, in the EC study the capacity factor seems rather high. The USA study uses a long lifetime but also calculates high losses (mainly due to high wake effects

Table 3.9 *Examples of calculated generating costs with assumptions*

	SERI/USDOE[60, 61]	*EC*[62, 63]	*Danish DEA*[64]
Total investment costs	US$400–500/m² = US$1,000–1,200/kWe	ECU400–600/m² = ECU900–1,100/kWe	DKK5,680/kWe (=US$770/kWe)
Annual mean wind speed	6.6 m/s at 25 m height	–	6.5 m/s at 30 m height
Annual gross energy capture	800–1,070 kWh/m² Great Plains site	–	1,000 kWh/m²
Capacity factor	–	28.5%	22.3%
Availability	95%	95%	–
Total losses	23%	–	–
O&M costs	1 US cent/kWh	2% of investment costs per year	1.4% year 1–2 2.0% year 3–5 2.5% year 6–20
Replacement costs after 8th and 20th year	US$27,000–40,000 for a 200 kWe wind turbine	–	–
Lifetime	30 years	20 years	20 years
Discount rate Fixed charge rate	0.061 0.102	–	–
Real rate of return	–	5% per year	7% per year
Total generating costs (1989) cents/kWh	6.8–10.3 US cents/kWh	ECU0.03–0.06/kWh = 3.5–7.0 US cents/kWh	DKK0.334/kWh = 4.5 US cents/kWh

in wind turbine arrays). The Danish study seems rather optimistic about the total investment costs. It can be concluded that in 1990 in general, for reasonably good sites, the total production costs were between 5 and 10 US cents/kWh.

The total production costs depend on a variety of assumptions. Two important assumptions are the figures for the wind speed and for the required rate of return of the investment costs. In a study done by the European Wind Energy Association (EWEA)[69] figures are presented about production costs as a function of these two variables, presented in Table 3.10. In this study also the range of investment costs is varied, because it is assumed that the windiness and the accessibility of a site are related. Therefore very windy sites are associated with higher investment costs than less windy sites. Again it can be concluded that the production costs are between 5 and 10 US cents/kWh.

Table 3.10 *Costs of electricity (in US cents/kWh) versus mean wind speed at hub height and required internal rate of return (net of inflation using a 20-year loan)*

Type of site	*Wind speed (m/s)*	*Investment costs (US$/kWe*	*Required internal rate of return*			
			5%	*8%*	*10%*	*15%*
Fairly good	6.5	1,035	7.0	8.5	9.5	12.4
Good	7.5	1,150	5.8	6.9	7.7	10.1
Very good	8.5	1,265	3.9	4.7	5.3	6.9

Wind Pumps

The economics of wind pumps can be evaluated using this expression[70, 71]:

$$\textit{water costs} = 0.114 \frac{(\alpha+\mu)\ SI <E_{cr}>}{\beta<V_{cr}>^3 <E_{an}>} \quad (US\$/kWh^h)$$

with:
- kWh^h the amount of energy needed to pump 367/z m^3 over a height of z metres;
- SI specific investment (US$$m^2$ rotor area);
- α annuity factor;
- μ annual maintenance and repair costs as fraction of investment costs;
- $<E_{cr}>$ mean daily hydraulic energy requirement in the critical month;
- $<E_{an}>$ annual mean daily hydraulic energy requirement;
- β quality factor indicating the efficiency of energy conversion;
- $<V_{cr}>$ mean wind speed in the critical month.

The expression takes into account various aspects which determine whether or not a wind pump is economic. The critical month is defined as the month in which the ratio (hydraulic energy requirement/win potential) is at a maximum. Depending on the type of application this ratio can vary between 1 (in the case of domestic use) and 3–4 (in the case of irrigation). The specific conditions and requirements determine the value of water.

As an example, consider the following case: $\alpha = 0.132$ (10% interest rate, 15 years lifetime), $\beta = 0.08$, SI = US$400/$m^2$ (classical wind pump), $\mu = 0.02$, and $<E_{cr}>$ $<E_{an}>$ (constant water demand). As a function of the mean wind speed in the critical month, the following costs are found:

$$\textit{water costs} = \frac{85}{<V_{cr}>^3} \quad (US\$/kWh^h)$$

At a value of $<V_{cr}> = 3$ m/s this results in water costs of US$3/$kWh^h$. At a value of $<V_{cr}> = 4$ m/s the water costs are more than halved. This means wind pumps are competitive with solar systems (cost figure about US$4/$kWh_h$). In general, operation costs of diesel pumps are US$0.7 to 1.1 /$kWh_h$, and those of kerosene pumps US$1.2 to 2.0 /$kWh_h$. The local situation will determine which system is the cheapest, depending on the size of the system, the cost and availability of fuel, and also on the cost of operation and maintenance.

Future prospects for modern wind pumps are that β will reach a value of 0.15 and SI of US$200/$m^2$, resulting in a cost reduction of about 60%. This cost reduction will make wind pumps more suitable for widespread application. One essential aspect is the reduction in production costs, that can be achieved if wind pumps are produced locally. Then, in many cases, the resulting water costs could be competitive with diesel or kerosene pumps. Table 3.11 shows which is the most economic way of pumping water as a function of the wind regime and the required application[71].

THE VALUE OF WIND ENERGY

The value of wind energy depends on the application of the energy and on the costs of alternatives to realize the same application. In this section, the value of electricity generated by wind turbines in grid-connected applications is discussed. As perceived by the utilities, this value can be defined as the costs of fuel, capacity and emissions that can be avoided if wind energy is applied. As perceived by the society as a whole, the value is equivalent to the net savings in the total social costs.

Fuel Savings

When wind energy is incorporated in an electricity production system it reduces the electricity produced by other types of units. Thus savings can be obtained on the consumption of fuel. How much fossil fuel and which type of fossil fuel are saved will depend on the electricity generating mix, now and in the near future. The savings also depend on the characteristics of the units, especially the heat rate as a function of capacity. However, the introduction of wind energy may also cause higher system heat rates due to fossil plants operating at lower loads and, potentially, cause some units to be operated near their minimum load point. Also it should be noted that the amount of fuel saved depends on the penetration level of wind power.

To calculate the avoided consumption of fuel it is necessary to analyze the electricity production system as a whole. Such an integral analysis has been carried out for the Netherlands, as shown in Figure 3.15[40]. It shows that over the years the specific amount of avoided fossil fuel is expected to decrease because of the assumed increase in wind generating capacity and because of the expected installation of more efficient fossil fuel power plants in the next decade.

Table 3.11 *Preliminary assessment of the attractiveness of using wind pumps**

Mean wind speed in critical month (m/s)	*Mean daily hydraulic power demand in critical month*					
	20–500 m^4/day[a]		*500–2,000 m^4/day*		*2,000–100,000 m4/day*	
	Rural water supply	*Irrigation*	*Rural water supply*	*Irrigation*	*Rural water supply*	*Irrigation*
>5	Wind best option	Wind best option	Wind best option	Wind best option	Wind best option	Wind best option
3½–5	Wind best option	Wind probably best option but check with kerosene and diesel	Wind best option	Wind probably best option but check with kerosene and diesel	Wind very good option but check with diesel	Wind very good option but check with diesel
2½–3½	Consider all options: wind, solar, kerosene diesel	Consider all options: wind, solar, kerosene, diesel	Consider wind, diesel, kerosene	Consider wind, diesel, kerosene	Consider wind and diesel	Consider wind and diesel
2–2½	Consider solar, kerosene, check wind	Consider solar, diesel, kerosene, wind doubtful	Consider diesel, kerosene, check wind	Consider diesel, kerosene, wind doubtful	Diesel best option, wind doubtful	Diesel best option, wind doubtful
<2	Consider solar, kerosene, diesel	Consider solar, kerosene, diesel	Consider kerosene, diesel	Consider kerosene, diesel	Diesel best option	Diesel best option

[a] Always consider hand pumps below 100 m^4.

Electric pumping is nearly always the best option if:

- a regular supply of electricity is guaranteed by the utility;
- the pumping location is close to a normal grid connection.

* This table only gives a very quick rough first indication. For example, the advice to consider various options does not necessarily mean that the options are equivalent; it means simply that none of the options can be disregarded beforehand. Wind pumps are indicated in a general way, without making a distinction between classical and modern wind pumps. When performing a detailed analysis each type of wind pump will prove to have different ranges of applicability, also depending on local conditions (loans, import duties, etc).

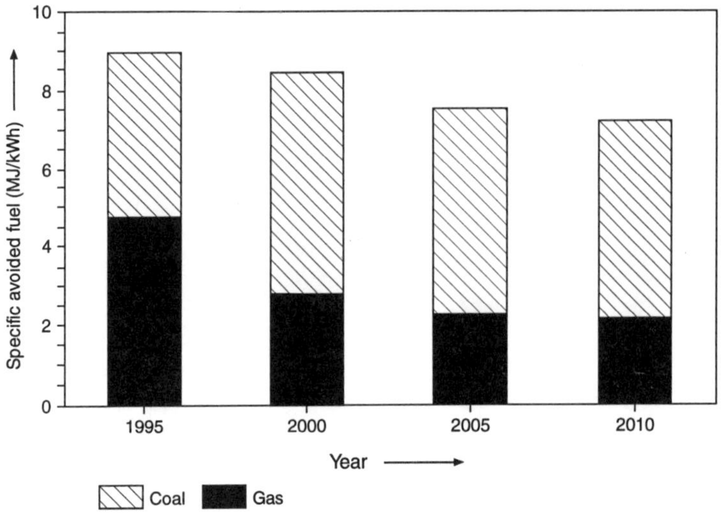

Figure 3.15 *Fuel avoided by wind energy (MJ/kWh) for the years 1995 (400 MWe wind power), 2000 (1,000 MWe), 2005 (1,500 MWe) and 2010 (2,000 MWe) in the Netherlands*

Capacity Savings

Because of the variability in wind speed, wind power is often regarded as a power source that has no capacity value. In reality, however, the contribution of wind power to the reliability of the total electricity production system is not zero. There is a probability that a wind power plant will be available at a time when conventional power plants are not available. Of course, this probability is smaller for wind power than for a conventional unit, but it indicates that wind power has a partial capacity credit. The capacity credit can be calculated by analyzing the total system reliability using statistical methods and by calculating the minimum required conventional production capacity with and without the inclusion of wind turbines in the power generating system.

The capacity credit of wind power has been determined for various power generating systems. For example, for the Netherlands it has been calculated that, in the year 2000, 1,000 MWe wind capacity might replace 165–184 MWe conventional capacity, so the figure for the relative capacity credit is 16.5–18.4%. For other countries, this figure might vary between minimal and 80%. Relatively high values are found for some areas in California (eg Solano County), where a good correlation between energy consumption and wind power production exists[70, 73]. Figure 3.16 clearly indicates that the relative capacity credit of wind power is decreased if the penetration level of wind power is increased.

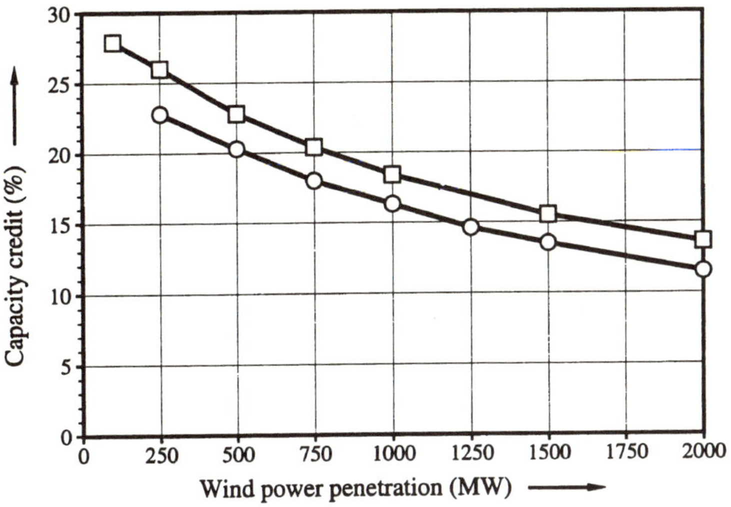

Figure 3.16 *Capacity credit of wind power in the Netherlands versus wind power penetration level (○ according to reference note 64, □ according to reference note 40).*

Emission Savings

During normal operation of wind turbines there are no emissions to the air, water or soil. These emissions occur in nearly all electricity generating systems, because of the burning of fossil fuels. This means, that each GWh generated by wind turbines that saves fossil fuels, prevents the emission of a certain amount of pollution. The amount of emissions prevented, depends on the mix of generating units in the region concerned, and on the measures that have been taken to reduce the emissions.

Several attempts have been made to calculate the emissions prevented. The journal *Windpower Monthly* publishes a table every month in which the pollution prevented is indicated for 60% of the Danish turbines registered in their statistics. The conversion factors used in *Windpower Monthly* are given in Table 3.12 and are valid for Danish coal-fired stations. In 1989 the estimated total output of 2,800 Danish wind turbines was 500 GWh, which prevented the emission of about 400,000 tonnes of pollution (most of which is CO_2). In the Netherlands two studies have estimated emissions prevented by wind turbines for the projected electricity production system in the year 2000[40, 66]. The results are given in Table 3.12 also.

It will be clear from the differences presented in Table 3.12, that the emissions prevented are not constant per GWh. They depend strongly on the mix of electricity production units and the assumptions about the emissions of each unit. This explains the differences in outcome for Denmark and the Netherlands. In the Netherlands, at least half the electricity is produced by gas-fired units, which give less pollution than coal-fired power plants. In the Netherlands, emission reduction techniques for

Table 3.12 *Comparison of prevented emissions (t/GWh produced by wind turbines) for Denmark and the Netherlands*

Emission component	*Denmark coal-fired stations 1989*	*The Netherlands total electricity system 2000*
Sulphur dioxide	5–8 t/GWh	0.25–0.40 t/GWh
Nitrous oxides	3–6 t/GWh	0.8–1.1 t/GWh
Carbon dioxide	750–1,250 t/GWh	650–700 t/GWh
Particles	0.4–0.9 t/GWh	na
Slag and ash	40–70 t/GWh	na

SO_2 and NO_x will be used on almost all units by the year 2000. Therefore, the figures for the Netherlands are much lower than 1989 numbers for Denmark. It can be expected that the emissions from fossil fuel-fired power stations will be reduced over the years under the influence of international discussions about global climate change and environmental pollution. As a result, the unit rate of emissions prevented by wind power (t/GWh) will decrease in time.

Avoided Fuel, Capacity, Operation, Maintenance and Emission Costs

On the basis of the savings on fuel, capacity and emissions the costs that are avoided by the application of wind energy can be calculated. This gives an indication of the value of wind energy as perceived by the utilities. Normally only the avoided fuel and capacity costs are analyzed. But the saving on emissions can also be translated into avoided costs. In some studies these avoided costs are estimated by investigating the cost of damage to flora, fauna, materials and mankind due to acid rain deposition, enhancement of the greenhouse effect, etc. In other studies the avoided costs are calculated by evaluating the costs of implementing extra emission reduction techniques in fossil fuel-fired power plants to realize the same amount of emission reduction as obtained by the introduction of wind power. Figure 3.17 presents the figures for the avoided costs as calculated for the Netherlands using the latter approach[40]. The outcome is only valid for the Netherlands, because the avoided costs depend very much on the regional electricity generating mix, the characteristics of each unit and the costs of fuel and capacity. As can be seen from Figure 3.17, the avoided costs by wind energy decrease over time. This is the result of several factors. The avoided capacity costs decrease due to an increase in the penetration of wind power over the intervening years. The avoided fuel costs decrease because of a changing electricity generating plant mix towards more coal-fired and more efficient power plants.

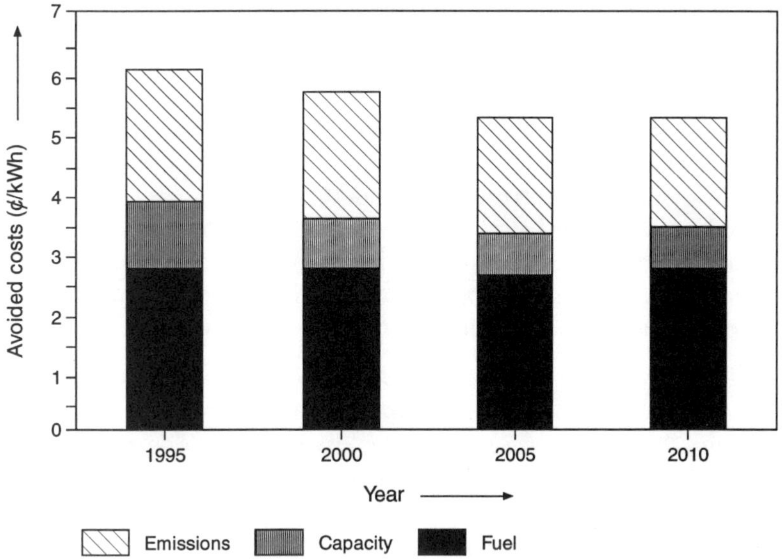

Figure 3.17 *The avoided costs of wind energy in the Netherlands for the years 1995, 2000, 2005 and 2010*

Total Social Costs

Several studies[74–81] have been carried out to compare wind energy with its alternatives on the basis of total "social" costs, which are all the costs society has to pay if a certain technology to generate electricity is applied. The total social costs can be divided into "internal" and "external" costs. Often, the following external costs, which have to be calculated for each option, are mentioned:

- R&D support;
- non-R&D government support (investment subsidies);
- tax provisions;
- loans (loan guarantees);
- environmental clean up costs;
- economic development costs;
- costs connected with the reliability and security of supply;
- system operation costs;
- resource depletion costs.

The first to present results of such a study was Hohmeyer in 1988[75]. At this moment many organizations are trying to estimate the social costs for their own country. Studies are being conducted in Germany and the United Kingdom, and by the EU together with USDOE. The latest results presented by Hohmeyer for Germany[82] indicate that the difference in costs not included in the prices of conventional electricity and electricity from wind turbines, ranges between 0.08 and 0.14 DM/kWh (5–9 US cents/kWh, 1989). From this analysis, it appears that wind energy is one of the cheapest electricity generation technologies now available.

A more comprehensive study about "externalities" has been carried out by Ottinger[78, 80, 81]. Besides making an analysis of the externalities of different energy conversion techniques, he also investigated how the different states of the USA have incorporated external environmental costs in their planning. Ottinger found 19 state commissions that have ordered some action to include environmental externalities in utility resource selection, and 10 more that have a plan under consideration. He characterizes the methodologies that are used to incorporate external environmental costs in the resource selection as follows:

- Quantitative, where the value of environmental costs control is determined or approved by the regulatory commission.
- Qualitative, where the commission gives the utilities a mandate to take environmental control and mitigation into account without specifying the means.
- Rate of return consideration, where the commission allows an increased profit for the use of environmentally friendly technologies.
- Avoided cost consideration, where the commission puts a premium on the avoided cost calculations for the use of "clean" sources in the resource selection procedures.

Ottinger concludes that "having decided which costs to include and exclude, the next major problem facing a commission or utility . . . is whether to use the cost of damages imposed on society by the resource, or the costs of control or mitigation of the pollutants emitted by the resource". The major advantage that Ottinger sees for using the control costs is that the figures are easier to determine. However, Ottinger believes that the damage costs are the relevant values to incorporate in the calculation, although these damage costs are much more complex and subject to disagreement.

Visual intrusion remains the most difficult cost to establish and to avoid. Individual perceptions and valuations of landscape vary and may be greatly modified where the individual concerned has a financial stake involved either through rent/cash receipts or employment. Nevertheless, where landscapes of great natural beauty or historic interest are involved, wind farm developments can impose very heavy social costs. This is likely to limit severely their location in such areas.

R&D PROGRAMMES AND MARKET INCENTIVES

Introduction

Before 1973, R&D activities to develop renewable energy sources were carried out in only a few countries, and primarily in areas of basic research. In the field of wind energy, some research was carried out in the mid-1940s and 1950s in Denmark, the United Kingdom and France. As a result of the marked increase in the oil price in 1973 and 1979, the total budget for renewable energy R&D increased dramatically. After 1981, the R&D expenditure for renewable energy has declined, as shown in Figure 3.18[83]. The decline is mainly due to major cuts in the United States. Outside the USA the R&D budgets were more or less stabilized. R&D expenditure for wind energy, however, was not influenced as dramatically as for example the expenditure for geothermal and solar thermal electric applications (see Figure 3.19).

For renewable energy R&D activities, the years between 1973 and 1979 were a definition phase: to develop analytical databases and methodologies, to implement research programmes and to develop and demonstrate components and systems. In the past few years, these activities have led to greater selectivity in R&D programmes. In the 1980s, government support continued to shift more and more from R&D activities towards market introduction, especially in the case of wind energy. In this section, a brief overview of some programmes that have been initiated to develop and implement wind energy is given.

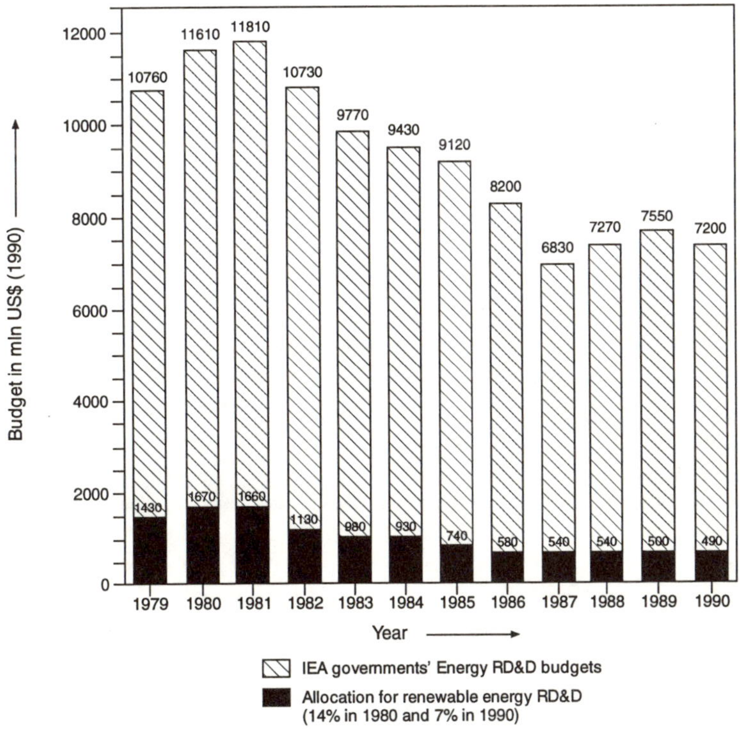

Figure 3.18 *IEA governments' energy RD&D (research, development and demonstration) budgets between 1979 and 1990*

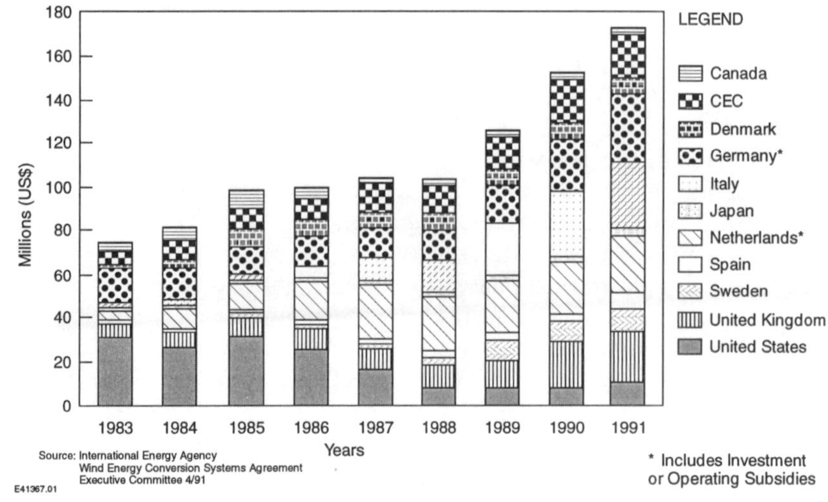

Figure 3.19 *IEA governments' RD&D expenditure for wind energy (1983–91)*

National Programmes

Since the end of the 1980s, there has been a renewed interest in developing renewable energy technologies, mainly because of growing environmental concern. Wind energy is especially of interest, since it is assumed that wind energy technology is close to being mature. Moreover, in economic terms wind turbines are nearly competitive with other grid-connected power plants. Especially in Western Europe, this renewed interest has led to government support, not only for R&D but also for market introduction. Countries such as Denmark, the Netherlands, Germany, Italy and the United Kingdom have decided to give investment support in order to develop a market for wind turbines which should reduce the production costs. Japan also has initiated an R&D programme[84]. For several countries the budgets for wind energy R&D between 1983 and 1991 are given in Figure 3.19[83].

Denmark initiated its first market development programme in 1979 by granting a 30% subsidy of the capital cost of each wind turbine. The subsidy percentage has been adjusted over the years in accordance with the improved economics of wind power, moving from 30% in 1979 to 10% in 1989. The subsidy was terminated in August 1989. The total economic support over this 10-year period amounts to about US$40 million. Another incentive was given by offering high payback tariffs for electricity produced by wind turbines and delivered to the grid[85].

The first market for wind energy in the USA was created in 1981 in California as a result of implementing a system of tax credits. However, this system was not especially developed for wind energy applications. At the end of 1985 the supply of tax credits was stopped and the market for wind turbines was reduced significantly. However, as a result of a favourable wind resource and rather high electricity payback tariffs, there is still a market of 100–200 MWe a year. In addition, an energy production incentive from the federal government of 1.5 US cents/kWh was begun in 1993, and at the state level a variety of property tax waivers, tax credits, and other incentives are being considered.

The Netherlands introduced a market stimulation programme in 1986. A subsidy of US$350/kWe was given in the first year. Since then this subsidy figure has remained fairly stable. Germany started its market incentive programme in 1989. It consisted of a subsidy on the electricity rates of DM0.08/kWh for a period of 10 years, or an investment subsidy of about 30–40% of the installation costs.

Other countries have also introduced market incentive programmes for wind energy. For example, Italy implemented a market stimulation programme in 1991 by subsidizing 30% of capital expenditure for plants over 3 MWe. Also a favourable selling price has been announced for electricity produced by renewable sources. The utility ENEL pays 13 US cents/kWh, fixed for a period of 5 years[12]. In Finland an investment subsidy of 40% is given, and no VAT is levied upon the investment costs[8]. Sweden gives a 25% investment subsidy on installations of minimum 60 kWe.

Over a 5-year period, ending 1996, around US$40 million will have been allocated for the introduction of wind energy to realize 100 MWe in 1996[84]. A summary of the budgets of these market incentive programmes is given in Table 3.13.

These programmes have met with some adverse criticism, in particular on grounds that they have encouraged large-scale schemes controlled by major corporations rather than smaller-scale developments at the community level, and that they have not conformed to solid local environmental criteria.

Table 3.13 *Overview of budgets for market development of wind energy*

Country	*Total budget*	*Period*
Denmark[35]	US$40 million (subsidies)	1979–89
USA[19]	No specific budget	1981–85
The Netherlands[35]	US$35 million (subsidies)	1986–90
Germany[35]	US$75 million	1989–94
Sweden[35]	US$40 million	1991–96

Besides setting up programmes to subsidize the installation costs, to provide tax credits or to implement favourable payback tariffs, governments have created other, direct or indirect, means to stimulate the development of wind energy. Some examples are[3]:

- **Subsidy for grid-connection costs.** Danish utilities offer a subsidy of 35% of the grid connection costs.

- **Legislation about planning issues.** One example is the incorporation of environmental externalities; about 19 states of the USA have implemented legislation about this issue. Another example is the recent Non-Fossil-Fuel Obligation Act in the UK which is used to force the building of wind power plants. In the Netherlands the government and the local authorities are working together in the framework of an intergovernmental accord to designate possible sites in order to reach the government goal of 1,000 MWe in 2000[87]. A legislative measure to the same effect has been taken in Italy[12].
- **General fuel taxes.** High taxes on the emission of carbon dioxide and partly also on the use of non-renewable energy sources are being debated in many countries and are already being implemented in Sweden and Norway.
- **Educational programmes.** Many countries have already instituted such programmes.

International Programmes

Besides national programmes there are also international programmes for the development of wind energy. For example, the World Bank is involved in applications of wind pumps in developing countries.

The EU started a renewable energy programme in the mid-1970s. It contains both a demonstration component as well as an R&D component. In the demonstration programme new wind turbine projects are supported. The total budget during the years 1978–88 amounted to about US$45million. The objective of the research and development programme is to identify the wind resources in the EU, and to develop methods to design and test wind turbines. This work should also create a basis for the development of European norms and standards essential for the post-1992 era.

The IEA has developed activities in the field of wind energy also. The IEA does not provide financial support but institutes working groups to co-ordinate research activities, to generate design and testing methods and to develop standards concerning wind energy. The IEA has implemented two agreements: a general R&D programme in the field of WECS and a special programme for the development of large-scale WECS[89]. As of January 1991 there is only one IEA agreement on wind R&D. The large-scale WECS agreement has been terminated. This activity is one of the tasks of the R&D agreement (Task XIII). R&D activities on wind energy carried out since 1978 have featured participation by Austria, Belgium, Canada, Denmark, Germany, Ireland, Italy, Japan, the Netherlands, New Zealand, Norway, Spain, Sweden, Switzerland, the United Kingdom and the United States. Cost and task-shared projects have included wind meteorology, wake effects, offshore WECS, wind-diesel systems and the development of recommended practices for testing and evaluating wind turbines. Nine countries were participating in the co-operative agreement for developing large-scale WECS.

CONSTRAINTS AND POSSIBLE SOLUTIONS

Grid-connected Wind Turbines

Although the wind energy resource has the potential to contribute substantially to the future electricity supply, there are several reasons why this contribution will remain limited in the near future. At this moment, wind turbines are only at an early stage of market introduction. The total amount of operating grid-connected wind power is about 2,200 MWe. Its contribution to world electricity production is small. However, in some areas the power of wind is significant. In California 1–2% of the state's electric power comes from wind, but during peak demand periods in the summer, Pacific and Gas Electric Company draws up to 10% of their electricity from wind power plants in their service area. The development of the market is being stimulated by national/state programmes (mainly in California, Denmark, the Netherlands, Germany, but also in other European countries such as Italy, the United Kingdom, Spain and Finland).

The market introduction has been limited so far because of technical, economic, environmental and institutional constraints. These are briefly discussed in this section. If it is not possible to relax these constraints to some extent, further penetration of wind energy will be difficult. An overview of the constraints as well as of the possible solutions is given in Table 3.14.

Technical Constraints

A major constraint is the technical lifetime of a wind turbine. At this moment it has not yet been proven that a technical lifetime of 20 years can be realized. An improvement in the reliability of all the components, especially the rotor, is necessary. This can be realized by choosing better materials, flexible components, variable speed rotors and inherently safe and easy control mechanisms. Such developments can lead to lower loads and therefore also to lower weights and costs of the different components of the wind turbine. If larger wind turbines are wanted it is necessary to improve the economics along these lines too, because development of larger wind turbines with the present state of technology will lead to higher costs due to the weight penalty. To realize this development more research is needed.

Economic Constraints

A major economic constraint is the difference between the cost of generating electricity with wind turbines and with other options. At present the generation costs with wind turbines are often higher. This might change when wind energy technology is developed further, when the market for wind turbines grows and when the price of fossil fuels increases (for example by imposing a carbon tax). Also it should be noted that at present only the direct costs for electricity production (fuel, capacity, operation and maintenance, and some emission reduction costs) are

Table 3.14 *Constraints and possible solutions for the development of grid-connected wind turbines*

Issue	*Constraints*	*Solutions*
Technical		
Lifetime	The lifetime of wind turbines has not yet proven to be more than 10 years.	Lifetimes up to 20 years by using better designs and better materials
Availability	The current availability of the best turbines is 95%; this figure has to be reached on average over the entire lifetime.	By maturing technology and by improving O&M schemes, a 95% availability can be achieved.
System efficiency	Performance factor is now about 3 for 100–300 kWe turbines; and the specific yield is about 1,000 kWh/m^2.	Performance factors can be improved; the specific yield can be improved by better design.
Testing	Designs must meet safety, availability, noise, lifetime, and performance goals.	Requirements must be standardized internationally for export purposes.
Quality control	There is no quality control to guarantee the quality of all wind turbines produced.	Standardized methods are needed to guarantee quality of all turbines.
MW-size development	MW-size turbines are necessary for a substantial energy contribution; now only prototypes are available. On the basis of present technology, these wind turbines are more expensive due to weight penalty.	By new design, better materials, flexible components, MW-size turbines must be available in 2000–10 with same production costs as present 100–500 kWe turbines.
Resources and siting	Wind climate detail not known; in densely populated countries good sites are scarce.	Wind atlases and handbooks with siting methods must be developed.

Table 3.14 *contd.*

Issue	*Constraints*	*Solutions*
Economic		
Conventional electricity prices	In most countries conventional electricity prices are low.	Wind turbine electricity production costs must be reduced and external costs must be included.
External costs	External costs are not included in electricity prices and a methodology of calculating external costs is not available.	External cost methodology and implementation strategy for these costs must be developed.
Wind turbine investment costs	Wind turbine costs are too high, especially for the larger ones. Also the additional costs (installation, land, infrastructure, siting) must be reduced.	Technology and mass production of wind turbines must be improved; installation and siting procedures must be optimized.
Operation and maintenance costs	Reduction of O&M costs is necessary.	Standardized O&M procedures, accessibility to spare components, agreements with eg insurance companies must be improved.
Grid connection costs	Many resource areas are far from main grid and therefore the grid-connection costs are too high.	The large resource areas and those with most economic grid-connection costs must be developed first.
Grid integration costs	Avoided costs depend on electricity production system. Control problems can cause extra losses.	Avoided costs must be identified, other control strategies must be developed and intermittent sources have to be taken into account in capacity expansion planning.

Table 3.14 *contd.*

Issue	*Constraints*	*Solutions*
Environmental		
Birds	Good sites for wind turbines are at times bird habitats or migration routes, so birds may be killed or disturbed.	Research has to indicate the potential danger to birds and can help in finding better sites.
Noise	In densely populated areas the siting of wind turbines can be restricted by noise problems.	Development of quieter wind turbines is necessary. Planning instruments to identify possible nuisance must be developed.
Visual impact	Sometimes unacceptable depreciation of landscape by placing wind turbines occurs.	Participation of public in planning process is necessary. Public feedback must be an important aspect in planning and R&D.
Telecommunication interference	There can be some interference with telecommunication transfer.	This aspect must be taken into account but will not be critical in most cases.
Safety	The probability of accidents with severe consequences should be sufficiently low.	Safety standards have to be developed world-wide. Wind turbine designs must meet these standards.

Table 3.14 *Contd.*

Issue	*Constraints*	*Solutions*
Institutional		
Financing	Wind turbines mainly need capital investment and don't have many annual costs. The need for capital can put limits on the expansion. In some countries this capital is not available.	Financing schemes for this type of energy investment must be developed. Investment in wind turbines will save foreign exchange for fuel import.
Training/education	For the expansion of wind turbines skilled engineers, public awareness about wind energy, political interest, etc, are needed.	Educational programmes are needed at all levels. Training engineers and transferring technology is also necessary.
Land use	There can be interference with other possible forms of land use.	Careful planning is required. In some cases there is a need for additional legislation.
Governmental and utility attitude	A positive attitude on the part of both government and utility is necessary for implementation.	Both governments and utilities must incorporate wind energy in their planning studies.
Manufacturing capacity	At this moment manufacturing capacity is insufficient to realize substantial growth.	Manufacturing capacity can expand only if market prospects are clear. Attention must be given particularly to the setting up of manufacturing capacity in developing countries.
Public and political acceptance	The "Not In My Back Yard" (NIMBY) syndrome.	Involvement beforehand of local people is needed.

included in the calculation of the kWh price. The figure does not include "external" cost components. This makes a comparison between the wind turbine generating costs and the overall electricity generating costs incomplete until the external costs are somehow internalized in electricity generating costs. This can be done for the emission of CO_2 by imposing a carbon tax. Another solution could be to compensate for the difference in external costs by providing a subsidy for wind turbines related to the value of avoided emissions.

Environmental Constraints

An important environmental constraint is the high number of wind turbines that will have to be installed to realize a significant contribution. Problems occur because the wind resource is often available in areas of outstanding natural beauty (eg the United Kingdom), far from the main grid (eg the USA), or close to inhabited areas (eg the Netherlands). On the one hand, this means higher costs due to the required infrastructure, and on the other hand it probably puts restrictions on siting because of bird damage, bird disturbance and noise problems. It is not only necessary to solve the various planning problems, but also research has to be carried out to reduce the noise level produced by wind turbines and to investigate how impact on birds can be minimized. There needs to be acceptance that many sites with high technical potential will be ruled out because they are environmentally unacceptable on visual, historical, noise, communications or ecological grounds.

Institutional Constraints

An important institutional constraint can be the integration of wind power into the electricity production system. At a low penetration level the integration of intermittent sources such as wind energy is not really a concern. At higher penetration levels it will be necessary to re-optimize the electricity production of the total system. At even higher penetration levels it will also be necessary to adjust the planning of new units in order to realize a sufficiently flexible and low-cost electricity production system, for example by integrating (new) storage units. This aspect needs further research and attention on the part of the utilities.

Another major institutional constraint concerns the financing of wind turbine projects. An expansion of wind turbine capacity will require a lot of capital, since almost all the generating costs are based on investments. This means that, especially in Third World countries, but also in East European countries, it might be more difficult to realize wind energy projects because of a lack of funding. This problem can only be solved by the creation of financing structures by, for example, the World Bank.

For the near future, another important constraint will be the need for trained personnel to construct, install, operate and maintain wind turbines. At this moment, human resources are limited. There must be adequate training, education and technology transfer to ensure a basis for further expansion of wind turbine capacity.

Table 3.15 *Constraints and possible solutions for wind pumps*

Category	*Constraints*	*Solutions*
Technical	New and lighter designs not yet fully developed	Use proven technology if possible. Set standards and certification procedures
	Piston pumps not yet able to fulfil requirements	Design pumps suitable for application in wind pumps
Economic	Classical wind pumps are expensive	High volume production in developing countries
	Lack of funds to set up local manufacturing industry	Incentive programmes and involvement of private sector
Environmental	No real constraints	None
Institutional	Market insufficiently developed	Market research, stimulation programmes and training are necessary
	Training of various targets groups is insufficient	Long-term involvement of donors (UN organizations, etc) is preferable to short-term involvement
		Policy decisions to disseminate wind pumps are necessary

Wind Pumps

The constraints for wind pumps are quite different from those for grid-connected wind turbines[71, 89, 90]. A list of constraints for implementing wind pumps is given in Table 3.15 and discussed in this section.

Technical Constraints

In newly developed wind pumps, manufactured locally in developing countries, there are often problems with the down-hole equipment – ie the pump, pump rod and rising main. Problems occur with the pumps due to shock forces (resulting from the higher operating speeds, in contrast with the older slow running designs), and due to heavy loading which occurs because of the higher output. In the near future attention should be focused on the development of better pumps. Proven technology should be used as much as possible.

The classical wind pumps, which are manufactured mostly in Australia, Argentina and the United States, represent proven technology and are highly reliable if maintained properly. The disadvantage of these wind pumps, however, is the high costs.

Economic Constraints

Wind pumps are increasingly used in developing countries. This means that the economic constraints are quite different from those for grid-connected wind turbines, which are mainly applied in industrialized countries. The greatest problem in developing countries is the lack of funds. Therefore, the market for wind pumps is still small, which makes it difficult to reduce the costs by mass production. This also means that there may not be adequate maintenance, infrastructure, spare parts and qualified personnel.

Institutional Constraints

Until now, the focus has been on research and development, and the industrialization and commercialization of developed wind pumps have been neglected. Policy decisions are needed to start the process of disseminating wind pumps. Attention should be paid to the development of the market and to matching supply and demand. The market can be stimulated by credit facilities for purchasers, or even subsidies in the initial phase. The private sector should be stimulated to participate in the dissemination process. Finally, all projects should be guided and evaluated thoroughly.

OPPORTUNITIES FOR GRID-CONNECTED WIND TURBINES UP TO THE YEAR 2020

Introduction

Using the world-wide resources map, which shows the mean wind speed and the mean wind power density throughout the world, it can be estimated that 27% of the earth's land surface is exposed to an annual mean wind speed higher than 5.1 m/s at 10 metres above the surface. Assuming that, on average (per square kilometre) 0.33 MWe wind turbine capacity can be placed, with an estimated mean production of 660 MWh/year, the global potential of onshore wind power production can be estimated to be 20,000 TWh/year. This is an enormous potential, but the question is: how much of this potential can be exploited in the next few decades? Therefore some estimates will be made for the wind power penetration up to the year 2020.

Table 3.16 *Energy and electricity demand in the year 2020 for the two scenarios*

Scenario	*Energy demand in 2020*	*Electricity demand in 2020*
Current policies	13,500 Mtoe (570 EJ)	25,625 TWh
Ecologically driven	11,560 Mtoe (485 EJ)	20,275 TWh

The amount of grid-connected wind power that can be realized up to the year 2020 will be given within the framework of two scenarios for the world energy demand. These scenarios are indicated by current policies and ecologically driven. The assumptions concerning the energy and electricity demand in the two scenarios are given in Table 3.16. The figures for the world energy demand in the current policies scenario and the ecologically driven scenario are chosen in relation to the WEC-M and WEC-L scenario, respectively, as presented in 1989 in Montreal[91].

Several factors will influence the development of wind power. The most important factors are, of course, the available wind resource and the wind power generating costs in comparison with grid electricity production costs. But financing aspects, grid integration of wind turbines, socio-economic aspects and governmental policy will also influence the development of wind energy. All aspects are very country-specific. Therefore the analysis of the possible penetration of wind power up to the year 2020 is carried out for each country/region.

Methodology

The methodology used for the analysis of the possible penetration of wind power up to the year 2020 is outlined in Figure 3.20. For each country/region the survey starts with a global investigation of the wind resource and a global inventory of the electricity production system[92, 93]. After this inventory, a cost comparison is made between wind power generation costs and the production costs of electricity from the national grid. The assumption is that substantial penetration of wind power will start when the long-run marginal costs (LRMC) for electricity production in that country are higher than the wind power electricity generation costs. If wind power penetrates substantially, it is assumed that the penetration of wind power will be limited by constraints, like financial limitations and growth rates of wind turbine manufacturing capacity and installation capacity. Finally, a more specific analysis of the wind resource, the electricity grid, and government policy is made to see if the penetration of wind power will be constrained still further.

For both the current policies and the ecologically driven scenarios, analyses have been made of the wind power penetration for each country up to the year 2020. The differences in wind power penetration for the two scenarios are caused by different assumptions shown, in Table 3.17. An important assumption in the current policies is that the wind power generation costs will reduce in the period 1990–2010 by 15%, whereas in the ecologically driven scenario the wind power generation costs will reduce by 30% in the period 1990–2010 and by 10% in the period 2010–20. Two important constraints restrict the penetration of wind power. The first constraint in the scenarios is that the investments in wind power in any year may not exceed 1% of GDP (1% = 0.1%) in the current policies scenario and 3% of GDP in the ecologically driven scenario[94, 95]. The second constraint in both scenarios is the limitation on the growth rate of the installed wind turbine capacity per year and per country, as defined in Table 3.17.

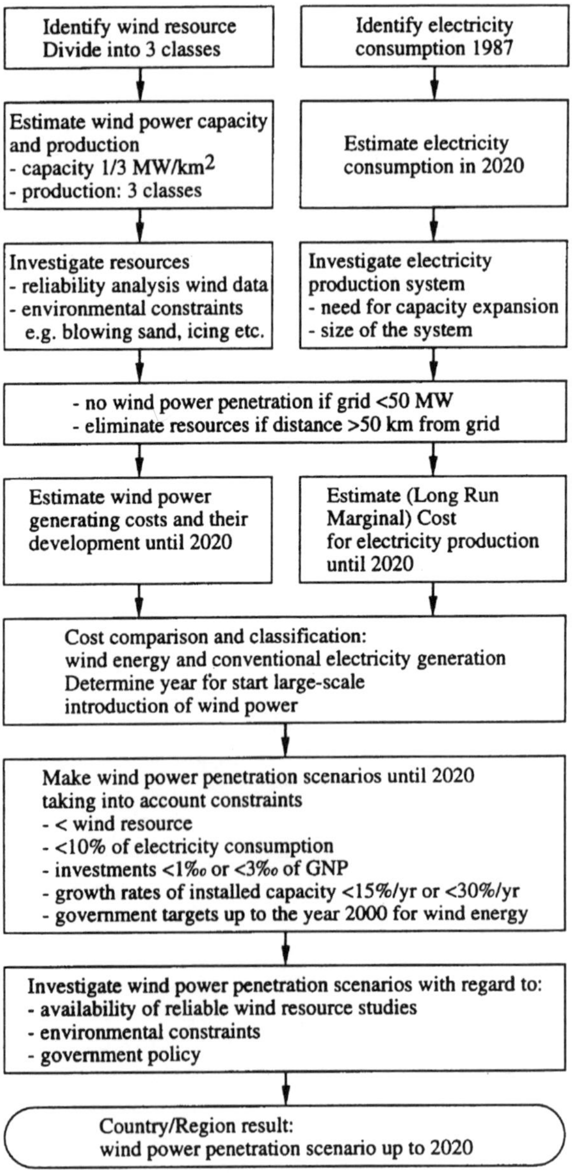

Figure 3.20 *The methodology for wind power penetration estimates to 2020*

Table 3.17 *The different assumptions in the current policies and the ecologically driven scenarios*

Assumptions	*Current policies*	*Ecologically driven*
Growth rate of world electricity consumption	2.8% per year	2.0% per year
Oil price	US$20–25 per barrel in 2000 US$30–35 per barrel in 2020	US$20–25 per barrel in 2000 US$30–35 per barrel in 2020
Carbon tax	No	Yes, starts in 2000 up to US$10 per barrel in 2005 and thereafter
Increase of efficiency fossil fuel plants	15% in the period 1990–2020	15% in the period 1990–2020
Wind power generation costs	15% reduction in the period 1990–2010	30% reduction in the period 1990–2010 10% reduction in the period 2010–20
Wind power investment costs	7.5% reduction in the period 1990–2010	15% reduction in the period 1990–2010 5% reduction in the period 2010–20
Financial constraint	Total annual investments per country/region not more than 1‰ of GDP	Total annual investments per country/region not more than 3‰ of GDP
Wind energy penetration into the grid	Up to 2020 not more than 10% of electricity consumption	Up to 2020 not more than 10% of electricity consumption
Growth rate of manufacturing capacity world-wide	Less or equal to 15% per year	Less or equal to 30% per year
Growth rate of installed wind turbine capacity per country	Less or equal to 15% per year	Less or equal to 30% per year

The Development of Wind Power Generation Costs

The state-of-the-art concerning the electricity generation costs of wind turbines has been discussed in the "Economic aspects" section above. The current costs of electricity produced by wind power for different wind speeds was presented in Table 3.10. The table shows that the wind power generation costs are between 4 and 7 US cents/kWh at sites with a mean wind speed of 8.5 m/s at hub height, between 6 and 10 US cents/kWh at sites with a mean wind speed of 7.5 m/s at hub height and between 7 and 12 US cents/kWh at sites with a mean wind speed of 6.5 m/s at hub height. The wide variety in costs is caused by differences in investment costs, O&M costs, availability, lifetime and interest rates for locations in different parts of the world.

The development of wind power generation costs that might be current policies scenario the R&D activities in wind energy will not be intensified. Therefore, the reduction in wind power generation costs will be modest. This reduction is estimated to be about 15% in the period 1990–2010, because of improvements in existing wind turbine designs and more efficient operation strategies.

In an ecologically driven scenario it is assumed that the R&D activities in wind energy are intensified. As a result, the generation costs might be reduced by about 30% in the next 20 years and by about 10% in the period 2010–2020. The cost reduction up to 2010 is mainly due to a reduction in wind turbine production costs and to more efficient operation of the wind turbines. The wind turbine production costs can be reduced by improvement of the turbine components, improvements in turbine design and by mass production of components and wind turbines. It is estimated that in the period 2010–2020 generation costs will be reduced by 10%. These cost reductions can be expected when large wind turbines become available based on innovative concepts, resulting in lower loads, and on the use of lighter weight materials and designs. These estimated reductions in wind turbine generation costs are in agreement with a study for the USA on the potential of renewable energy[20].

Economic Competition with Conventional Electricity Production

One of the important constraints for the penetration of grid-connected wind power is the relative economic performance of wind power compared to conventional electricity production. The relative economic performance of wind power can be estimated by using the avoided cost principle. In such an analysis, the avoided costs are evaluated if wind power were to be introduced into an electricity production system. These costs can be compared with the wind power electricity production costs. When these production costs are lower than the avoided costs, wind power is a favourable option.

Earlier the avoided costs were presented for the Dutch situation in Figure 3.17. These costs contain avoided fuel costs, avoided capacity costs and avoided emission reduction costs for NO_x, SO_2 and CO_2. The total avoided costs change from 6 US cents/kWh in 1995 to 5 US cents/kWh in 2010. When only the avoided fuel and capacity costs are evaluated, the figures are 4 US cents/kWh in 1995 and 3 US cents/kWh in 2010[40]. For other countries, these costs might differ significantly because of a different electricity production system, different wind regime, different cost figures for fuel, and capacity of conventional fossil and nuclear power plants.

Usually the relative economic performance of wind power is calculated by comparing the wind power production costs with the marginal costs (fuel and variable O&M costs) of conventional power production. Such a comparison is easy to make, but unfavourable for wind power, because it excludes the capacity savings and the environmental benefits of wind power. The results of two studies are mentioned here. Figure 3.21 gives a comparison for the USA up to the year 2030[20], and shows that at outstanding wind sites, the economic crossover might occur in 1995 for gas displacement and in 2020 for coal displacement[96]. In a study designed to identify the market potential for wind power in developing countries[4], the long-run marginal costs for future conventional capacity expansion and for wind energy have been estimated for almost all developing countries. In the calculation of the LRMC, an oil price of US$23 per barrel was assumed, and also a coal price of US$35 per tonne. Table 3.18 presents the LRMC figures, together with estimates for the costs of electricity production with wind turbines in these countries.

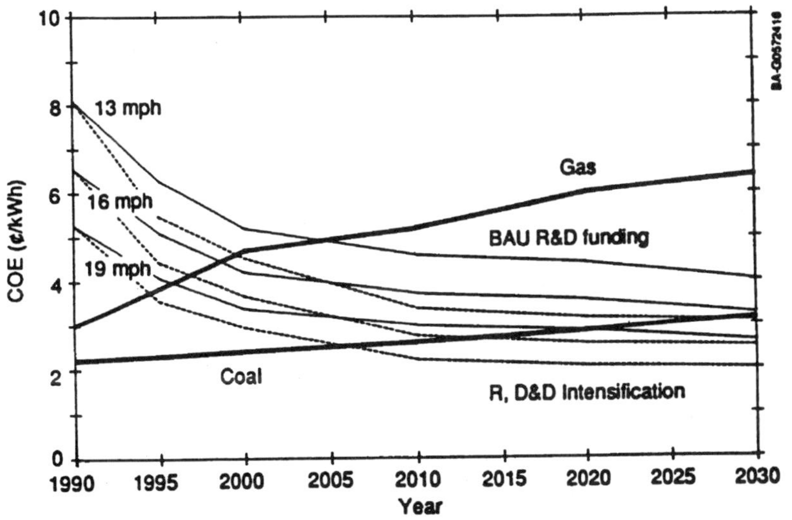

Figure 3.21 *Comparison of COE for wind power plants' costs versus fuel and variable O&M costs for natural gas and coal power plants*

From the table on the following page, it can be seen that the LRMC can be as low as 3 US cents/kWh for a country like Colombia, and as high as 33 US cents/kWh for parts of China. These differences in electricity production costs show that in some countries wind power is much more attractive than in other countries. Using these results, it can be concluded that wind power is already a favourable option in developing countries like China and India, whereas it will stay unfavourable for many years to come in other countries. The results of this study are used to select the developing countries in which large-scale penetration of wind power before the year 2020 is possible.

For other countries a more generalized approach is followed. On the basis of the development of the fossil fuel prices, development of the efficiency of the fossil-fired power plants and the introduction of a carbon tax, as indicated in Table 3.17, a range for the development of the fuel costs of fossil-fired power plants is identified, for both the current

Table 3.18 *Long-run marginal cost comparison of wind in developing countries*

Country	*LRMC of future marginal plant (US$/kWh)**	*LRMC of wind energy (US$/kWh)***	*Wind energy perspective*
Argentina	0.04–0.08	0.05–0.21	Unlikely
Brazil	0.04–0.08	0.15–0.23	No
Chile	0.10–0.20	0.06–0.23	Yes
China	0.33	0.04–0.23	Yes
Colombia	0.03–0.06	0.13–0.23	No
Costa Rica	0.03–0.06	0.12	No
Cyprus	0.10	0.15	Unlikely
Egypt	0.06–0.07	0.13–0.18	Unlikely
India	0.15	0.09–0.23	Yes
Jamaica	0.09–0.12	0.09–0.10	Yes
Jordan	0.09	0.10	Likely
Kenya	0.06–0.09	0.13–0.19	No
Mauritania	0.15	0.10–0.23	Yes
Morocco	0.06–0.12	0.08–0.19	Likely
Pakistan	0.08–0.12	0.09–0.20	Likely
Peru	0.08–0.09	0.09–0.15	No
Portugal	0.09	0.08–0.15	Unlikely
Romania	0.09–0.12	0.04–0.13	Yes
Sri Lanka	0.06–0.13	0.10–0.18	Unlikely
Syria	0.09–0.12	0.07–0.13	Yes
Tanzania	0.06–0.10	0.08–0.09	Likely
Tunisia	0.09	0.16–0.21	No
Turkey	0.07	0.13–0.23	No
Uruguay	0.05	0.07	No
Venezuela	0.04	0.07–0.14	No
Yemen, Arab	0.04–0.05	0.06	No
Yemen, PDR	0.11–0.22	0.07–0.23	Yes
Zambia	0.04	0.18–0.23	No
Zimbabwe	0.05	0.19	No

* The LRMC ranges above for the future marginal plant represent alternative least cost expansion plans, as it cannot be determined which expansion path will be taken in the future.
** The ranges above for the LRMC of wind reflect wind stations with differing wind resources.

policies scenario and the ecologically driven scenario, as shown in Figures 3.22 and 3.23. Also shown is the development of the wind power generation costs is identified. The year in which the wind power generation costs are lower than the electricity production costs is taken as the start of large-scale wind power penetration. As can be seen in Figures 3.22 and 3.23, this year is not easy to determine. Therefore best estimates are made using additional information per country, if available.

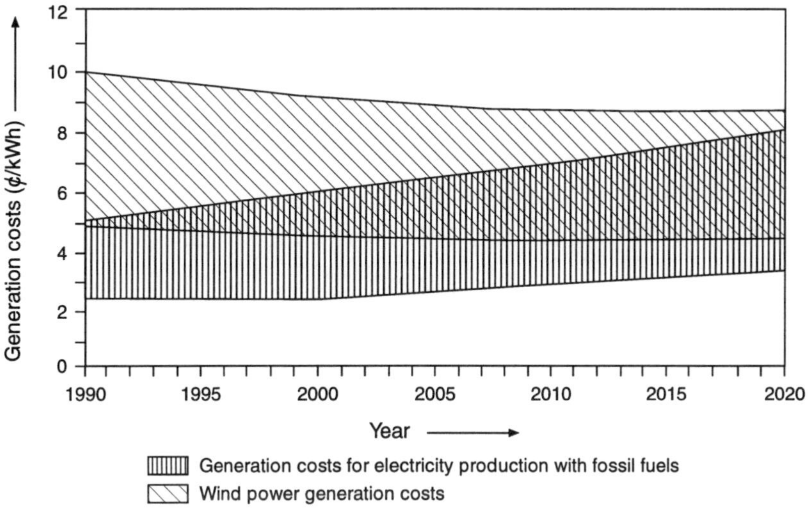

Figure 3.22 *Wind power costs and costs for electricity production with oil up to 2020 – current policies scenario*

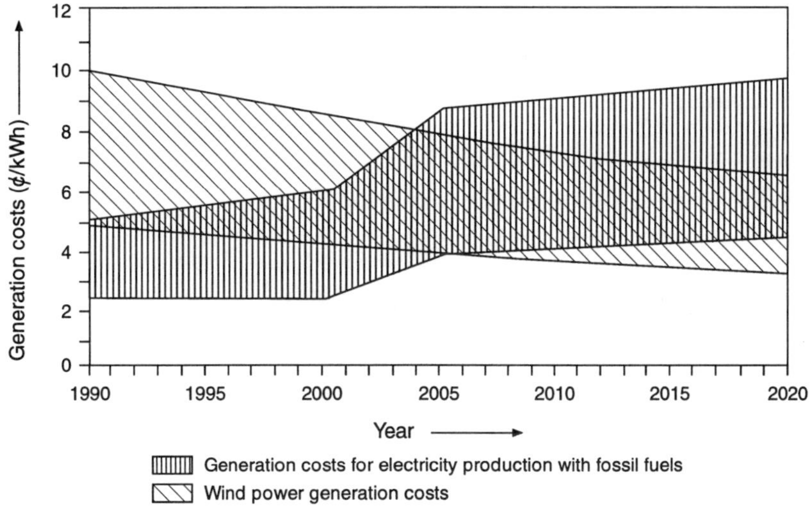

Figure 3.23 *Wind power generation costs and costs for electricity production with oil up to 2020 – ecologically driven scenario*

Grid Integration of Wind Power

Wind energy is an intermittent power source. There are variations in wind speed (and therefore in wind power output) on a yearly, seasonal, daily and hourly scale. A utility has to serve the varying load of its customers by a dispatch of the power plants. As wind power is a varying power source which cannot really be dispatched, conventional power plants or storage facilities have to deal with these variations. This means that output variations in wind power can give rise to extra costs due to extra power changes or start/stops of conventional units.

If the penetration of wind power is increased, it might reach a level where the economics of the total electricity production is affected in a negative way. This will limit the penetration of wind power into the electricity production system, especially if no storage capacity is available. Several studies have been carried out to investigate the optimum penetration level of wind power with regard to this problem[38, 96]. The results indicate that no universal figure can be given for the optimum penetration level. The optimum can be as low as 10% but also as high as 50% and depends on the specific circumstances and characteristics of the utility system[3]. We conclude that for wind power penetration levels less than 10% of the total electricity production no severe problems will occur. Moreover, we assume that below this level the penetration causes no practical economic disadvantage for wind power. Of course, a higher penetration level of wind power is feasible, but it requires re-optimization of the total electricity production system[97]. This might result in, for example, the integration of more peak load units or storage capacity plants. In this study we assume that in the year 2020 this reoptimization has only just started. Therefore, we assume that the wind power penetration level will not exceed the figure of 10% in all electricity production systems.

Two other aspects concerning the grid integration of wind power have to be taken into account. First, the existing capacity of a utility must have a certain minimum size. It has been pointed out that the size of the grid must be at least 50 MWe[4]. Secondly, the distance of the wind resource from the grid might be a limiting factor, as it influences the economics of wind power, especially in developing countries. In reference note 4 the resources that can be exploited have been limited to those resources which are within 50 km of a transmission line. Here, we have taken these two constraints into account also in estimating the wind power penetration up to the year 2020. A third consideration, the erection of overhead transmission lines and towers between the wind resource and grid substation, has been assumed. Such overhead transmission lines, perhaps particularly in rural areas of great natural beauty, are visually intrusive and cause opposition. No allowance has been made for placing lines underground.

Other Aspects Concerning the Introduction of Wind Power

The implementation of wind power, even below a penetration level of 10%, will not only be determined by an economic comparison between marginal costs of electricity production and wind power generating costs. Other aspects will influence the realization of wind power as well. These aspects will be discussed here under the heading of financial, technical, environmental, institutional and potential market factors. Some of these factors are translated into constraints that limit the penetration of wind power in a specific country. It should be noted that the translation to constraints is based on global investigations only.

An important financial factor for the development of wind power will probably be the amount of money that can be invested in wind power. Not only developing countries but probably also economies in transition will have limited financial resources to invest in wind power. Therefore, we constrain the development of wind power per country by limiting the investments. In the current policies scenario we limit the investments in wind power to a maximum of 1‰ of the GDP. This criterion has been chosen by looking at the investments in wind power in Denmark and the Netherlands for the early 1990s as a percentage of GDP. For Denmark this figure will be about 2‰, and for the Netherlands about 1‰. It is not reasonable to expect that other countries, especially developing countries, will invest more. Therefore a criterion of 1‰ has been set. In an ecologically driven scenario we assume that at maximum 3‰ of GDP will be used for investments in wind power. This will most probably need an intensive international support programme, especially as far as developing countries are concerned.

There are of course many technical factors, especially connected with the wind turbine technology, that can constrain wind power development. In this study we assume that MW-size wind turbines will become a mature technology. We also assume that the lifetime of wind turbines will be extended to at least 20 years. Another technical factor concerns the availability of wind resource studies for a country and the reliability of these resource data. The actual decision to set up a wind farm in a specific country cannot be based on a global estimate of the wind resource. More detailed analyses of the wind resource have to be made per country or per region. This can be very difficult because of a lack of data or methods to describe the wind climatology. Practically no developing countries have this information. Therefore, we assume that in almost all developing countries (except India and China) wind power development will not start before the year 2000. We also assume that initiatives will be taken to obtain the required data in the coming decade.

Environmental factors will also constrain the implementation of wind power, due to noise problems, telecommunication interference, bird damage or hindrance and impact on landscape. In our study these aspects are taken into account globally. We also assume that limitations due to impact on landscape are less severe within the ecologically driven scenario than within the current policies scenario. This assumption is somewhat arbitrary, given that beyond certain limits it would suggest local environmental considerations were being set aside in a manner inconsistent with the title of this scenario.

Institutional factors include elements like the existence of special programmes for wind energy. It is evident that until now the development of wind power has been strongly influenced by the policy of governments and utilities on wind energy. For the next few years it is also clear that only countries with a certain commitment to wind energy will realize a certain amount of wind power. Therefore, we assume that the development of wind power up to the year 2000 will take place according to the government goals for wind power.

Potential market factors include questions like: is the potential for wind capacity large enough to expect economies of scale, are there national industries involved in wind energy, and are there enough trained people to realize and maintain wind farms? These aspects have been taken into account by limiting the growth rate for the installed wind turbine capacity per country and the growth rate in world-wide manufacturing capacity. We assume that for each country the growth rate for the installed wind turbine capacity will not exceed 15% per year in the current policies scenario and 30% in the ecologically driven scenario. We also assume that the growth rate for manufacturing capacity world-wide will not exceed 15% per year in the current policies scenario and 30% per year in the ecologically driven scenario. The latter percentage has been derived from the growth rate of the world sales of gas turbines in the period 1960–70. In this period this growth rate was about 30% per year[98].

Wind Power Penetration in the Current Policies Scenario

In the current policies scenario the total energy consumption is estimated to be 13,500 Mtoe (570 EJ) in 2020 (see Table 3.16). In 1987 the total energy consumption was about 8,100 Mtoe. The total electricity consumption in 1987 was 10,500 TWh. It is assumed that the electricity growth rate will be higher than the energy growth rate, because of the penetration of electricity in more market segments[5]. Also, it is assumed that the regional growth rates will differ considerably. Developing countries will have higher growth rates than the industrialized countries. In Table 3.19 the assumed growth rates are indicated per WEC region. Overall, this would mean that in the year 2020 electricity consumption will be about 25,625 TWh.

For each country, the amount of wind turbine capacity and production up to the year 2020 can be estimated, using the methodology, the assumptions and the constraints as described earlier. For the year 2020 the results are summarized in Table 3.19, aggregated for the 8 WEC regions. Figure 3.24 shows the development of installed wind capacity up to the year 2020 within the current policies scenario. For the years 2000 and 2010 the total installed wind capacity is estimated to be 13.6

GWe and 61.9 GWe respectively. The total amount of wind capacity in operation in the year 2020 is estimated to be about 180 GWe, which produces about 375 TWh per year. This is approximately 1.5% of the estimated electricity consumption in the year 2020. This will save 3.5 EJ (83 Mtoe), which is 0.6% of the projected primary energy consumption in 2020. This could reduce the annual emission of CO_2 by about 330 Mt.

Table 3.19 *Electricity consumption and wind energy production by region in the current policies scenario*

WEC region	*Electricity use 1987 (TWh)*	*Annual growth (%/yr)*	*Electricity use 2020 (TWh)*	*Potential wind capacity (GWe)*	*Projected wind capacity 2020 (GWe)*	*Projected wind capacity 2020 (TWh)*
North America	2,990	2.0	5,740	2,360	90	187
Latin America and Caribbean	525	4.0	1,916	1,005	3	5
Western Europe	2,101	2.0	4,035	635	38	80
Eastern Europe and CIS	2,046	3.0	5,421	2,295	6	16
Middle East and North Africa	271	5.9	1,798	930	3	5
Sub-Saharan Africa	210	2.7	506	2,170	2	4
Pacific	1,972	3.0	5,222	975	28	57
(China)	(497)	(4.0)	(1,815)	(290)	(20)	(40)
Central and South Asia	271	4.0	988	80	10	22
Total	10,386	2.8	25,626	10,450	180	376

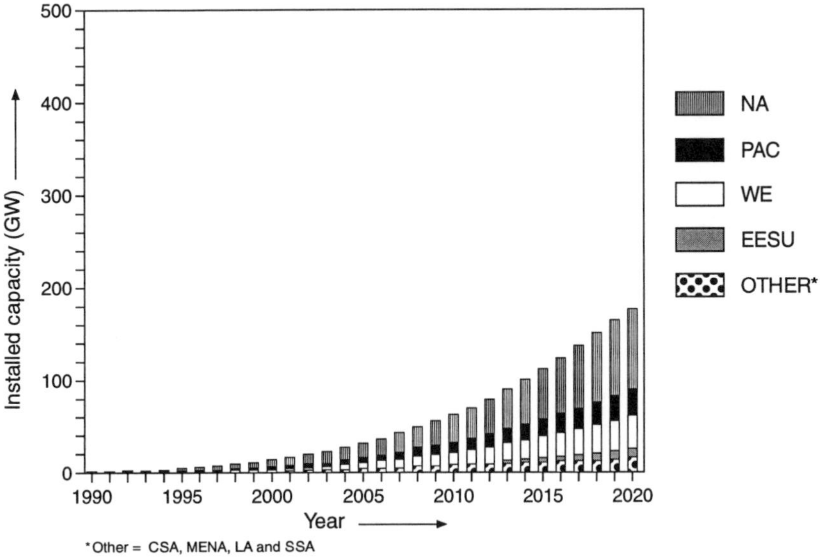

Figure 3.24 *Installed wind capacity up to 2020 per WEC region for the current policies scenario*

Wind Power Penetration in the Ecologically Driven Scenario

In the ecologically driven scenario, the total energy consumption is estimated to be 11,560 Mtoe (485 EJ) in 2020 (see Table 3.16). In 1987 total energy consumption was about 8,100 Mtoe. The total electricity consumption world-wide in 1987 was 10,500 TWh. It is assumed that the electricity growth rate is higher than the energy growth rate, because of the penetration of electricity into more market segments[5]. Again, it is also assumed that the regional growth rates will differ. Developing countries will have much higher growth rates than the industrialized countries. In Table 3.20 the assumed growth rates are indicated per WEC region. Overall, this would mean that in the year 2020 electricity consumption will be about 20,125 TWh.

Table 3.20 *Electricity consumption and wind energy production by region in the ecologically driven scenario*

WEC region	*Electricity use 1987 (TWh)*	*Annual growth (%/yr)*	*Electricity use 2020 (TWh)*	*Potential wind capacity (GWe)*	*Projected wind capacity 2020 (GWe)*	*(TWh)*
North America	2,990	1.5	4,887	2,360	212	423
Latin America and Caribbean	525	3.0	1,530	1,005	11	21
Western Europe	2,101	1.5	3,435	635	78	165
Eastern Europe and CIS	2,046	2.0	3,932	2,295	54	120
Middle East and North Africa	271	4.8	1,302	930	12	25
Sub-Saharan Africa	210	2.0	404	2,170	9	19
Pacific	1,972	2.2	4,067	975	80	160
(China)	(497)	(3.0)	(1,319)	(290)	(51)	(102)
Central and South Asia	271	3.0	718	80	18	34
Total	10,386	2.0	20,275	10,450	474	967

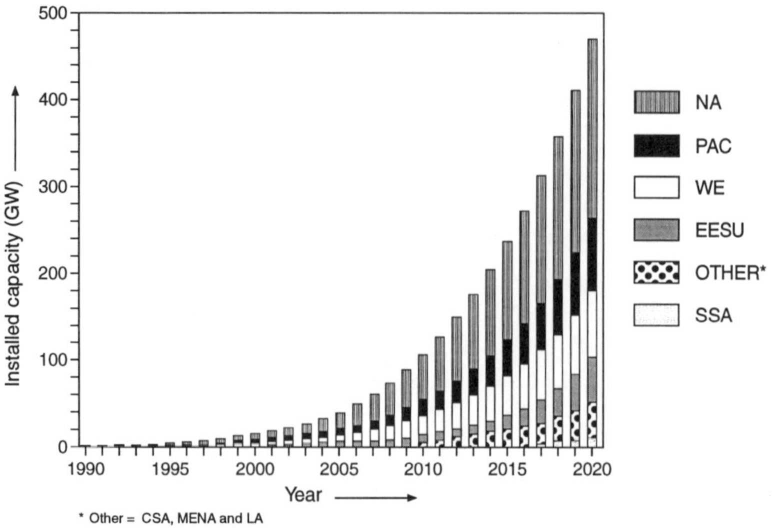

Figure 3.25 *Installed wind capacity up to 2020 per WEC region for the ecologically driven scenario*

For each country, the amount of wind turbine capacity and production up to the year 2020 can be estimated, using the methodology, the assumptions, and the constraints for the ecologically driven scenario as described above. For the year 2020 the results are summarized in Table 3.21, aggregated for the 8 WEC regions. Figure 3.25 shows the development of the installed wind capacity up to the year 2020 within the ecologically driven scenario. For the years 2000 and 2010 the total installed wind capacity is estimated to be 13.7 GWe and 83.3 GWe respectively. The total amount of wind capacity in operation in the year 2020 is estimated to be about 470 GWe, which produces about 970 TWh per year. This is approximately 4.8% of the estimated electricity consumption in the year 2020 according to this scenario. It will save about 9.0 EJ (215 Mtoe). This is 1.8% of the projected primary energy consumption in 2020. This could reduce the annual emission of CO_2 by about 840 Mt.

Comparison Between the Two Scenarios

The possible development of wind power in a current policies scenario and an ecologically driven scenario was outlined in the previous sections. In Table 3.21 the wind turbine capacity and electricity production per WEC region for the years 2000, 2010 and 2020 are summarized for both scenarios. Figure 3.26 shows the development of the electricity production by wind power up to the year 2020 for the current policies scenario, and Figure 3.27 for the ecologically driven scenario. The figures show that up to the year 2000 the development of wind power in the two scenarios will be the same, whereas after the year 2000 the development will be different. The difference is mainly caused by two assumptions. In the ecologically driven scenario, the larger decrease in wind power generation costs in combination with the imposition of a substantial carbon tax on the use of fossil fuels stimulates the application of wind power. Also, the less severe financial constraint in the ecologically driven scenario (investments up to 3‰ instead of 1‰ of the GNP) and the less severe growth rate constraint for both the manufacturing capacity and the installed capacity per country, allow a greater increase in the development of wind power. The other assumptions, such as the restriction that the electricity production of wind turbines should not exceed 10% of the electricity production, have a minor impact.

One aspect worth mentioning separately here, is the growth rate of the world-wide manufacturing capacity of wind turbines. In 1990 the manufacturing capacity world-wide was approximately 250 MWe/year, whereas the installed capacity was about 2000 MWe. To realize the amount of wind power as indicated in the current policies and ecologically driven scenarios, the wind turbine manufacturing industry will have to be expanded considerably. For the current policies scenario the development of manufacturing capacity is shown in Figure 3.28, and for the ecologically driven scenario in Figure 3.29. The growth rates for the

Table 3.21 *Electricity production (in TWh) and primary fuel savings (in Mtoe) by wind turbines for 2000, 2010 and 2020 for the two scenarios*

	Current policies scenario						*Ecologically driven scenario*					
	2000		*2010*		*2020*		*2000*		*2010*		*2020*	
	Electricity production						*Electricity production*					
WEC region	*TWh*	*Mtoe*	*TWh*	*Mtoe*	*TWh*	*Mtoe*	*TWh*	*Mtoe*	*TWh*	*Mtoe*	*TWh*	*Mtoe*
North America	14	3	61	14	187	42	14	3	107	24	423	94
Latin America and Caribbean	0.2	0	1	0	5	1	0.2	0	2	0	21	5
Western Europe	7	2	29	6	80	18	7	2	47	10	165	37
Eastern Europe and CIS	0.7	0	5	1	16	4	0.7	0	10	2	120	27
Middle East and North Africa	0.3	0	1	0	5	1	0.3	0	2	0	25	6
Sub-Saharan Africa	0.2	0	1	0	4	1	0.2	0	2	0	19	4
Pacific	4	1	20	4	57	13	4	1	34	8	160	36
(China)	(3)	(1)	(15)	(3)	(40)	(9)	(3)	(1)	(25)	(6)	(102)	(23)
Central and South Asia	3	1	11	2	22	5	3	1	12	3	34	8
Total	29	6	129	29	376	85	29	6	216	48	967	215

Note:

Primary electricity is converted to thermal equivalents assuming 1 TWh_e = 0.222 Mtoe.

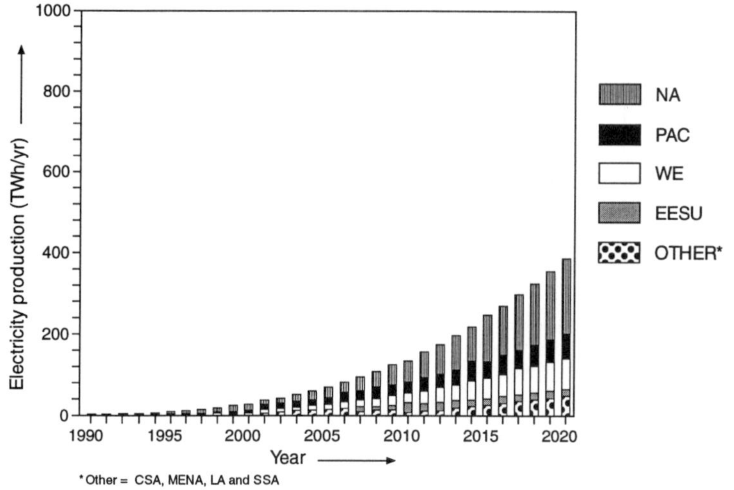

Figure 3.26 *Electricity production by wind turbines per WEC region for the current policies scenario*

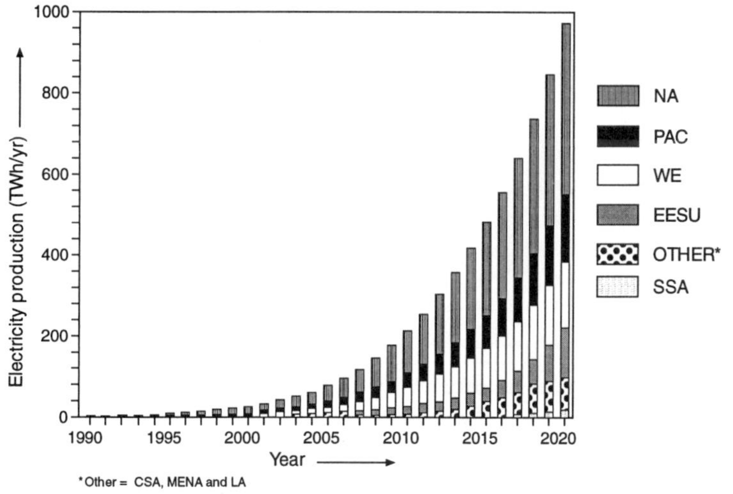

Figure 3.27 *Electricity production by wind turbines per WEC region for the ecologically driven scenario*

manufacturing capacity and the installed capacity per country are never higher than 15% per year in the current policies scenario and 30% per year in the ecologically driven scenario. The resulting industrial manufacturing capacity in 2020 will be sufficient for the necessary replacement of wind turbines and for the realization of moderate growth in the years after 2020.

The cumulative investments in wind turbines up to the year 2020 in both scenarios are given in Table 3.22. It can be seen that in the current policies scenario the cumulative investments will be about US$178,000

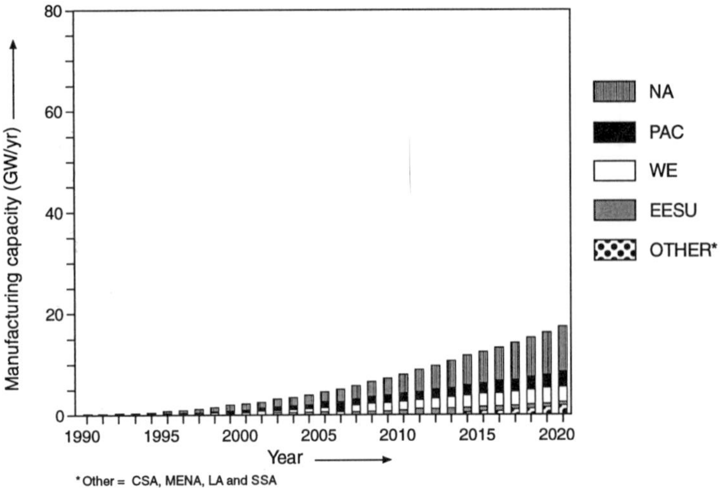

Figure 3.28 *Development of the wind turbine manufacturing capacity per WEC region for the current policies scenario*

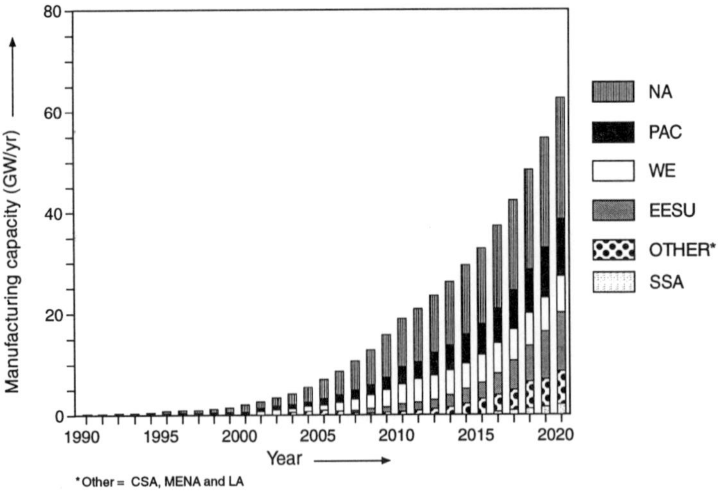

Figure 3.29 *Development of the wind turbine manufacturing capacity per WEC region for the ecologically driven scenario*

million and in the ecologically driven scenario about US$407,000 million (1990 dollars). The development of the investments per WEC region over the years for the current policies scenario is shown in Figure 3.30, and for the ecologically driven scenario in Figure 3.31.

Finally, it should be noted that 180 GWe wind turbine capacity in the current policies scenario in the year 2020 (producing about 375 TWh per year) is equivalent to about 1.7% of the global potential of wind energy, whereas 470 GWe wind turbine capacity in the ecologically driven scenario in the year 2020 (producing about 970 TWh per year) is equivalent to about 4.5% of the global potential of wind energy.

Table 3.22 *Cumulative investment costs in wind energy for 2000, 2010 and 2020 for the two scenarios*

	Cumulative investment costs (in US$ billion, 1990)	
Year	*Current policies*	*Ecologically driven*
2000	11	11
2010	59	93
2020	178	407

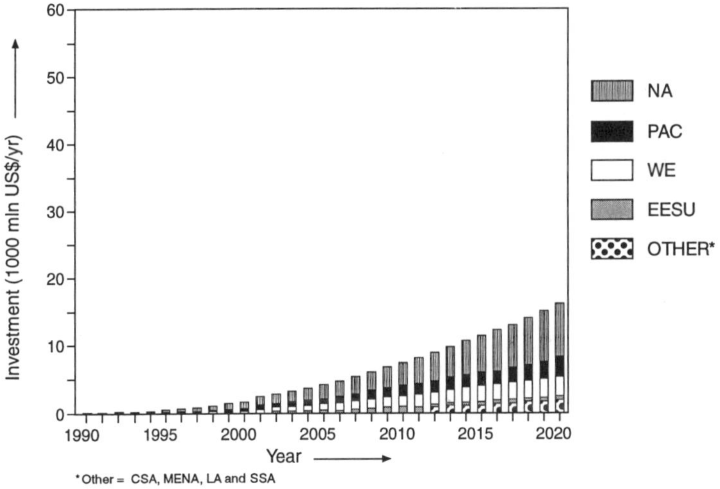

Figure 3.30 *Investments in wind turbine capacity per WEC region for the current policies scenario*

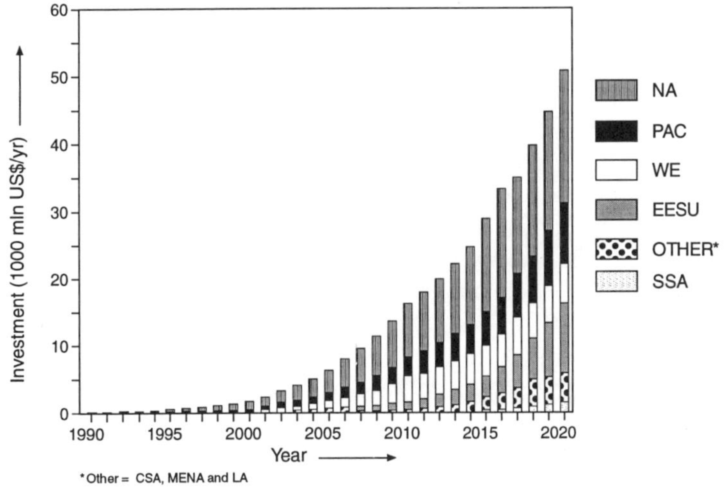

Figure 3.31 *Investments in wind turbine capacity per WEC region for the ecologically driven scenario*

The contribution of electricity produced by wind turbines to the total electricity supply for the year 2020 in both scenarios is very modest. Also, only a few per cent of the total estimated wind resource will be used in the year 2020. This means that the resource and the grid integration of wind energy will not really limit a further expansion of electricity production by wind turbines after the year 2020. Even factors such as financial constraints, economics and manufacturing capacity will not really limit a further expansion on a global scale. Therefore, it is expected that beyond 2020 the contribution of wind energy will continue to grow.

OPPORTUNITIES FOR WIND PUMPS, BATTERY CHARGERS AND WIND-DIESEL SYSTEMS UP TO THE YEAR 2020

About 80% of the world population live in rural areas, 80% of which do not have a grid connection. Estimates show that in 2025 about 2 billion people will not have a grid connection[99]. If the population density is less than 2 people per km^2, the costs of the electricity cable will be more than US$5000 per connection. This means that there is a huge potential for small and off-grid electricity production and for wind pumps.

For example, in China there are 196 million farmers and herdsmen without an electricity supply. In Inner Mongolia about 83,000 battery chargers are in use by herdsmen and in the rest of China about 20,000. It is estimated that farmers and herdsmen in China, who live in areas with a reasonable wind resource, represent a demand for 1,000,000 battery chargers[6]. This implies that there will be an annual market for at least 50,000 battery chargers during the next 20 years in China alone. Other countries have also indicated that there is a market for wind pumps and battery chargers. For example, in Colombia it is estimated that in the next few years there will be a market for 20,000 wind pumps and 30,000 battery chargers[100].

Using the figures for China as a rough estimate for the world it can be expected that there will be a market for at least 100,000 battery chargers over the next 20 years. A similar approach can be followed for wind pumps. The World Bank indicates that the present rate of installations is just over 10,000 wind pumps per year[28]. A conservative estimate of the total world-wide potential of wind pumps would result in a market of 100,000 installations per year. The market for wind-diesel systems is also very promising. A study by SERI estimated that in 1990 there were over 130,000 diesel generators world-wide[101]. This study also shows a retrofit market of at least 15 GWe, and another 26 GWe for new applications.

CONCLUSIONS AND RECOMMENDATIONS

Conclusions – Status and Potential

- It is estimated that at a height of 10 m, about one-quarter of the world's surface (3×10^7 km^2) is exposed to mean wind speeds above

5.1 m/s. Assuming that in this area it is possible to install a wind power capacity of 0.33 MWe per km^2 on mean, it would be possible to generate 20,000 TWh per year. This is twice current global electricity consumption. The possibilities for electricity production by wind turbines placed offshore have not been taken into account, but are probably much smaller.

- The development of modern wind turbines began about 20 years ago. A commonly used technology today is the horizontal axis wind turbine with two or three blades. Wind turbines with a rating of 50 kWe to 500 kWe are commercially available. The mean rating per wind turbine installed has increased from 50 kWe in 1981 to 160 kWe in 1990.
- Wind turbine technology has improved remarkably, resulting in a better efficiency, availability and capacity factor and a significant reduction in costs for manufacture and installation. Despite these improvements, wind turbine technology is still in an early stage of commercial development. Fatigue life and reliability are identified as key problems. Advanced airfoils using better and more flexible materials, advanced control systems and a variable-speed or stall-controlled rotor are the main areas for improvement.
- In regions with a very good wind regime, the economics of modern wind turbines are about break-even with conventional power plants. Current electricity production costs by wind turbines are in the range of 5–10 US cents/kWh.
- Large wind turbines (>500 kWe) are still under development. With the present status of technology the application of large wind turbines would lead to higher electricity production costs due to the weight penalty. Research is required to develop cost-effective large wind turbines.
- World-wide about 2,000 MWe wind turbine capacity has been installed up to 1990, mainly in the USA (especially California) and Europe (especially Denmark).
- The main planning issues related to the installation of wind farms concern visual impact and noise. Avoidance of the most visually sensitive sites would seem to be a basic necessity regardless of technical potential and economics, but to date some schemes have been created or proposed without due regard for this consideration. This stricture is particularly applicable to the United Kingdom. Improvements in technology to reduce the noise production are remarkable, but research and development are still needed to improve performance further. Both issues require careful planning procedures, incorporating local communities, utilities and environmental groups. Impact on birds may also be important for some sites.
- Because of the fluctuating nature of the wind, the integration of wind power in an electricity supply system requires attention. Integral planning of the electricity supply system including wind power will be necessary.

- Apart from grid-connected applications, wind turbines are also used for purposes like water pumping and battery charging. It is estimated that over one million wind turbines for water pumping were in operation in 1990. Although the contribution of these small-scale systems to the global energy supply will always be small, their application is important, especially for the development of rural areas. These systems should not be judged in terms of megawatts, but in terms of the number of people who benefit from the energy produced that meets their basic needs.

Conclusions – Future of Wind Energy

Two scenarios for the development of wind power up to the year 2020 have been presented: a current policies scenario and an ecologically driven scenario. In these scenarios different assumptions are made about R&D efforts and the effect of these efforts on the wind turbine electricity production costs, the growth rate of the capacity to manufacture and to install wind turbines, financial constraints and the introduction of an energy or carbon tax. In Table 3.23 the wind turbine capacity and electricity production per WEC region for the years 2000, 2010 and 2020 are summarized for both scenarios.

It is concluded that the contribution of wind energy to the world electricity supply in the year 2020 can be about 375 TWh in a current policies scenario and about 970 TWh in an ecologically driven scenario, which might be approximately 1.5% and 5% respectively of the total electricity consumption in these scenarios in the year 2020. It is assumed that there is no difference in wind power penetration in the year 2000 because of the times involved to obtain a difference in results if policies are changed.

Recommendations

The implementation of wind energy has just started. In many cases major technical, economic, environmental and institutional constraints will tend to limit future implementation of wind energy. Therefore the following recommendations have been formulated:

- Although significant progress has been made in the past 20 years in developing modern grid-connected wind turbines, it is necessary to increase R&D in order to develop wind turbines which are more reliable, have a better performance and lower generation costs.
- A better understanding should be achieved of the wind resources and the characteristics of the wind on a local, regional and global scale. Together with the WMO (World Meteorological Organization), the World Energy Council could initiate a study to identify the main resource areas.

Table 3.23 *Electricity production (in TWh) and primary fuel savings (in Mtoe) by wind turbines for 2000, 2010 and 2020 under the two scenarios*

	Current policies scenario						*Ecologically driven scenario*					
	2000		*2010*		*2020*		*2000*		*2010*		*2020*	
	Electricity production						*Electricity production*					
WEC region	*TWh*	*Mtoe*	*TWh*	*Mtoe*	*TWh*	*Mtoe*	*TWh*	*Mtoe*	*TWh*	*Mtoe*	*TWh*	*Mtoe*
North America	14	3	61	14	187	42	14	3	107	24	423	94
Latin America and Caribbean	0.2	0	1	0	5	1	0.2	0	2	0	21	5
Western Europe	7	2	29	6	80	18	7	2	47	10	165	37
Eastern Europe and CIS	0.7	0	5	1	16	4	0.7	0	10	2	120	27
Middle East and North Africa	0.3	0	1	0	5	1	0.3	0	2	0	25	6
Sub-Saharan Africa	0.2	0	1	0	4	1	0.2	0	2	0	19	4
Pacific	4	1	20	4	57	13	4	1	34	8	160	36
(China)	(3)	(1)	(15)	(3)	(40)	(9)	(3)	(1)	(25)	(6)	(102)	(23)
Central and South Asia	3	1	11	2	22	5	3	1	12	3	34	8
Total	29	6	129	29	376	85	29	6	216	48	967	215

Note:
Primary electricity is converted to thermal equivalents assuming 1 TWh_e = 0.222 Mtoe.

- It is essential that all (new) designs of wind energy conversion systems fulfil minimum requirements for safety, availability, lifetime, performance and noise. Also, control of the quality of manufacturing and installing wind turbines is important. Therefore, it is recommended to standardize procedures for testing the minimum requirements and the quality internationally. This should be realized within an international framework. The work of the IEC (International Electrotechnical Committee) in this field can be intensified.
- The realization of large amounts of wind power can only be achieved if the technology is also applied in developing countries. Therefore, it is necessary to disseminate knowledge and technology among countries. How the transfer of knowledge and technology to developing countries can be realized should be investigated further, not only for wind energy but also for other renewable energy technologies as well as energy conservation methods.
- Wind energy is a source that by its nature has a fluctuating output. Therefore it is necessary that utilities develop strategies and methods for power system planning such that an optimum use of wind power is made.
- In many cases resources of wind energy are not located near urban areas with a large electricity demand. It is recommended to investigate the possibilities to trade energy from these resources internationally.
- Electricity prices do not reflect the full social and environment costs of electricity production. As a result, not enough credit is given to the relative benefits of using wind energy. It is recommended that the WEC, in co-operation with other international bodies, tries to establish a methodology for determining the total costs of producing electricity from different sources of energy. Also, it should be investigated how these costs should influence the decision making on power production expansion.
- Wind energy is a diffuse source of energy, which means that a large amount of wind turbines need to be installed in order to make a significant contribution to the total electricity supply. Also, the use of wind energy can have some direct environmental impacts, such as disturbance of landscape, noise and bird hindrance. Therefore, decision-making procedures have to be worked out in which all relevant parties are involved. All parties should work to ensure that wind farm developments do not take place in areas of great natural beauty or historic interest, because of visual intrusion, or in very sensitive natural habitats, where significant damage to flora and fauna can be anticipated.
- For the expansion of wind turbine capacity, skilled engineers are needed, as well as public awareness and political interest. Therefore, it is necessary to establish adequate training and education about wind energy at all levels.

- Small wind turbines for water pumping and battery charging can play a vital role in improving the quality of life of people in developing countries. It is recommended to set up financial arrangements and to fund research to disseminate these technologies.

ACKNOWLEDGEMENTS

Under the responsibility of the WEC study group on Wind Energy, which is part of the WEC Study Committee on "Renewable Energy Resources: Opportunities and Constraints 1990–2020", this report has been prepared by Dr. A.J.M. van Wijk, Dr. J.P. Coelingh, and Prof.dr. W.C. Turkenburg from Utrecht University in the Netherlands (Department of Science, Technology and Society, Padualaan 14, 3584 CH Utrecht, The Netherlands).

The WEC study group on Wind Energy is grateful to them and to all persons and organizations who supplied information or comments, including P. Smulders and J. de Jongh (The Netherlands), K. Averstad (Sweden), L. Tallhaug (Norway), N. Meyer (Denmark), E. Sesto (Italy), D. Ancona, E. Miller, J. Darnell and M. Skowronski (USA), J.-R. Frisch (France), M. Grubb (United Kingdom) and D. Strange (Canada).

The NOVEM (Netherlands Organization for Energy and Environment) is thanked for financial support.

REFERENCES

1. World Meteorological Organization. *Meteorological Aspects of the Utilization of Wind as an Energy Source.* Technical Note 175, Geneva, Switzerland, 1981.
2. D.L. Elliott, C.G. Holladay, W.R. Barchet, H.P. Foote, and W.F. Sandusky. *Wind Energy Resource Atlas of the United States.* Pacific Northwest Laboratory, prepared for USDOE, SERI, Golden, Colorado, USA, 1986.
3. W.C. Turkenburg. On the Potential and Implementation of Wind Energy, *Proceedings Amsterdam EWEC '91,* Elsevier, Amsterdam, The Netherlands, 1992, Part II, pp. 171–180.
4. *The Potential for Wind Energy in Developing Countries.* SERI/STR-217-3219, Golden, Colorado, USA, 1987.
5. *Energy in a Finite World, A Global System Analysis.* IIASA, Ballinger, Cambridge, Massachusetts, USA, 1981.
6. Chen Zhixin, Liu Fang, Yang Xiaosheng, He Dexin, and Shi Pengfei. Status, Programme and Policy of Chinese Wind Power. *Proceedings Amsterdam EWEC '91*, Elsevier, Amsterdam, The Netherlands, 1991, Part I, pp. 155–157.
7. Association of Danish Windmill Manufacturers. *Wind Power in the 90s – Pure Energy.* Herning, Denmark, 1991.
8. P.D. Lund and E.T. Peltola. Large-scale Utilization of Wind Energy in Finland: An Updated Scenario. *Proceedings Amsterdam EWEC*

91, Elsevier, Amsterdam, The Netherlands, 1991, Part I, pp. 164–168.
9. P. Mann, N. Stump, R. Windheim, W. Sandtner and S. Scheller. The 250 MW Wind Energy Programme in Germany. *Proceedings Amsterdam EWEC '91*, Elsevier, Amsterdam, The Netherlands, 1991, Part I, pp. 169–171.
10. P.L. Tsipouridis. PPC's Wind Energy Experience. *Proceedings Amsterdam EWEC '91*, Elsevier, Amsterdam, The Netherlands, 1991, Part I, pp. 182–186.
11. Department of Non-Conventional Energy Sources, *Annual Report 1988–89*, Ministry of Energy, Government of India.
12. G. Ambrosini, U. Foli, E. Sesto and R. Vigotti. The Wind Energy Research and Demonstration Programme in Italy. *Proceedings Amsterdam EWEC '91*, Elsevier, Amsterdam, The Netherlands, 1991, Part I, pp. 143–147.
13. Ministry of Energy and Mineral Resources. *Wind Energy Project in Jordan, Final Report*. Jordan Electric Authority, May 1989.
14. Ministry of Economic Affairs. *Memorandum on Energy Conservation: Strategy for Energy Conservation and Renewable Energy Resources*. The Hague, The Netherlands, 1990.
15. Institutt For Energiteknikk. *Oppdatering av Vinddraftpotensialet i Norge*. Kjeller, Norway, 1990 (in Norwegian).
16. A. Ceña-Lazaro. Situation of Wind Energy in Spain. *Proceedings Glasgow EWEC '89*, Peter Peregrinus Ltd., London, United Kingdom, 1989, pp. 42–45.
17. Ministry of Housing and Physical Planning. *Locations for Wind Energy Development*. Stockholm, Sweden, 1989.
18. World Energy Council. *1989 Survey of Energy Resources*. Holywell Press Ltd., Oxford, United Kingdom, 1989.
19. R.R. Loose and D.F. Ancona. The US Wind Energy Programme. *Proceedings Amsterdam EWEC '91*, Elsevier, Amsterdam, The Netherlands, 1992, Part II, pp. 191–195.
20. *The Potential of Renewable Energy: An Interlaboratory White Paper*. US-DOE/SERI, Golden, Colorado, USA, March 1990.
21. D.L. Elliott, L.L. Wendell and G.L. Gower. Wind Energy Potential in the United States Considering Environmental and Land-use Exclusions. *Proceedings Solar World Congress*, Denver, Colorado, USA, August 1991, Vol. 1, Part II, pp. 576–581.
22. L. Arkesteijn, G. van Huis and E. Reckman. *Space for Wind* (in Dutch), Rijksplanologische Dienst, The Hague, The Netherlands, 1987.
23. J.C. Berkhuizen, E.T. de Vries, J.C. van den Doel and H. Muis. The Potential of Wind Energy in the Netherlands: A Modelling Approach (in Dutch), Hoek van Holland, The Netherlands, 1986.
24. *Prospects for Foreign Applications of Wind Energy Systems – Final Report*. DOE Wind/Ocean Technologies, PL 96-345, 1986.
25. M.J. Grubb and N.I. Meyer. Wind Energy: Resources, Systems and Regional Strategies, In T.B. Johansson, H. Kelly, A.K.N. Reddy and

R.H. Williams (eds.). *Renewable Energy: Sources for Fuels and Electricity*. Island Press, Washington DC, USA, 1993.
26. R. Viljoen. The Diffusion of Wind Energy Technology in South Africa. *Proceedings Amsterdam EWEC '91*, Elsevier, Amsterdam, The Netherlands, 1991, Part I, pp. 187–192.
27. R. Swisher and D.F. Ancona. Wind Energy Developments in the Americas. *Proceedings Madrid ECWEC '90*, H.S. Stephens & Associates, Bedford, United Kingdom, 1990, pp. 43–48.
28. The World Bank/UNDP. *Global Windpump Evaluation Programme, Preparatory Phase: Overall Country Study Report, Main Report*. The Netherlands, 1989.
29. T.F. Jaras. *Wind Energy 1987, Wind Turbine Shipments and Applications*. Wind Data Center, Stadia, Inc., Great Falls, Virginia, USA, 1987.
30. T.F. Jaras. *Wind Energy 1988, Wind Turbine Shipments and Applications*. Wind Data Center, Stadia, Inc., Great Falls, Virginia, USA, 1988.
31. P. Nielsen. Wind Energy Activities in Denmark. *Proceedings Amsterdam EWEC '91*, Elsevier, Amsterdam, The Netherlands, 1991, Part I, pp. 177–181.
32. International Energy Agency. *Wind Energy Annual Report 1991* Paris, France, 1992.
33. *WindStats Newsletter*. Winter 1991, Vol. 4, No. 1.
34. Wang Lan. The Development of Wind Energy and Wind Pumps in China. *Proceedings Glasgow EWEC '89*, Peter Peregrinus Ltd., London, United Kingdom, 1989, pp. 51–53.
35. B.T. Madsen. *Evaluation of the Stimulation of Wind Energy in Europe: European Strategy Document on Wind Energy Utilization*. Status report B, September 1990.
36. D. Lindley and P. Musgrove. *Evaluation of the Supply Side of the Market: Status Report on Wind Energy in Europe*. European Wind Energy Association, 1991.
37. International Energy Agency. *Renewable Sources of Energy*. Paris, France, 1987.
38. *Wind Energy Technical Information Guide*. SERI, Golden, Colorado, USA, December 1989.
39. Expert Group Study on Recommended Practices for Wind Turbine Testing and Evaluation. *Power Performance Testing*, 2nd edition. International Energy Agency, 1990.
40. A.J.M. Van Wijk. *Wind Energy and Electricity Production*. Ph.D. Thesis, Utrecht University, The Netherlands, 1990.
41. S.M. Hock, R.W. Thresher and P. Tu. Potential for Far-term Wind Turbines: Performance and Cost Projections. *Proceedings Solar World Congress*, Denver, Colorado, USA, August 1991, Vol. 1, Part II, pp. 565–570.
42. H.J.M. Beurskens and B. Pershagen. The International Energy Agency: Collaboration in Wind Energy. *Proceedings Amsterdam*

EWEC '91 Elsevier, Amsterdam, The Netherlands, 1991, Part I, pp. 148–154.

43. E. Hau. Study on the Next Generation of Large Wind Turbines, Part 4. Energy, Cost Analysis and Conclusions, *Proceedings Madrid ECWEC '90*, H.S. Stephens and Associates, Bedford, United Kingdom, 1990, pp. 443–447.
44. J.A. de Jongh. *Wind Electric Battery Chargers*, Course Material prepared under contract of the Dutch Ministry of Development Co-operation, October 1991.
45. H.J.M. Beurskens and E.H. Lysen. *Perspective of Wind Energy: Status Report.* European Wind Energy Association, 1991.
46. R.L. San Martin. Environmental Emissions From Energy Technology Systems: The Total Fuel Cycle, *Proceedings IEA/OECD Expert Seminar on Energy Technologies for Reducing Emissions of Greenhouse Gases*, Paris, France, 12–14 April 1989, Volume 1, pp. 255–272.
47. J. Schmid, H.P. Klein and G. Hagedorn. How Renewable is Wind Energy?, *Windirections*, Vol. X, No. 2, 1990, pp. 24–29.
48. J. Winkelman. *Birds and Wind Farms.* Institute for Nature Management, RIN report 89/15 (in Dutch).
49. Seldom do birds fly into houses, *Windpower Monthly*, Vol. 5, No. 10, October 1989.
50. California Energy Commission. *Wind Turbine Effects on Avian Activity, Habitat Use, and Mortality in Altamont Pass and Solano County Wind Resource Areas 1989–91*, no. P700-92-001, March 1992.
51. Luke, A. Watts Hosmer and L. Harrison. Bird Deaths Prompt Rethink on Wind Farming in Spain, *Windpower Monthly,* Vol. 10, No. 2, February 1994.
52. Expert Group Study on Recommended Practices for Wind Turbine Testing and Evaluation, 4. *Acoustics Measurements of Noise Emission from Wind Turbines,* (2nd ed.) International Energy Agency, 1988.
53. N.J.C.M. van der Borg and W.J. Stam. Acoustic Noise Measurements on Wind Turbines. *Proceedings Glasgow EWEC '89,* Peter Peregrinus Ltd., London, United Kingdom, 1989, pp. 453–457.
54. T.R. Lee, B.A. Wren and M.E. Hickman. Public Responses to the Siting and Operation of Wind Turbines. *Proceedings Glasgow EWEC '89*, Peter Peregrinus Ltd., London, United Kingdom, 1989, pp. 434–438.
55. M. Wolsink. Attitudes, Expectancies and Values about Turbines and Wind Farms. *Proceedings Glasgow EWEC '89*, Peter Peregrinus Ltd., London, United Kingdom, 1989, pp. 439–443.
56. Dutch Judge Wind Parks to be Ugly, *Duurzame Energie*, Vol. 4, September 1989 (in Dutch).
57. Expert Group Study on Recommended Practices for Wind Turbine Testing and Evaluation, 5. *Electromagnetic Interference: Preparatory Information*, International Energy Agency, 1986.

58. D.T. Swift-Hook (ed.). *Wind Energy and the Environment.* IEE, Peter Peregrinus Ltd., London, United Kingdom, 1989.
59. E.T. De Vries. The Safety of Wind Energy. *Proceedings Glasgow EWEC '89*, Peter Peregrinus Ltd., London, United Kingdom, 1989, pp. 355–357.
60. H.J.M. Beurskens. Certification and Quality Standards: A Must for the Development of Markets for Wind Turbines. *Proceedings Herning ECWEC '88*, H.S. Stephens & Associates, Bedford, United Kingdom, 1988, pp. 440–444.
61. Expert Group Study on Recommended Practices for Wind Turbine Testing, 6. *Structural safety: Preparatory Information*, International Energy Agency, 1988.
62. J.M. Cohen, T.C. Schweizer, S.M. Hock and J.B. Cadogan. A Methodology for Computing Wind Turbine Cost of Electricity Using Utility Economic Assumptions. *Proceedings Windpower '89,* American Wind Energy Association, Arlington, Virginia, USA, 1989, pp. 168–172.
63. H.N. Nacfaire and K. Diamantaras. The European Community Demonstration Programme for Wind Energy and Community Energy Policy. *Proceedings Glasgow EWEC '89*, Peter Peregrinus Ltd., London, United Kingdom, 1989, pp. 1–5.
64. Danish Energy Agency. *Wind Energy in Denmark: Research and Technological Development, 1990.* Ministry of Energy, 1990.
65. J.A. Kuipers and J.B. Molendijk. Co-operation of the Dutch Utilities to Promote the Increased Application of Wind Energy. *Proceedings Amsterdam EWEC '91,* Elsevier, Amsterdam, The Netherlands, 1991, pp. 540–544.
66. N. Halberg. *Wind Energy Penetration Study.* CEC DG XII, Arnhem, The Netherlands, July 1989.
67. N.C. van de Borg, A. Curvers and W.J. Stam. The Energy Production of Wind Turbines (in Dutch). *Proceedings National Wind Energy Conference,* Noordwijkerhout, The Netherlands, 1988, pp. 265–274 and personal communication.
68. H. Lawaetz. The Successful Story of Wind Energy in Denmark. *Proceedings Solar World Congress*, Denver, Colorado, USA, August 1991, Vol. 1, Part II, pp. 571–575.
69. European Wind Energy Association. Time for Action! Wind Energy in Europe. *EWEA Strategy Document*, October 1991.
70. P.T. Smulders. Wind Energy for Water Pumping in Developing Countries. *Proceedings Rome EWEC '86*, A. Raguzzi, Rome, Italy, 1986, pp. 99–105.
71. J. van Meel and P. Smulders. *Wind Pumping: A Handbook.* World Bank technical paper number 101, Industry and Energy Series, July 1989.
72. S. Rashkin. 1988 CEC Wind Performance Results and External Costs. *Proceedings Windpower '89*, American Wind Energy Association, Arlington, Virginia, USA, 1989, pp. 147–167.

73. S. Hock and T. Flaim. Wind Energy Systems for Electric Utilities: A Synthesis of Value Studies. *Proceedings 6th Biennial Wind Energy Conference and Workshop*, USA, 1983, pp. 857–866.
74. M. DeAngelis and S. Rashkin. *The Social Benefits and Costs of Electricity Generation and End-use Technologies*. California Energy Commission, Sacramento, California, USA, 1990.
75. O. Hohmeyer. *The Social Costs of Energy Consumption: Social Effects of Electricity Generation in the Federal Republic of Germany*. Springer Verlag, Berlin, Germany, 1988.
76. R. Friedrich and A. Voss. Die Sozialen Kosten der Elektricitätserzeugung (in German: The Social Costs of Electricity Generation), *Energiewirtschaftliche Tagsfragen*, 39 (1989) 10, pp. 640–649.
77. A.J.M. van Wijk and W.C. Turkenburg. The Value of Wind Energy. *Proceedings BWEA '91*, Swansea, United Kingdom, Mechanical Engineering Publications Ltd., London, 1991.
78. Survey of State Actions to Factor Environmental Costs in Utility Planning, *Solar Industry Journal*, third quarter 1990, Vol. 1, Issue 3, pp. 11–12.
79. H.M. Hubbard. The Real Cost of Energy. *Scientific American*, Vol. 264, No. 4, April 1991, pp. 18–23.
80. R.L. Ottinger. *Environmental Costs of Electricity*. Oceana Publications Inc., Washington, USA, 1990.
81. O. Hohmeyer and R. Ottinger. *External Environmental Costs of Electric Power: Workshop Report*. Ladenburg, Germany, 1990.
82. O.H. Hohmeyer. Latest Results of the International Discussion on the Social Costs of Energy: How Does Wind Compare Today?, *Proceedings Madrid ECWEC '90*, H.S. Stephens & Associates, Bedford, United Kingdom, 1990, pp. 718–724.
83. International Energy Agency. *Energy Policies of IEA Countries: 1990 Review*, Paris, France, 1991.
84. H. Matsumiya and I. Ushiyama. Wind Energy Activities in Japan. *Proceedings Amsterdam EWEC '91*, Elsevier, Amsterdam, The Netherlands, 1991, Part I, pp. 172–176.
85. E.J. van Zuylen, A.J.M. van Wijk and C. Mitchell. Comparison of the Financing Arrangements and Tariff Structures for Wind Energy in European Community Countries. Paper C11, *Proceedings Travemünde ECWEC '93,* to be published.
86. K. Averstad, Swedish State Power Board, *Personal Communication*, October 1991.
87. F. Vlieg and E.T. de Vries. Intergovernmental Accord Regarding the Siting of Wind Energy Projects. *Proceedings Amsterdam EWEC '91*, Elsevier, Amsterdam, The Netherlands, 1991, Part I, pp. 927–930.
88. IEA LS WECS Executive Committee, *IEA Large-scale Wind Energy, Annual Report 1990*, Stockholm, Sweden, 1991.
89. Office of Energy. *New Directions for AID's Renewable Energy Activities: Renewable Energy Applications and Training Project*. Bureau for Science and Technology, US Agency for International Development, 1987.

90. P.T. Smulders. Summary, Conclusions and Recommendations (draft). *International Workshop on Wind Energy for Rural Areas, WERA '91*, Bergen, The Netherlands, October 1991, Proceedings to be published by ECN, The Netherlands.
91. World Energy Council. *Global Energy Perspectives 2000–2020.* 14th Congress, Montreal, Canada, 1989.
92. International Energy Agency. *World Energy Statistics and Balances, 1971–1987.* Paris, France, 1989.
93. E. Taşdemiroğlu. The Energy Situation in OIC Countries: The Possible Contribution of Renewable Energy Resources. *Energy Policy*, December 1989, pp. 577–590.
94. Commission of the European Communities. *Energy in Europe, Energy for a New Century: the European Perspective.* Luxembourg, July 1990.
95. United Nations. *Annual Bulletin of Electric Energy Statistics for Europe, 1980, 1985, 1986, 1987, 1988.* New York, USA, 1989.
96. J.O.G. Tande and J.C. Hansen. Determination of Wind Power Capacity Value. *Proceedings Amsterdam EWEC '91*, Elsevier, Amsterdam, The Netherlands, 1991, Part I, pp. 643–648.
97. W.J. Steeley. *Requirements and Experience with Interconnecting Wind Power Plants to PG&E's Power System.* PG&E, San Ramon, California, USA, 1989.
98. F.A.W.H. van Melick. Kenmerken van een gecombineerde gas/stoomturbine-installatie (in Dutch; Characteristics of a Combined Cycle Installation), *Elektrotechniek*, 52 (1974) 3, pp. 158–164.
99. *Prospects for Industrial Co-operation on Joint Manufacture and Marketing of Wind Turbines and Wind/Diesel Systems in China.* February 1990.
100. A. Pinilla. Global Windpump Evaluation Programme: Country Study of Columbia. *Proceedings Amsterdam EWEC '91*, Elsevier, Amsterdam, The Netherlands, 1991, Part I, pp. 871–875.
101. W.R. King and B.L. Johnson III. *World-wide Wind/Diesel Hybrid Power System Study: Potential Applications and Technical Issues.* SERI/TP-257-3757, Golden, Colorado, USA, 1991.

CHAPTER 4

Geothermal Energy

OVERVIEW

General Characteristics

Geothermal energy is quite widely used for a variety of applications in many areas of the world. Old technologies are still viable; new technologies have matured and better developments are in view. The potential for further development is substantial, both in energy resources and technology transfer. Geothermal energy is independent of weather, has inherent storage and is used both in base-load and on-demand basis. National and international mechanisms for promoting the development have proven their effectiveness in the past and can be expanded.

Geothermal energy refers to heat stored within the solid earth. The quantity is much larger than man's current use, with the largest amounts concentrated along the boundaries of the tectonic plates, in areas also known for volcanism and earthquakes. It is renewable if the heat extraction rate does not exceed the replenishment rate. It is most useful as hot, pressurized water or steam that can be tapped with wells, but mainly exists in hot dry rock.

Utilization of Geothermal Energy

Low-temperature resources have been used from time immemorial for bathing and space heating, and recently also for heating of greenhouses, heat pumps and some process heat. Dry steam and high-temperature water have been in commercial use for electric power production for decades. In the past decade considerable progress has been made in using medium temperature geothermal water (as low as 100°C) for power generation through binary (dual-fluid cycle) plants which are now a matured technology. Significant progress has also been achieved lately in geothermal heat pumps (GHP). This has led to higher growth rate in the use of these resources and has increased the potential of economically available resources in many countries. In the long run, development of effective means of extracting useful energy from hot dry rock and from geopressured and magma resources may further increase the useful potential of geothermal energy.

Technologies for exploration and extraction have relied on petroleum industry experience, which had to be modified in order to meet specifics of geothermal resource (eg, high-temperature environments and salinity of the resource). Exploration and extraction costs represent a substantial portion of the total energy costs. These technologies have been successfully transferred to developing nations by a concerted effort of the United Nations (UNITAR and UNDP).

Environmental Impact

During the early years of development, some environmental problems were created by the direct release of steam into the atmosphere and hot water into rivers. The steam often contains H_2S and may contain CO_2, and the brines may be saturated in dissolved minerals. Modern abatement systems and reinjection techniques have substantially reduced the environmental impact of geothermal energy.

Present Utilization

Installed electric capacity in 1990 reached almost 6,000 MWe, and direct uses of geothermal resources reached the equivalent of 4 Mtoe[8]. The leading countries for application of geothermal energy for electric power generation are the United States, the Philippines, Mexico, Italy, Iceland, New Zealand, Japan, El Salvador, Nicaragua and Kenya. Power is also produced in Turkey, China, Indonesia, Thailand, Taiwan and Russia, while Iceland and Hungary are leading the world in the direct use of geothermal resources.

Future Trends

Although they are concentrated in certain areas, geothermal energy reserves of significant quantity exist in all regions of the world, and they represent an indigenous energy resource for many countries. The world geothermal resource is estimated at 10^9 Mtoe (the estimated demand for world primary energy for 2020 is about 10^4 Mtoe per year[1].

Development of geothermal resources world-wide also means a significant contribution to developing countries, which possess these resources. Apart from reducing their dependence on energy import, it also yields a meaningful contribution to local technical infrastructure and creation of jobs in exploration and drilling operations.

By 2020 in the "current policy scenario", the annual world-wide contribution of geothermal energy will reach 40 Mtoe, and in the "ecologically driven scenario" the figure could reach 90 Mtoe. The more favourable scenario can be achieved through a better accounting of external costs and/or the reactivation of national mechanisms proven in the past (in the USA, loan guarantees and energy tax credits), as well as increased international support for exploration, technology transfer and demonstration projects (UNITAR and UNDP programmes).

INTRODUCTION

Source of Geothermal Energy

Geothermal energy is renewable heat energy from deep in the earth. It originates from the earth's molten interior and the decay of radioactive materials. Heat is brought near to the surface by deep circulation of groundwater and by intrusion into the earth's crust of molten magma originating from great depth. In some places this heat comes to the surface in natural streams of hot steam or water, which have been used since prehistoric times for bathing and cooking. By drilling wells this heat can be tapped from the underground reservoirs to supply pools, homes, greenhouses, and power plants.

The quantity of this heat energy is enormous; it has been estimated that over the course of a year, the equivalent of more than 100 PWh of heat energy is conducted from the earth's interior to the surface. But geothermal energy tends to be relatively diffuse, a phenomenon which makes it difficult to tap. In fact, if it were not for the fact that the earth itself concentrates geothermal heat in certain regions (typically those regions associated with the boundaries of tectonic plates in the earth's crust) geothermal energy essentially would be useless as a heat or a source of electricity using today's technology.

In the strictest sense, geothermal energy is not a "renewable" resource, but rather a minable resource like oil, where the amount that is ultimately recoverable will depend on the technology available to recover it. Reinjection of the water (the heat transportation media) enhances the renewable characteristic by avoiding exhaustion of the aquifer. There is some ambiguity on this issue, however, because certain geothermal sites may be developed in a manner in which the heat withdrawn equals the heat being replaced naturally. In any case, even if it is not technically a renewable resource, potential global geothermal resources represent such a huge amount of energy that, practically speaking, the issue is not the finite size of the resource but the availability of technologies that can tap the resource in an economic manner.

Nature of the Geothermal Energy Resource

On average, the temperature of the earth increases by about 3°C for every 100 metres in depth. This means that at a depth of 2 km the temperature of the earth is about 70°C, increasing to 100°C at a depth of 3 km and so on. However, in some places tectonic activity allows hot or molten rock to approach the earth's surface, creating pockets of higher temperature resources at accessible depths.

The extraction and practical utilization of this heat requires a carrier which will transfer the heat towards the heat extraction system. This carrier is provided by geothermal fluids forming hot aquifers inside permeable formations. These aquifers, or reservoirs, are the hydrothermal fields. Hydrothermal sources are distributed widely, but unevenly, across the earth. High-enthalpy geothermal fields occur within well-defined belts of geologic activity, often manifested as earthquakes, recent volcanism, hot springs, geysers and fumaroles. The geothermal belts are associated with the margins of the earth's major tectonic or crustal plates, and are located mainly in regions of recent volcanic activity or where a thinning of the earth's crust has taken place.

One of these belts rings the entire Pacific Ocean, including Kamchatka, Japan, the Philippines, Indonesia, the western part of South America running through Argentina, Peru, Ecuador, Central America, and Western North America. An extension also penetrates across Asia into the Mediterranean area. Hot crustal material also occurs at mid-ocean ridges (eg, Iceland and the Azores) and interior continental rifts (eg, the East African rift, Kenya and Ethiopia).

Low-enthalpy resources are more abundant and more widely distributed. They are located in many of the world's deep sedimentary basins – for example along the Gulf Coast of the United States, Western Canada, in Western Siberia, and in certain areas of Central and Southern Europe, as well as at the fringe of high-enthalpy resources.

There are four types of geothermal resources: hydrothermal, geopressured, hot dry rock, and magma. Although they have different physical characteristics, all forms of the resource are potentially capable of electric power generation if sufficient heat can be obtained for economical operation.

Hydrothermal

The only commercially used resources at the present time, these contain hot water and/or steam trapped in fractured or porous rock at shallow to moderate depths (from approximately 100 m to 4,500 m). Hydrothermal resources are categorized as vapour-dominated (steam) or liquid-dominated (hot water) according to the predominant fluid phase. Temperatures of hydrothermal reserves used for electricity generation range from 90°C to over 350°C, but roughly two-thirds are estimated to be in the moderate temperature range (150°C–200°C). The highest quality reserves contain steam with little or no entrained fluids, but only two sizeable, high quality dry steam reserves have been located to date – Larderello in Italy and The Geysers field in the United States.

Recoverable resources available for power generation far exceed the development to date. Many countries are believed to have potential in excess of 10,000 MWe[2] which would fulfil a considerable portion of their electricity requirements for many years (eg, the Philippines and Indonesia).

Important low-enthalpy hydrothermal resources are not necessarily associated with young volcanic activity. They are found in sedimentary rocks of high permeability which are isolated from relatively cooler near-surface ground water by impermeable strata. The water in sedimentary basins is heated by regional conductive heat flow. These basins (eg Pannonian Basin) are commonly hundreds of kilometres in diameter at temperatures of 20–100°C. They are exploited in direct thermal uses or with heat pump technology.

Geopressured

Geopressured geothermal resources are hot water aquifers containing dissolved methane trapped under high pressure in sedimentary formations at a depth of approximately 3 km to 6 km. Temperatures range from 90°C to 200°C, although the reservoirs explored to date seldom exceed 150°C. The extent of geopressured reserves are not yet well known world-wide, and the only major resource area identified to date is in the northern Gulf of Mexico region where large reserves are believed to cover an area of 160,000 km^2. This resource is potentially very promising because three types of energy can be extracted from the wells; thermal energy from the heated fluids, hydraulic energy from the high pressures involved, and chemical energy from burning the dissolved methane gas.

Hot Dry Rock

These resources are accessible geologic formations that are abnormally hot but contain little or no water. The hot dry rock (HDR) source has no practical limit. The basic concept in HDR technology is to form a man-made geothermal reservoir by drilling deep wells (4,000–5,000 m) into high-temperature, low-permeability rock and then forming a large heat exchange system by hydraulic or explosive fracturing. Injection and production wells are joined to form a circulating loop through the man-made reservoir, and water is then circulated through the fracture system. Figure 4.1 illustrates this concept[2].

Magma

Magma is molten rock at temperatures ranging from 700°C–1,200°C. Magma chambers represent a huge potential energy source, the largest of all geothermal resources, but they rarely occur near the surface of the earth. Extracting magma energy is expected to be the most difficult of all resource types. The accessible depths are thought to be between 3,000 and 10,000 m. This resource has not been developed as yet, and since the USA research programme was terminated in 1991, it awaits further research in the future.

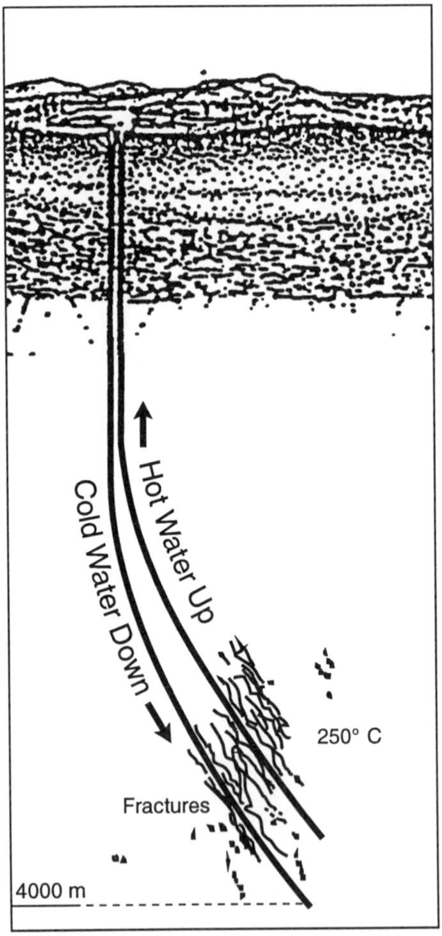

Figure 4.1 *Hot dry rock concept*

CURRENT GEOTHERMAL ENERGY UTILIZATION

Brief History

Geothermal energy has had a long history of use in applications such as therapeutic hot baths, space and water heating, and agriculture. It was not until 1904 that the power of natural geothermal steam was first harnessed to produce electricity, by Prince Piero Ginori Conti at Larderello, Italy. During these early decades there was little growth of this application due to cheap competing sources of electric power and it was 1958 before the next large-scale geothermal power station was commissioned, at Wairakei, in New Zealand. Only those sources which were easy to exploit such as The Geysers in California, United States and some liquid-dominated geothermal resources in Japan and New Zealand were developed to the commercial power stage before 1970.

Of the geothermal resource types, hydrothermal energy is the most advanced and cost-competitive and the only one presently being used commercially. Magma, geopressured and hot dry rock systems are still in experimental stages, although the latter two types have been successfully technically demonstrated, and energy extraction has been experimentally verified. Hot dry rock technology development has been carried out in the United Kingdom, Japan, France, Germany, the former Soviet Union and the United States. The latter is the only country currently investigating geopressured resources.

Present State-of-the-Art

For geothermal energy utilization a number of technology solutions have been introduced. Several of them are still under development, some are already commercial but undergoing continuous improvement activities. The following is an overview of the technology solutions and their developmental status, thus establishing a basis for subsequent discussion.

Exploration and Extraction

Hydrothermal development begins with exploration, to locate and confirm the existence of a reservoir with economically exploitable temperature, volume and accessibility. Geosciences (geology, geophysics and geochemistry) are used to locate reservoirs, to characterize their condition and to optimize the location of wells. Drilling technology used for geothermal development historically derives from the petroleum industry. Certain critical components, such as drilling muds, were modified to work in high-temperature environments, which proved only marginally adequate. Materials and equipment capable of dealing not only with increasing temperatures, but also with hard, fractured rock formations and saline, chemically reactive fluids were needed. As a result, a specialized part of the drilling industry devoted to geothermal development has evolved. Efforts are under way to develop improved methods and materials to deal with the temperatures, salinity and reactive nature of the geothermal environment and to develop better procedures for forecasting geothermal reservoir performance.

Most known reservoirs were discovered from surface manifestations such as hot springs, but exploration now relies increasingly on techniques such as volcanological maps, gravity meters for assessing variations in the density of the rock, electrical methods, seismic, chemical geothermometers, sub-surface mapping, and temperature and heat flow measurements. Gravity surveys are not of much use in initial exploration, although they may help to elucidate the geological structure in an area where it is not well known. Their main use is in underground fluid movement monitoring. Electrical resistivity (and, to a growing extent, electromagnetic) survey is the main method, followed by chemical geothermometry and heat flow. Magnetic prospecting is becoming more important.

In hydrothermal fields, the main advantage of using the electrical resistivity method is that it depends on changes in the electrical properties of the actual resource that is being sought – the hot water itself. Most other methods look at the geological structure – and not all geothermal reservoirs actually conform to any structural model.

Exploratory drilling and production testing is used to establish reservoir properties. If a suitable reservoir is confirmed, field development follows. The geometry and physical properties of the reservoir are modelled, changes in reservoir fluids and rock are analyzed, long-term behaviour of the reservoir is predicted by numerical simulation, and the siting of production wells and injection wells for disposal of spent fluid is determined. (Reinjection is also considered to recharge the reservoir and extend its life.)

Hydrothermal fluids may be produced from wells by artesian flow (ie fluid forced to the surface by ambient pressure differences), or by pumping. In the former case, the fluid may "flash" into two phases (steam and liquid), whereas under pumping, the fluid remains in the liquid phase. The choice between production modes depends on the characteristics of the fluid and the design of the energy conversion system.

Geothermal fields generally lend themselves to "staged" development, whereby a modestly sized plant can be installed at an early stage of field assessment. It can be small enough to operate with confidence on the basis of what is known of the field. Its operation provides the opportunity for obtaining reservoir information which can lead to the installation of future stages.

Other types of geothermal energy have special requirements in the exploration phase. For example, the forces that drive fluids from geopressured brine reservoirs differ greatly from those in conventional oil and gas reservoirs and require a special technology for forecasting geopressured reservoir performance. Better sensing techniques besides seismic methods are needed for exploring magma deposits. Drilling technology requirements and costs increase as the geothermal environment becomes hotter, deeper and more abrasive to drill. Recovery of geopressured energy requires high-pressure technology and the use of heavy drilling muds. Hot dry rock requires the drilling of deep wells in very hard rock and the creation of artificial heat exchange fractures through which fluid can be circulated, with entering and leaving facilitated through one or more deep wellbores.

Successful magma drilling technology has not been established. Magma technology will require special drilling technology to deal with the interaction of the drill bit with molten rock, the effects of dissolved gases, and mechanisms of heat transport in molten magma.

Direct Heat Use

The abundant low and moderate-temperature hydrothermal fluids can be used as a direct heat source for space and water heating, for industrial processes, and for agricultural applications. Some of the major uses have been for balneology, space heating and hot water supply for public institutions; district heating systems for groups of buildings (the predominant use); greenhouse heating; warming fish ponds in aquaculture; crop drying; and for various washing and drying applications in the food, chemical, and textile industries. In regions where high-temperature resources occur, the combination of electricity production with the above uses (eg in Iceland) is possible.

In direct-use geothermal systems, fluids are generally pumped through a heat exchanger to heat air or a liquid; although the resource can be used directly if the salt and solids content is low. These systems are the simplest applications, using conventional off-the-shelf components.

In most of the above uses the hydrothermal source is above 40°C. By use of heat pump technology a hydrothermal source of 20°C or less can be used as a heat source, (eg in the USA, Canada, France, Sweden and other countries). The heat pump operates on the same principle as the home refrigerator, which is actually a one-way pump. The geothermal heat pump, however, can move heat in either direction. In the winter, heat is removed from the earth and delivered to the home or building (heating mode). In summer, heat is removed from the home or building and delivered for storage to the earth (air conditioning mode). In either cycle water is heated and stored, supplying all or part of the function of a separate hot water heater. Because electricity is used only to transfer heat, not to produce it, the GHP will deliver 3–4 times more energy than it consumes. It can be effectively used over a wide range of earth temperatures. Current growth rates for GHP systems run as high as 20% per year in the USA and the outlook for continued growth at double-digit rates is good. The US Department of Energy Information Administration (EIA) has projected that GHPs could provide up to 68 Mtoe of energy for heating, cooling and water heating by 2030.

Power Generation

If temperatures are high enough, the preferred use of geothermal resources has been the generation of electricity which could either be fed into the utility grid or be used to power industrial processes on site. It is normally used for base-load power production, and peak-load generation can be obtained in particular situations. The reason is connected with difficulty of control for peak load and with the scaling/corrosion problems which can occur if the various vessels and turbines are not kept full of fluid and air is allowed into them. Cycling load may be necessary for remote locations (eg islands or non-connected areas) or may be useful to increase the value of the produced energy. (This concern about scaling/corrosion does not apply necessarily to dry steam fields.)

Power Generating Plant Options

There are several types of energy conversion processes for generating electricity from hydrothermal resources; dry steam and flash steam systems, which are traditional processes, and binary cycle and total flow systems, which are newer processes with significant advantages.

Dry Steam Plants

Conventional steam-cycle plants are used to produce energy from vapour-dominated reservoirs. As shown in Figure 4.2, steam is extracted from the wells, cleaned to remove entrained solids and piped directly to a steam turbine[3]. This is a well-developed, commercially available technology, with typical size units in the 35–120 MWe capacity range. Recently, in some places, a new trend of installing modular standard generating units of 20 MWe has been adopted. In Italy, smaller units in the 15 to 20 MWe range have been introduced.

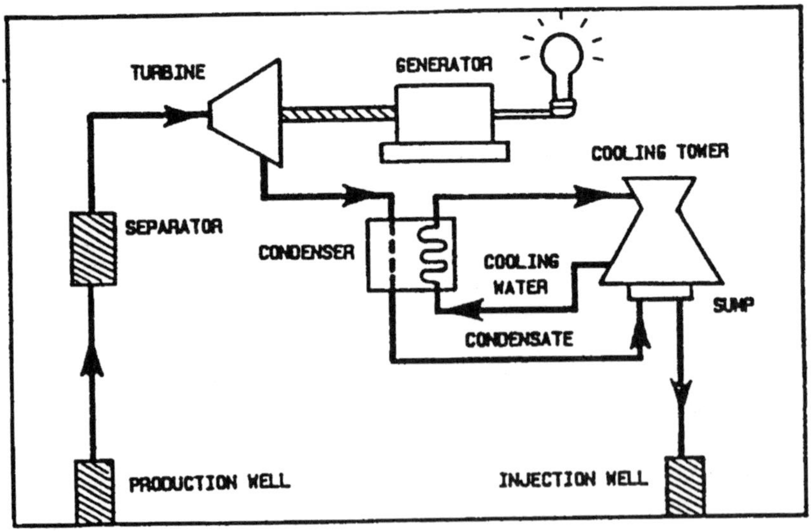

Figure 4.2 *Dry steam system*

Flash Steam Plants

More complex cycles are used to produce energy from liquid-dominated reservoirs which are hot enough (typically above 160°C) to flash a large proportion of the liquid to steam. As shown in Figure 4.3, single-flash systems evaporate hot geothermal fluids into steam (by reducing the pressure of the entering liquid) and direct it through a turbine[3]. In dual-flash systems, steam is flashed from the remaining hot fluid of the first stage, separated and fed into a dual-inlet turbine or into two separate turbines. In both cases the condensate can be used for cooling and the brine is re-injected into the reservoir. This technology is economical at many locations and is being developed with turbogenerators in capacities of 10–55 MWe. Here also a modular approach using standardized units of 20 MWe is being implemented in the Philippines and Mexico.

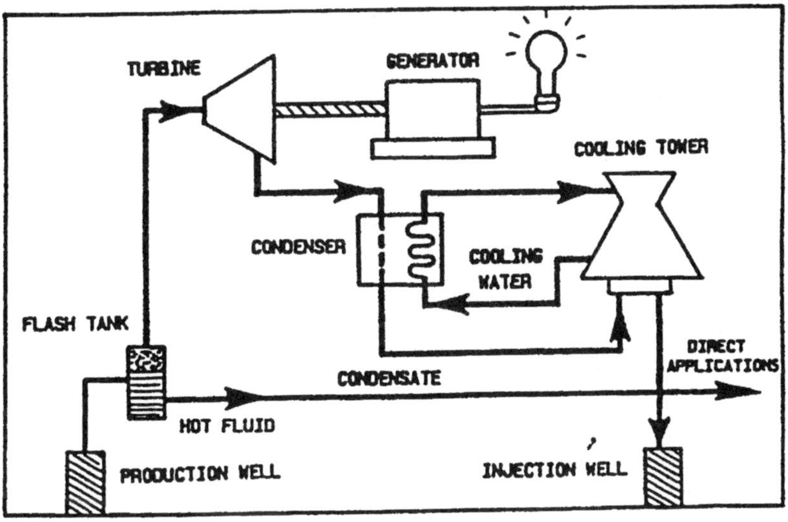

Figure 4.3 *Flash steam system*

Binary-cycle Plants

These are most appropriate for liquid-dominated resources which are not hot enough for efficient flash steam production and for making use of the heat remaining in water separated from steam in flash steam plants (eg Kawerau, New Zealand). They are also useful for saline brine that cannot be flashed because of the resulting deposition of scale, and geothermal fluids with high dissolved (non-condensible) gas content. In this conversion process, the geothermal hot fluid, in the temperature range of 90°C–200°C, is maintained under pressure by a down-well pump and is sent through a heat exchanger where it vapourizes a secondary working fluid (eg, isobutane). The working fluid is expanded through a turbine, condensed and reheated for another cycle. Spent geothermal fluids are usually disposed of by sub-surface injection. This system is depicted in Figure 4.4[3].

Binary systems offer the advantage of enabling utilization of moderate-temperature resources. Binary systems almost avoid corrosion, scaling, and environmental problems through the use of a closed-cycle system where all the geothermal fluid is reinjected. Moreover, in many cases, they are capable of higher conversion efficiencies than flash steam plants and consequently require smaller amounts of geothermal fluids per unit of electricity generated. Because of the lack of steam condensate, they must, however, rely on external sources of cooling water or air cooling. Generating electricity with binary cycle technology is the primary candidate for use with the hot dry rock resource, due to the moderate temperature of the circulated fluids. Typical unit size is 1 to 3 MWe. For example, in East Mesa in southern California, a 30 MWe Binary Power Plant consists of 26 modular units of 1.2 MWe each. Already some 200 MWe binary power plants are in operation world-wide.

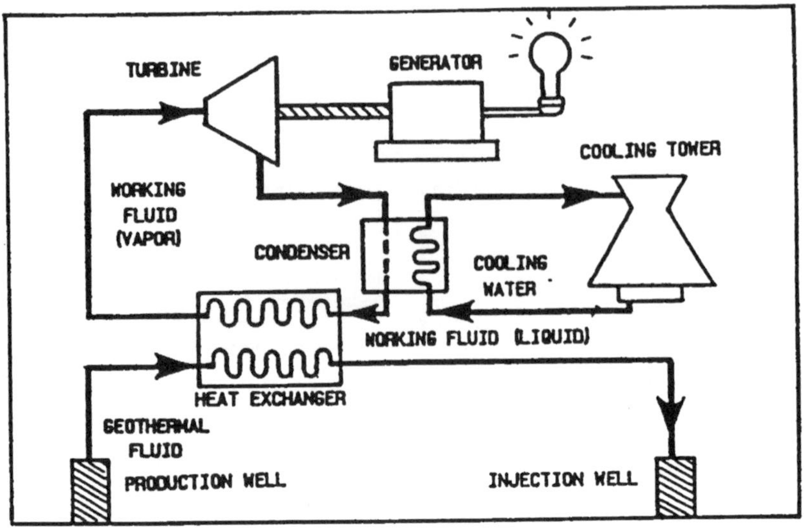

Figure 4.4 *Binary system*

Total Flow Turbines

This is an experimental process, based on using concurrently the steam, hot water, and pressure of geothermal resources (ie, the total resource), thereby eliminating energy losses associated with the conventional method of flashing and steam separation. These systems usually channel a mixture of steam and hot water into a rotating conversion system and capture the kinetic energy of the mixture to power an electric generator. Work on total flow technology was conducted under the auspices of an IEA Implementing Agreement and resulted in testing of a prototype system (the Helical Screw Expander) in Italy, Mexico, New Zealand and the United States. One example of this concept is shown in Figure 4.5[3].

Hybrid Cycle

In this power conversion concept, electricity is generated from two or more sources of energy. In such plants geothermal heat is supplemented with biomass or fossil fuel. Such plants may be appropriate for reservoirs which are too cool to generate electricity economically from the hot water alone but which can play a fuel-saving role.

Potential Utilization of Other Resources

Geopressured resources, which contain recoverable dissolved methane in deep reservoirs under very high pressures but moderate temperatures, can be utilized through three principal conversion technologies. The heat can be converted to electricity in a binary turbine generator. The methane can be sold to a pipeline or converted to electricity in a gas turbine or reciprocal engine. In the early years of the useful life of the reservoir, energy in the wellhead pressures can be converted by a hydraulic turbine generator. A hybrid cycle where exhaust heat from a gas turbine is added

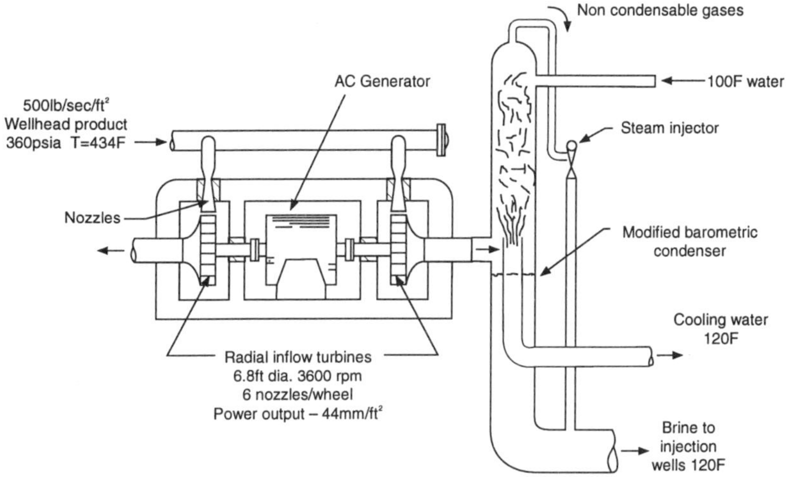

Figure 4.5 *Total flow system*

to the binary working fluid stream was tested for a year at Pleasant Bayou, Texas, and produced about 15% more electricity than would have been generated if the combustion process and binary module had not been interconnected[4].

Potential magma energy applications include the possibility of electric power production and of using the chemical reactivity and/or heat of the magma to evolve hydrogen from water or synthesize gas from water/ biomass mixtures.

Geothermal Energy Project Cost Structure

A typical geothermal energy project has several well-defined cost components. In order to facilitate analysis of technology improvements, the objections are structured in a hierarchy that reflects those components. Four major cost components are recognized:

- resource analysis – finding and defining a geothermal energy resource;
- fluid production – producing geothermal fluid and maintaining that production;
- energy conversion – extracting energy in a useful form from the geothermal fluid;
- other operations – any other application-specific cost factors.

The cost tree, illustrated in Figure 4.6, is generic and it can be applied to any resource type (hydrothermal, geopressured, hot dry rock or magma), conversion system (eg flash or binary) or final use (ie electricity or direct heat)[5].

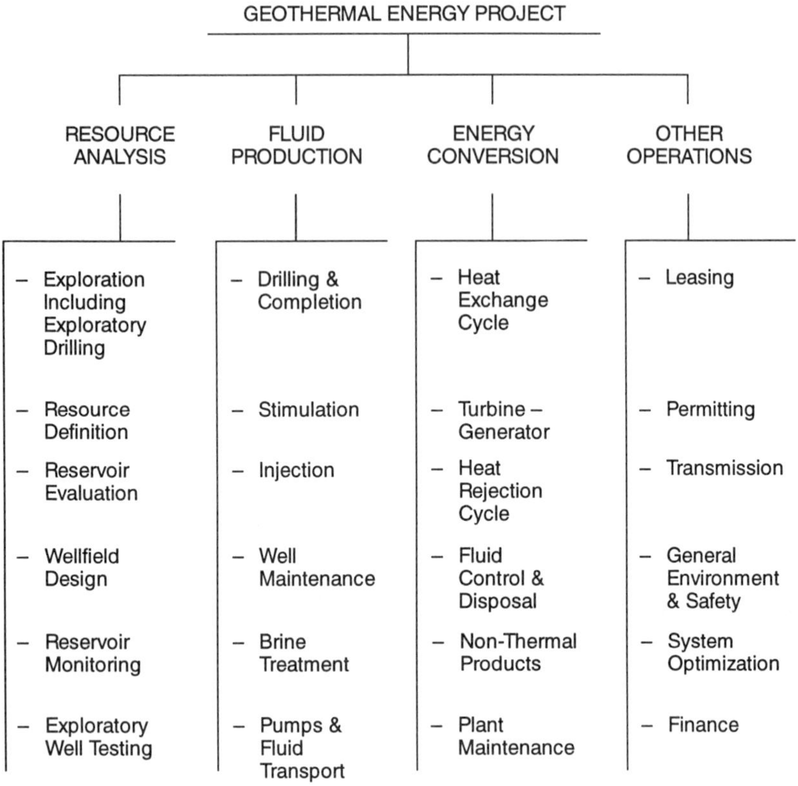

Figure 4.6 *Geothermal energy project cost structure*

Current Use and Commercial Status

Although the exploitation of geothermal energy is not a new notion, and since the beginning of this century attempts have been made to convert it economically in order to produce electricity, the real "push" towards geothermal energy utilization occurred only in connection with the energy crisis after 1973–74, as shown in Figure 4.7[6].

Electricity from geothermal energy has been generated in Italy for more than 80 years, but until 1974 the total installed capacity converting geothermal energy into electricity was some 770 MWe (in Italy, Japan, New Zealand, the United States and Mexico). After the second oil shock the installed capacity world-wide achieved its highest growth of 17.2%. In this period the number of geothermal power producing countries increased from 10 to 17. Emphasis has been placed on power production using the liquid hydrothermal resource since power production with dry steam has been commercially viable for several decades. In 1990 a total of about 6,000 MWe were produced from geothermal resources in more than 20 countries. Substantive market penetration has thus far occurred only in hydrothermal technology.

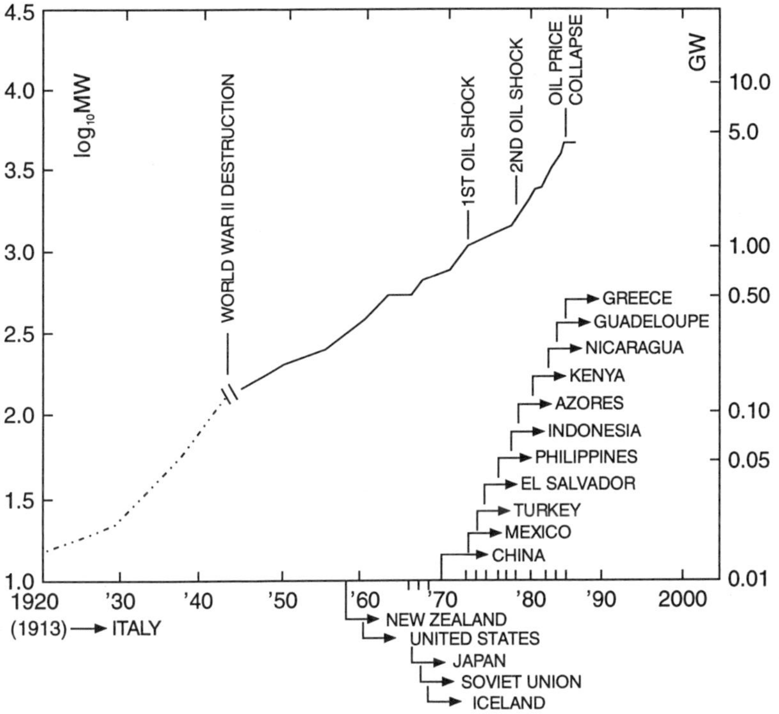

Figure 4.7 *Growth pattern for geothermal power*

Direct Application of Hydrothermal Energy

During the last decade, the greatest growth has occurred in the electricity sector, but considerable activity has also taken place in direct use applications, involving a wide variety of markets. Direct applications utilize well-known technology and straightforward engineering. However, in some cases the technology is complicated by dissolved solids or gases in the geothermal fluid. The technology, reliability, economics and environmental acceptability of direct use of geothermal have been demonstrated throughout the world. Some of its advantages, as compared to electric plants are high utilization efficiency, shorter development time, and less capital investment.

Iceland has been the world's leading country in geothermal district space heating. About 80% of the population use geothermal space heating. Recently France has become a significant user of low-enthalpy geothermal energy. Hungary is the leading country in the world in the use of geothermal for horticulture with about 2 Mm^2 of greenhouses heated by geothermal water.

Up to 1988, the total installed capacity for direct use of geothermal energy world-wide was some 14,000 MWt. The primary users of direct geothermal heat and 1988 capacity are listed in Table 4.1[7]. The total installed capacity in 1975 was about 3,100 MWt (excluding balneology) – thus the growth in direct uses in the past decade has been approximately 6.8% per year.

Table 4.1 *Estimated capacity of direct geothermal use*

	Countries	Space heating (cooling heat pumps[1] warm water) (MW_t)	Green-houses (breeding aqua-culture) (MW_t)	Industry (oil recovery[3]) (MW_t)	Balne-ology (pool, spa) (MW_t)	Multi-purpose (or undefined uses) (MW_t)
NA	United States	936[2]	129	427[3]	284	–
WE	France	660[2]	16	–	na	–
	Iceland	945	77	75	209	–
	Italy	131	94	30	376	–
	Turkey	34	4	–	97	65
EE	Bulgaria	na	na	–	na	135
	Hungary	75	565	70[4]	581	289
	Romania	30	85	na	136	22
	Yugoslavia	14	111	–	2	–
CIS	ex-Soviet Union[4]	429	395[4]	220[3]	360	–
PAC	Japan	49[2]	50	38	4,475	152
	New Zealand	42	2	105	26	1
China		133	146	91	25	–
Others (below 40,000 TOE)[5]		68[2]	55	4	103	59
Roundings and na data		54	60	40	76	27
Peak capacity MW_t		3,600	1,789	1,100	6,750	750

[1] Calculated with references to the heat actually used
[2] Heat pumps (groundwater, earth): France 50 MW_t; Switzerland 15 MW_t; USA 725 MW_t; Germany (W), Japan: data not available.
[3] Oil recovery (OR): load factor 0.8; Hungary: 10 wells = 50 MW_t; CIS: 200 (?)MW; USA: 395 MW_t. Other countries data not available.
[4] Commonwealth of Independent States (CIS) former Soviet Union: some 1,400 hectares (approx. 2,000 MW_t) of hot beds, permafrost de-icing, warm irrigation not included.
[5] Others (28 countries, below 40,000 TOE/yr): Algeria, Australia, Belgium, Brazil, Canada, Chile, Czechoslovakia, Denmark, Germany (W), Greece, India, Indonesia, Israel, Kenya, Korea (N.S.), Mexico, Netherlands, Philippines, Poland, Spain, Sweden, Switzerland, Thailand, Tunisia, UK, Zimbabwe.

Hydrothermal Electric

World-wide installed geothermal electric capacity amounted to almost 6,000 MWe in 1991. Table 4.2 shows the countries with installed power plants and the energy saved[8]. From 1978 to 1985, the installed electrical capacity grew at an average annual rate of about 17.2%. The cause of this surge was the two oil shocks (1973 and 1979) and expectation of further oil price rises. In this period most of the known profitable resources were exploited and much work was devoted to exploration of new hydrothermal resources. After the oil price collapse, the growth rate fell to about 4% a year. Taking into account that most of the subsidies for renewable energy and mainly that for geothermal energy were almost completely stopped, this growth rate is not negligible.

Table 4.2 *World geothermal energy update*

Region	*Countries*	*Installed capacity* Electric uses MW_e	Direct uses MW_t
NA	USA	2,837	463
	Canada	5.1	2
LA	Argentina	0.6	–
	Colombia	–	12
	El Salvador	95	–
	Guatemala	–	10
	Mexico	700	–
	Nicaragua	70	–
WE	Azores (Port.)	3	–
	Austria	–	4
	Belgium	–	93
	Denmark	–	1
	Germany	3	8
	Greece (Milos)	2	18
	France & Antilles	4.2	337
	Iceland	45	774
	Italy	548	329
	Switzerland	–	23
	Turkey	20	246
EE	Bulgaria	–	293
	Czech & Slovak FR	–	105
	Hungary	–	1,276
	Poland	–	9
	Romania	1	251
	Yugoslavia	–	113
CIS	ex Soviet Union	11	1,133
MENA	Algeria	–	13
	Ethiopia	–	38
	Jordan	0.36	–
	Tunisia	–	90
SSA	Kenya	45	–
PAC	Australia	0.02	11
	Japan	270	3,321
	New Zealand	264	258
	Philippines	888	–
	Taiwan	3.3	–
	Indonesia	143	–
	Thailand	0.3	–
China	China	25	2,154
	TOTAL	5,984	11,385

Over the last 15 years geothermal technology (mainly hydrothermal) has made a big step forward, from mainly balneological uses to widespread industrial, agricultural and district heating usage, and from the use of dry steam resources only to power production from a wide spectrum of resources. The energy conversion technology has become a mature and commercially viable technique. Binary power plants, which reached maturity with almost 200 MWe commercially installed capacity, operate as closed-loop geothermal power plants with almost zero pollutants. Plants of a few hundred kilowatts up to tens of megawatts can be installed in 1–2 years and provide a clean, reliable indigenous energy source.

Research Activities

Hydrothermal Research

The major elements of hydrothermal research have been reservoir definition, fluid injection technology, heat cycle research, emissions abatement, permeability enhancement, reservoir engineering, geothermal materials development, deep drilling and development of reservoir simulators.

Considerable progress has been made in refining geological, geophysical and geochemical technologies as a result of the research of the past 10 years. Exploration technologies designed especially for detecting hydrothermal reservoirs have been developed. Reservoir definition and engineering have been advanced by new methods for characterizing fractured reservoirs and realistically predicting their response to production and injection. Reservoir predictive models have been improved and can be used with greater confidence. A number of reservoir confirmation projects have been conducted, and the state of the art of reservoir characterization has been advanced through the development of instruments for well testing. Improved drilling systems and components are now available, but drilling costs must be lowered and research continued.

Extensive materials development has been undertaken in response to serious problems encountered with material failures. Many materials typically used for mechanical components suffer thermal degradation on exposure to the hot fluids, and the chemicals present in the fluid create costly problems due to corrosion and scaling. The materials research has resulted in a number of solutions to these problems through high temperature elastomers and polymer concrete, leak-tight metallic seals, high-temperature down-hole cable, steels for improved drill bits and pitting-resistant alloys. In addition, standardized fluid sampling and analysis procedures have been developed, and precipitators/clarifiers and scale inhibitors to handle saline and corrosive brine are now commercially available.

Thus, major strides have been made in improving the viability of hydrothermal technology. Numerous technical advances have been made, but further improvements are needed, particularly in the areas of exploration techniques and resource assessment, reservoir behaviour, improved drilling techniques and exploitation of the chemicals in the brine.

A great part of research activities in the geothermal resources and their utilizations are performed by USA agencies and academic institutions. Yet, in the past two decades an international effort in geothermal R&D, through transfer of information and education, has been promoted. The International Geothermal Association (IGA), the Geothermal Resources Council (GRC), and the UNDP lead in the dissemination of information; three geothermal educational institutes (in Italy, Iceland and New Zealand) are educating new specialists in geothermal activities.

Geopressured Resource

Research has been directed toward determining the economic feasibility of extracting the resource as well as gaining a better understanding of the quantity, productivity and longevity of the reservoirs. For the past 15 years, the United States Department of Energy has funded extensive long-term field tests on a number of wells on the Gulf Coast of Louisiana and Texas as well as laboratory research. This research has established that the quantity of geopressured brine contained in the Gulf Coast is very large. The area's geopressured resources are estimated to contain from 23 to 240 GWe for 30 years. Louisiana alone has the potential of from 4.1 to 43 GWe for 30 years[5]. The large quantity of methane saturated in the brine can be extracted from the brine by a simple and economical gravity separation technique. Petroleum industry drilling and production technology can be successfully adapted to geopressured brine production. Researchers have also concluded that, because of their relatively benign nature, spent brines may be disposed of by injection into saline reservoirs at relatively shallow depths.

A major problem revealed during field tests was calcium carbonate scaling which can severely restrict brine flow in a well. A scale inhibitor has been successful in achieving an effective solution; a test well produced about 30,000 barrels of brine per day for more than one year with no sign of scaling.

The temperature of the brine has been the cause of some concern. While it ranges between 120°C and 205°C in situ, the temperature in test wells seldom exceeds 150°C, which is marginal compared to other geothermal resources. However, advances in binary conversion, as well as the possibility of co-production of methane, may make the low-temperature resource profitable. A hybrid geopressured conversion unit (1 MWe) was successfully operated in the United States in 1990, as a demonstration unit.

Hot Dry Rock

A major long-term hot dry rock (HDR) research effort in the USA led by Los Alamos National Laboratory has been under way since 1972 at Fenton Hill in New Mexico. Early investigation demonstrated the technical feasibility of extracting heat at reasonable rates from a hydraulically stimulated region of low-permeability hot crystalline rock. In 1979, work was initiated to develop a larger and hotter Phase II reservoir at the same

HDR site, partly in collaboration with German and Japanese scientists under the IEA Hot Dry Rock Implementing Agreement. The primary objective of this latter effort on the Phase II reservoir has been to determine whether sustainable electric power can be generated from hot dry rock resource. The initial 30-day flow test of the Phase II reservoir in 1986 produced water at 190°C, corresponding to about 10 MWt. Since then, work has focused on the installation of a pilot plant to demonstrate the commercial feasibility of producing energy from hot dry resources. The continuous operation of the plant was initiated in March 1992 with the start of the planned long-term flow test. It is expected that the test will yield important information on thermal lifetime, reservoir impedance, thermal drawdown, energy output, water consumption rates, economics, and system operating parameters.

A complementary approach to the Fenton Hill work has been adopted in the HDR experiments at the Camborne School of Mines in Cornwall, in the United Kingdom. Funded by the United Kingdom Department of Energy with some support from the Commission of the European Communities, the project has aimed to reveal the complex processes which take place when rocks at depth under stress are fractured. Differences between the Fenton Hill and United Kingdom projects include British use of explosive fracturing and high viscosity fracturing fluids. The Cornwall reservoir is less than 100°C.

A limited interconnection was achieved between the first two wells in Cornwall. The hydraulic connection was poor, with only about 60% of the water being recovered. In an attempt to improve the reservoir connection, a third well was deviated into the fracture at a different angle. Recent tests have shown a great improvement in circulation characteristics. Heat extraction tests have been conducted, and researchers are considering working at greater depths. However, the UK has suspended its work at the Cornwall site.

The former Soviet Union has also begun a HDR experimental programme with the drilling of a 4,300 m well in the western Ukraine. Recently the EU and Japan have begun major HDR projects.

Magma

In almost 20 years since interest first developed in the possibility of extracting heat from shallow magma sources, the scientific feasibility for doing so has been established. During a United States Department of Energy-funded multi-year research project conducted by Sandia National Laboratories, experimental boreholes were drilled to 74 m into molten rock at a temperature of up to 1,100°C, and heat extraction experiments were successfully demonstrated at the Kilauea Iki crater in the Hawaii Volcanoes National Park. The approach used was to cool the lava to solidify it before drilling rather than directly contacting the liquid rock. The technology required to drill into molten or semi-molten magma at depth is being evaluated, since the western continental United States has a number of relatively young volcanoes that probably overlie magma chambers.

A site in Long Valley, California has been selected for intensive investigations of drilling and completion technology, characterization of the magma environment and energy extraction studies. The first two phases of the drilling were conducted in 1989 and 1991 respectively, followed by scientific experiments in alternate years. The United States suspended its magma research programme in 1991 in order to focus on hydrothermal research.

DEVELOPMENT TRENDS IN GEOTHERMAL ENERGY

Examples of Past Approaches to Development

Geothermal heat and power generating plant development (as well as that of other renewable resources) flourished over the past 15 years due to the sharp rise in oil prices, expectations of further rises and to government incentives. The USA promoted more than 45% of the world's installations in the geothermal domain, thus the impact of the USA development trend was significant to the whole geothermal market. In the USA the combination of federal incentives with the sharply rising oil price, spurred development of this new technology beyond most earlier expectations. The surge in hydrothermal development provided an opportunity to test and advance many emerging technologies, resulting in significant cost reduction and technological improvements.

While the use of dry-steam resources was practical and economical many years ago, lately the use of water-dominated resources became an economically viable technology under the right conditions. The use of the binary system with moderate temperature, high salinity and high gas content resources, which at the beginning of the 1980s was in a demonstration stage, became a mature technology.

The high energy price combined with the government incentives in the USA introduced a significant change in the geothermal industry. Historically most of the non-utility geothermal resource developers viewed their business as exploring for, developing and selling geothermal steam to utilities. In the past decade this image changed, as developers began to build their own generating plants and sell the electricity generated to utilities under the Public Utility Regulatory Policies Act (PURPA). Similar trends can be found today in other countries including developing countries. Companies active in geothermal field development expanded their business to include power plant development and electricity sales. Contracting structures such as Build, Operate & Transfer (BOT) or Build, Operate & Own (BOO) were often helpful for project development.

At one end of the scale in the US geothermal industry, there are large oil and resource companies, which in the past started to diversify into geothermal exploration. At the other end there are the many small, and new, companies that entered the geothermal field in response to changes in the business environment.

The upsurge of the geothermal market was not solely due to the rise in energy price and government incentives. The incentives can be seen as an internalization of external societal costs of power generation and not a subsidy. This was a correction of an existing built-in bias against renewables in favour of fossil fuel burning technologies. It did nothing more than correct an inequity by introducing proper accounting of certain costs of power generation. In the USA this amounted to about 10% of power costs.

In the world there were many geothermal sites where the surface exploration test drilling and even the pre-feasibility study had already been performed. These sites were not exploited due to the energy economy before the energy crises. Use of these sites considerably reduced the cost of the geothermal power plant as well as shortened the time lapse between the decision to utilize geothermal energy to the actual building of a geothermal power plant (by 1–4 years).

In developing countries, the preliminary investigations leading up to a pre-feasibility report were commonly financed by local governments with heavy support from international agencies or bilateral aid agreements. The power plant construction is commonly financed by the World Bank or regional development banks, and by loans provided by suppliers of plant equipment from export credit agencies and commercial banks. The appraisal study and drilling are costly and project-based financing is difficult to obtain.

The UNDP initiated an exploration programme for developing countries which materialized in many cases into geothermal projects. This was the case in Kenya, Tibet and Latin America. For example in El Salvador, 30% of the country's total installed electricity generation capacity is from geothermal energy (within 15 years from the start of the exploration programme). In the Philippines 890 MWe (ie 15% of the country's total electricity capacity) is from geothermal energy and nearly 25% on the island of Luzon, the most populated area.

During the past few years the Philippines, and to a limited extent Indonesia, leased geothermal fields to international companies in a fashion similar to oil fields. The development and operation of the steam fields are in the hands of the international companies and the local power company buys the geothermal steam. The main idea of the above strategy is on the one hand to make the geothermal energy more attractive and more economically appealing to private developers, and on the other hand to enhance the use of indigenous energy resources, reduce conventional fuel use and at the same time reduce environmental pollution.

The dynamics of government support for geothermal R&D have also significantly affected recent development. For example, the USA federal assistance to geothermal grew from US$39 million in 1975 to US$158 million in 1979, but since then the direct assistance to geothermal research and development has fallen to about US$28 million in 1992. The technological achievements of the above productive years resulted in the maturation of some hydrothermal technologies such as small binary power plants using organic working fluids.

Recent Situation

In recent years, the operating environment faced by the developers of power plants fueled by renewable resources has deteriorated markedly in the USA, affecting the entire geothermal market. The most important federal and state tax incentives for many of these technologies either have expired or soon will. Petroleum and natural gas prices, while remaining volatile, have receded to levels of the late 1970s, to less than 50% at their peak in 1982. US federal research and development support for renewable energy resources has fallen by 82% since it peaked in 1980, a victim of the budget deficit and a general sense of complacency about energy issues on the part of lawmakers and the public. Finally, utility managers and regulators have reacted to the electric power industry's more competitive operating environment by scrutinizing costs, often rescinding the favourable power purchase contract terms that enabled independent power developers to finance generating plants or forcing these developers to bid against developers of competing technologies, such as fossil fuel cogeneration plants. Similar patterns can be found in other countries.

Ironically, the slackening pace of renewable energy development is occurring at a time when interest in non-polluting sources of energy is growing rapidly. While skyrocketing oil prices provided the main impetus for energy policy making in the 1970s, it is the environmental costs of energy production that account for much of the renewed interest today. A spate of environmental disasters in 1988 and 1989, most of them linked to conventional energy supply, focused the public's attention, sparking media coverage and legislative proposals reminiscent of the energy crises of the 1970s. Major continuing policy debates, including increasing concern over global warming resulting from the emission of carbon dioxide and other "greenhouse" gases, a sweeping multi-billion dollar clean-air initiative by the US government that seeks to curb the industrial and auto emissions that cause acid rain, and a far-reaching plan to address worsening air quality in southern California that would require significant lifestyle changes, have once again brought energy and environmental issues to the forefront of attention. Power plant emissions such as carbon dioxide, sulphur dioxide and nitrogen oxide are the primary focus of the new environmental concern. Recent series of major oil spills off Europe's and North America's coastlines as well as natural gas pipeline explosions in the former Soviet Union focused the attention of public opinion on the risks associated with the use of fossil fuels.

But none of these have been translated into practical substantial incentives for use of clean energy. In the near term, current market forces, especially low energy prices and uncertainties about the availability of favourable utility power purchase contracts, will inhibit the outlook for renewable technologies in some developed countries, especially the United States. The inherent advantages of geothermal energy, minimal environmental impact, short lead times, modular design characteristics, inflation proof fuel costs, and their indigenous nature are, however, still appealing to some countries.

Factors Affecting the Growth of the Geothermal Industry

In the present market situation, the geothermal power plant development is limited to a narrow "niche" in the electricity generation area. Whenever proven resources exist, geothermal energy can compete with small thermal or internal combustion power plants. This is one of the reasons for the rapid growth of hydrothermal usage in some of the developing countries. An additional "niche" exists in countries where foreign exchange problems favour reliance on indigenous resources (in many cases with foreign funding). In the USA exploration was done mostly by oil companies (after the oil crisis); in the rest of the world exploration is sponsored mainly by international institutions such as UNDP and the World Bank.

The level and pace of growth of the geothermal power plant market appears to depend largely on four key factors:

1. **The prices of competing fuels, especially oil and natural gas.** Fuel prices will continue to have an enormous influence on the business climate for renewable energy resources affecting everything from utility power purchase rates to the availability of private investment capital and government R&D support.
2. **Accounting for environmental costs.** Many of the environmental costs associated with conventional energy technologies are externalities. That is, they are not fully incorporated into the market prices of these technologies. Increasingly, however, the environmental costs of energy supplies are being internalized as a result of public pressure, domestic regulatory actions and international treaties. Since renewable energy technologies tend to have significant advantages over conventional power generating technologies in areas such as air pollution impacts, hazardous waste generation, water use and pollution, and carbon dioxide emissions, they will benefit from policies that internalize environmental costs. Geothermal fields usually occur in remote regions, some of them with great natural beauty, but some of them in the middle of the desert. Almost in every case there are people who oppose the installation of a new power plant. Today there are well-head units as well as binary closed-cycle units without any effluent but still they are regarded as conventional power plants or even worse. It is ironical that geothermal energy is seen by these opponents as being more environmentally damaging than coal or fuel oil.
3. **The rate of future technological development.** Research and development will reduce the energy cost and probably the uncertainty of the hydrothermal reservoir performance which is still a deterrent to rapid growth (at the moment too high a financial risk). The technological improvements should address all parts of geothermal energy projects.

4. **Administrative permits.** One of the advantages of geothermal energy is its short lead time (due to modularity and prefabricated units like the binary power plants), but often the main hindrance is permitting. Some authorities are still treating the geothermal energy as a threat to the environment and not as a means of reducing the environmental hazards.

The attitude towards geothermal energy should be changed, with some terms of reference created in order to distinguish energy sources from one another and thus prevent them from being collectively regarded as an environmental foe. Such a change can be accomplished through a comprehensive public information programme directed toward cognizant regulatory officials and/or the public as to the nature of the impact of geothermal development and its ability to co-exist with competing interests such as scenic national forests.

The governmental incentives must be different in different parts of the world. In the industrial countries there could be a "negative incentive" to fossil fuel burning, such as a CO_2 tax, which will make renewable energy resources economically more attractive and thus enhance their use. In developing countries such a tax might penalize their economy or divert them to non-commercial fuels and be much more harmful to the world ecology. Therefore it should be in the interest of the industrialized world to divert future energy development in developing countries as much as possible towards clean, renewable energy resources by providing positive incentives to use of renewable energy.

Ultimate Potential

The growth rate of the geothermal market is not limited by lack of resources. In the era of the oil crises, intensive investigation revealed many geothermal reservoirs for electricity generation, some of which are in operation, and some 11,000 MWe proven resources are not yet tapped. In the near future the growth rate will most probably be 3–4% annually, as it has been in the past few years, exploiting the actual proven potential.

But if the real value of this technology, such as its superior environmental characteristics and its being a local resource are taken into account, it will render the geothermal market more profitable and will enhance geothermal explorations and R&D. In such a situation the growth rate can reach 6–7% and more. Such an outlook will encourage the development of the other geothermal resources. The hot dry rock and the geopressured technologies may reach maturity towards 2010.

In 1978 the Electric Power Research Institute (EPRI) published a report on the ultimate potential of geothermal energy of a global basis[9]. The results, summarized in Table 4.3, show the regional breakdown of geothermal resources for each of four temperature ranges, and demonstrate the widespread distribution of this energy resource. The global total is very significant compared to today's usage, but most of the total resource is contained in hot dry rock.

Table 4.3 *Regional geothermal resource estimate*

*Region** *Temperature*	*NA*	*LA*	*WE*	*EE/CIS*	*MENA*	*SSA*	*PAC*	*China*	*CSA*	*Total*
<100°C	160	130	44	160	42	110	71	62	88	870
100–150°C	23	27	4.8	5.8	2.1	7.4	6.2	13	5	95
150–250°C	5.9	28	0.8	1.5	0.5	2	4	3.3	0.6	47
>250°C	0.4	0.5	0.01	0.11	0.1	0.1	0.2	0.2	0.04	1.7
TOTAL	189	186	49.6	167	44.7	119	81.2	78.3	93.6	1,000

* Regional abbreviations:
NA North America;
LA Latin America;
WE Western Europe;
EE/CIS Eastern Europe and Commonwealth of Independent States (former USSR);
MENA Middle East and North Africa;
SSA Sub-Saharan Africa;
PAC Pacific (excluding China);
CSA Central and South Asia.

In the EPRI report, the prospects for world-wide geothermal energy utilization were investigated from two approaches. The first was based on projections made by professionals and organizations in various countries of the world, as reported in the literature and as reported in response to the EPRI questionnaire that was sent to member nations of the World Energy Council and to many non-member nations. The second approach was to estimate the potential for utilization based on geothermal, geologic, and geophysical parameters and engineering estimates of recovery and conversion efficiencies.

The geothermal resource base underlying the continental land masses of the world to a depth of 3 km and at temperatures higher than 15°C was calculated to be 1.2×10^{13} GWh_t or 1.03×10^9 Mtoe[10]. It therefore appears that geothermal energy is one of the most abundant resources we have. If we are able to exploit only 1% of this energy we will have enough for several hundred years. But to tap the majority of this energy resource we need to invest money to improve the existing technology – especially extraction of energy from hot dry rock.

A recent study by the US DOE's Office of Renewable Energy, argued that geothermal energy is by far the most abundant energy source in the US, accounting for nearly 40% of the total energy resource base, shown in Figure 4.8[11]. The total resource base is defined as: "concentration of naturally occurring solid, liquid, or gaseous materials in or on the Earth's crust in such a form that economic extraction of the commodity is currently or potentially feasible". In this study the resource base includes geothermal reservoirs with a minimum temperature of 80°C at a maximum depth of 6 kilometres, except for geopressured resources which are included to seven kilometres. Also included are low-temperature resources in the 40–80°C range to a depth of 2–3 kilometres.

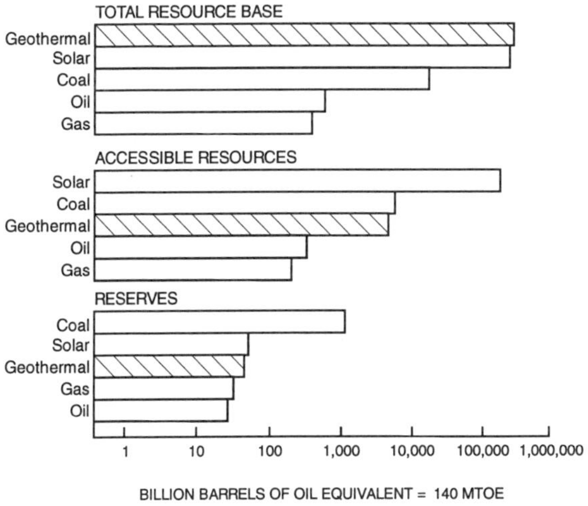

Figure 4.8 *US resources comparison*

FUTURE TRENDS

Introduction

The technology to harness the hydrothermal energy, as mentioned in previous sections, is already available and the geothermal industry has proved its ability to deal with substantial growth rate in parallel with technical developments. The Committee explored two possible scenarios for geothermal energy development:

1. **Current policy scenario,** which will continue the geothermal market at the recent pace of development. It should be noted that this is not a "business as usual" scenario, because the world is becoming more environmentally conscious every year. With increasing emphasis on renewables and building on continued success, a steady growth in geothermal use is predicted for this scenario.
2. **Ecologically driven scenario,** which would stress the importance of environmental considerations in energy industry and thus create incentives which would make geothermal energy even more attractive, leading to a maximum practical contribution by 2020, with increased usage in later years.

The world-wide growth trend is estimated in Table 4.4 and is divided into respective world regions.

Current Policy Scenario

In the USA, which is the leader in geothermal usage, at present there is no significant incentive for use of this resource. The bidding system in use

Table 4.4 *Geothermal outlook to 2020 for two scenarios*

	Ecologically driven scenario			*Current policy scenario*		
	*Electric Use**		*Direct thermal use*	*Electric Use**		*Direct thermal use*
	MWe	*Mtoe/ year*	*Mtoe/ year*	*MWe*	*Mtoe/ year*	*Mtoe/ year*
1990						
North America	2,842	4.17	0.05	2,842	4.17	0.05
Latin America	866	1.27	–	866	1.27	–
Western Europe	625	0.92	1.36	625	0.92	1.36
CIS and East Europe	12	0.02	0.89	12	0.02	0.89
Mid East/No. Africa	0.3	–	–	0.3	–	–
Sub-Saharan Africa	45	0.07	–	45	0.07	–
Pacific/China	1,594	2.34	1.7	1,594	2.34	1.7
Central/South Asia	–	–	–	–	–	–
Total	5,984.3	8.79	4.0	5,984.3	8.79	4.0
2000						
North America	6,000	8.80	1.0	4,200	6.16	0.8
Latin America	1,700	2.49	0.1	1,320	1.93	0.1
Western Europe	1,400	2.05	1.6	940	1.38	1.6
CIS and East Europe	350	0.51	1.6	80	0.12	1.3
Mid East/No. Africa	–	–	0.1	–	–	0.1
Sub-Saharan Africa	200	0.29	–	80	0.12	–
Pacific/China	3,120	4.57	2.1	2,350	3.45	1.9
Central/South Asia	50	0.07	–	–	–	–
Total	12,820	18.78	6.5	8,970	13.16	5.8
2010						
North America	12,000	17.60	1.6	6,300	9.24	1.1
Latin America	3,000	4.40	0.1	2,000	2.93	0.1
Western Europe	2,500	3.67	2.4	1,400	2.05	2.0
CIS and East Europe	1,400	2.05	2.8	200	0.29	1.8
Mid East/No. Africa	100	0.15	0.1	–	–	0.1
Sub-Saharan Africa	500	0.73	–	120	0.20	–
Pacific/China	6,400	9.38	3.5	3,500	5.13	2.3
Central/South Asia	200	0.29	–	–	–	–
Total	26,100	38.27	10.5	13,520	19.82	7.4
2020						
North America	24,000	35.20	2.0	9,300	13.64	1.4
Latin America	6,000	8.80	0.2	3,000	4.40	0.2
Western Europe	4,500	6.60	3.9	2,100	3.08	2.5
CIS and East Europe	3,000	4.40	4.6	400	0.59	2.4
Mid East/No. Africa	300	0.44	0.2	100	0.15	0.1
Sub-Saharan Africa	1,000	1.47	–	200	0.29	–
Pacific/China	11,000	16.13	5.7	5,200	7.63	3.1
Central/South Asia	500	0.73	–	100	0.15	–
Total	50,300	73.77	16.6	20,400	29.93	9.7

* Conversion efficiency = 0.3846 (1 TWh = 9.36 PJ = 0.223 Mtoe)
Capacity factor = 0.75 (6,580 hr/year)
Therefore 1GWe = 6,580 GWh/year ≈ 1.47 Mtoe

for power generation and supply fails to assign proper value to the positive social attributes of renewable technologies – such as their superior environmental characteristics and the fact they reduce dependence on imported fuels – because these factors are not incorporated into market-based energy prices. Competitive bidding tends to favour technologies with low capital (but uncertain fuel) costs over capital-intensive technologies, because of overoptimistic projections about fuel costs and because some programmes emphasize first-year costs in their price criteria. Although first generation bidding systems appear likely to work to the disadvantage of renewable power development, some signs already exist indicating that state regulators may try to achieve a more balanced result – either by assigning greater weight to non-price criteria in the bidding process or by reserving minimum quotas for certain technologies.

In other parts of the world the rate of construction of new geothermal power plants is mainly driven by the financing ability from international institutions such as the World Bank and industrialized countries' export facilities.

In 1990 there were almost 6,000 MWe operational in power plants with another 2,000 MWe planned or under construction. If these planned plants become operational by 1995, it will mean an annual growth rate of about 4%. With a continuation of this growth rate, by 2020 there will be about 20,000 MWe generated in geothermal power plants. A lower growth rate is expected for direct use based on limited opportunity to use heat in close proximity to many resource areas.

Ecologically Driven Scenario

The US Energy Information Administration evaluated the potential of future contributions of renewable energy in the US, in the framework of two fundamental questions:

1. How much renewable energy might be derived by the year 2030 under existing laws and regulations?
2. By how much might that contribution be reasonably increased through cost reductions attributed largely to R&D and regulatory changes?

The conclusions are rather encouraging for renewables in general and geothermal in particular. In the production of electricity, renewable energy in the US was found to have the potential to increase from 161 Mtoe in 1990 (8% of the US energy consumption) to 458 Mtoe with incentives in 2030. *Geothermal electric power makes the earliest and largest contribution* with over ten-fold increase by 2010. By 2030, geothermal energy could provide almost 100 Mtoe, more than any other renewable source except hydropower. The growth rate in the US between 1990–2030 is estimated by the Agency at 5.6% under existing laws or 7.3% with incentives[12].

A report prepared by the US Department of Energy estimated geothermal power production potential by 2030, for the US only, to be nearly 44,000 MW. This would constitute the equivalent of 6.5% of current US total generating capacity, and nearly half the projected increase required by the turn of the century[13].

An important element to consider in the development of the geothermal resource is the lead time required. A steady development of geothermal fields is required in order to establish foundations for positive developments in the future. As the time between initial discovery of the resource and first power production has decreased over the past 2 decades from about 20 years to about 6 to 8 years[14], without an enhanced, steady investment in exploration of geothermal fields, it will not be possible to develop the geothermal plants in the future when needed.

Assuming that the growth rate until 1995 remains 4% per year, and 6–8% thereafter, the total power plant installed capacity will reach over 50,000 MW by the year 2020. The growth potential of direct use is lower due to siting limitations, but can be increased considerably by R&D investment. With only 3% growth a year over the present energy contribution of 4.1 Mtoe/year, by 2020 the direct use contribution will reach 16.6 Mtoe.

These estimates have not taken into account the potential of geopressured, hot dry rock and magma resources. Under other market conditions there would be great impetus for investment in the technology development and maturation process of these resources.

Ecological and Environmental Considerations

The carbon dioxide problem has been selected by the authors of the *World Energy Horizons 2000–2020*[15] as a significant example of the environmental effects arising from energy use. The same approach is adopted here. The burning of fossil fuels releases carbon dioxide into the atmosphere and, whatever the result of the many climatic studies being carried out all over the world, it can already be assumed that the resulting political pressure is likely to lead to constraints on the growth of this form of energy.

In 1985 the industrialized countries' share of world primary energy consumption was 70% while its contribution to the atmospheric CO_2 was 75%. By 2020 the first figure will have fallen to 57%–58% while the latter will approach 60%.

The problem of carbon dioxide is a global problem. Therefore any single nation that attempts to mitigate changes in climate through any unilateral programme of energy conservation, fuel switching or scrubbing carbon dioxide from smoke-stacks, in the absence of some international rationing or compensation arrangement, pays alone for the cost of its programme while sharing the benefits with the rest of the world. Even the United States, the largest energy consumer, phasing in a one-third cut-back in fossil fuel consumption over the next 20 years at a cost perhaps equivalent to US$150 or 200 billion per year at today's prices would reduce emissions world-wide by less than 10%[8].

Any significant effort to curtail emissions would require an international agreement regarding policies. Two main approaches are under discussion:

1. rationing consumption of fossil fuels according to their contribution to carbon dioxide pollution (as shown in Figure 4.9);
2. charging the producer of the emission the social value of the specific emission, and using this "tax" for mitigation of environmental problems.

The second approach will actually raise the cost of electricity from fossil fuels and will enhance the use of non-polluting systems like most renewable energy sources including geothermal energy.

It is very difficult to define the social cost of a pollutant. A reliable value for social benefits or costs should use marginal costs and benefits, but at the moment there is insufficient data on which to value them. However, in California,[16] preliminary proposed values of different pollutants have been published as follows:

NO_x	SO_x	*PM 10*(*)	*ROG*(**)	CO_2
$11,600/ton	$11,500/ton	$7,800/ton	$3,300/ton	$7/ton

(*) Particulate matter less than 10 microns
(**) Reactive organic gases

The carbon dioxide value is based on average rural reforestation. A recent report published by MIT Energy Laboratory has claimed that the complete disposal of power plant emission by means of air

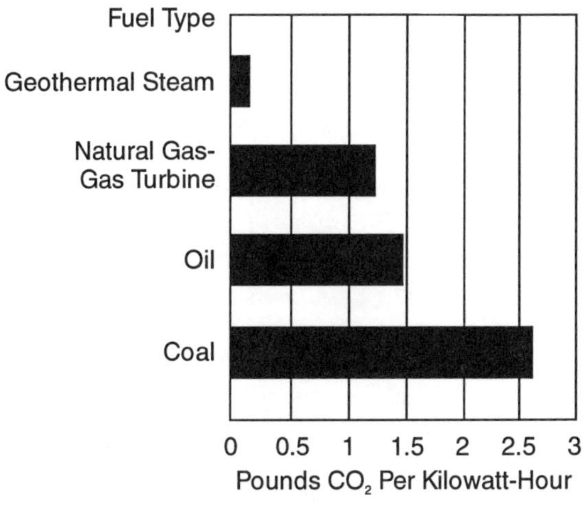

Figure 4.9 *Relative CO_2 emissions*

separation/flue gas recycling and disposing of it in the deep ocean would double the cost of electricity. Since pollution knows no political borders, it is in the best interests of developed countries, if they want to get the maximum return on their investment in pollution control, to direct their investment where it provides them with the maximum return in reduction of CO_2, SO_x, NO_x, and particulate emissions. As an example, a Dutch company obtained 7 times more emission reduction per dollar for investing in Poland than in Holland.

There are other environmental advantages to geothermal energy, as power plants using it consume less land area than other energy sources:

Technology	*Land occupied (sq. metres per GWh/year over 30 years)*
Coal (including coal mining)	3,642
Solar thermal	3,561
Photovoltaics	3,237
Wind (land with turbines and roads)	1,335
Geothermal	404

CONCLUSIONS AND RECOMMENDATIONS

Conclusions

- Geothermal energy resources are substantial. The quantity of this heat energy is enormous; it has been estimated that over the course of a year, the equivalent of more than 100 PWh of heat energy is conducted from the earth's interior to the surface, which is orders of magnitude larger than the estimated demand for world primary energy in 2020 (1.16 to 1.35×10^4 Mtoe).
- Geothermal energy is used for a variety of applications in many areas of the world. Old technologies are still viable; new technologies have matured and better developments are in view. The potential for further development is substantial, both in energy resources and technology transfer.
- Properly implemented, geothermal energy is renewable and non-polluting. Initially some environmental problems were created by the direct release of steam into the atmosphere and hot water into rivers. Modern reinjection techniques have very much reduced and even eliminated environmental impact of geothermal energy.
- Because geothermal energy potential often exists in areas of great natural beauty, and in remote areas, extreme care will be required in the siting, design and size of plant to avoid or minimize adverse environmental impact. Carbon dioxide, hydrogen sulphide and mercury emissions need to be minimized.
- It has an inherent energy storage capability, is independent of weather, and can be used both for base-load or peaking power plants. In most cases it will be more economical to run these plants as base-

load generators, and result in fewer operating concerns such as turbine blade corrosion.

- The low-temperature resources have been used from time immemorial for bathing and space heating, and recently also for greenhouses and some process heat. Dry steam and high-temperature water have been in commercial use for electrical power production for decades. In the last decade considerable progress has been made in using medium-temperature geothermal water (as low as 100°C) for power generation through binary plants which are now a matured technology. This has led to higher growth rate in the use of these resources and has increased the potential of economically available resources in many countries.
- Proven mature technology for direct use and power production for second and third generation equipment have widened the useable resources from dry steam to high-temperature hydrothermal (steam flash plants) to moderate temperature (binary). Progress in binary-cycle plants enables utilization of moderate temperature fluid as well as aggressive fluids, greatly increasing the exploitable resources in many countries.
- Direct use of hydrothermal energy is economically viable, for many applications. There is consistent growth of this market.
- Installed electric capacity in 1990 reached almost 6,000 MWe. The direct use of geothermal resources reached the equivalent of 4.1 Mtoe. The leading countries for application of geothermal energy for electric power generation are the United States, the Philippines, Mexico, Italy, Iceland, New Zealand, Japan, El Salvador, Nicaragua, Kenya. Power is also produced in Turkey, China, Indonesia, Thailand, Taiwan and Russia, while Iceland and Hungary are leading the world in direct use of geothermal resources.
- Geothermal technology offers important benefits of flexibility and modularity, which allow additional capacity to be added in comparatively small increments requiring comparatively small capital outlays. Lead times for installation of a plant are relatively short (between one and two years), although initial reservoir exploration and development takes significantly longer.
- Geothermal energy use began in developed countries (Italy and the USA), where deployment was fostered by regulations (such as PURPA in the USA). Later, through intensive international support programmes (UNITAR, UNDP) geothermal technology was disseminated to developing countries by technology transfer in exploration, exploitation and training.
- Development of geothermal resources world-wide also means a significant contribution to developing countries, which possess these resources. Apart from reducing their dependence on energy import, it also means a meaningful contribution to local technical infrastructure and creation of jobs in exploration and drilling operations.
- By 2020, in the "current policies" scenario, the annual world-wide contribution from geothermal energy will reach about 40 Mtoe. In

the "ecologically driven" scenario, the figure could reach more than 90 Mtoe. The more aggressive policy scenario can be achieved through a better accounting of external costs and/or the reactivation of national mechanisms proven in the past (eg, in the USA, loan guarantees and energy tax credits) and international support for exploration, technology transfer and demonstration projects (UNITAR and UNDP programmes).

- There will be a larger relative growth rate in primary energy consumption in developing countries, because of the high rate of population growth and labour productivity (per capita GDP), relatively high income flexibility and relatively low price flexibility. Many of these countries are endowed with renewable resources, including geothermal. Diverting a large part of the new energy demand towards the clean renewable energy resources will have a larger effect on world pollution than just conserving energy in the developed countries. Such a diversion will occur only if substantial incentives are given.
- In their industrialization, developing countries do not necessarily need to pass through increasing dependence on fossil fuels. They can leapfrog to clean renewable energies such as geothermal.
- Apart from reducing their dependency on fossil fuels (which they import), it means a successful transfer of modern technology to those countries and also a meaningful contribution to the increase of local technical infrastructures by leaving the great bulk of engineering and resource development to local companies.
- Accessible potential can be increased by further R&D in extracting useful energy from hot dry rock, and from geopressured and magma resources, with probable contributions in a significant way to global energy supplies after 2020.

Recommendations

To maximize the useful contribution from geothermal energy by 2020, the global community should intensify the support for mechanisms which have demonstrated their effectiveness in the past. In developed countries this means: internalize societal and other external costs in the power generation costs so as to give the true value to this renewable, non-polluting energy resource; and promote new technologies by incentives such as tax credits and direct R&D support. In developing countries, expand existing support mechanisms of international organizations in exploration, project demonstration, technology transfer and training.

Because of the long lead time of new projects, exploration of geothermal resources is a prerequisite to further development. Countries with resources and international organizations should continue or initiate investment today in geothermal field exploration. Incentives in field development (made mostly by governments or international organizations) are necessary to ensure rapid development of geothermal energy.

In addition, we should internationalize the emission reduction problem by assisting projects in developing countries either with international financing or insurance of BOT or BOO schemes, to promote geothermal projects where the increased investment above the fossil fuel plants is justified.

ACKNOWLEDGEMENTS

This chapter was produced by the Geothermal Subcommittee of the WEC Study Committee on Renewable Energy Resources, under the direction of Chairman L.Y. Bronicki of Israel. Other authors include B. Doron and M. Lax of Israel, and contributions were made by R. DiPippo, J.E. Mock, and K.G. Pierce of the U.S.A., J.T. Lumb of New Zealand, G. Allegrini of Italy, G. Cuellar of El Salvador, D.L.P. Strange of Canada, and I.B. Friedleifsson of Iceland. There were also many other contributors and reviewers, and grateful recognition is extended to them all.

REFERENCES

1. *Global Energy Perspectives 2000–2020.* World Energy Conference, 14th Congress, Montreal, Canada 1989.
2. *Renewable Sources of Energy.* International Energy Agency, OECD, Paris, France, 1987.
3. *Energy Technology Status Report.* California Energy Commission, Report P500-88-003 October 1988.
4. *Geothermal Energy Programme, Multiyear Plan FY 1992–1997* (Draft), Geothermal Division, US Department of Energy, June 1991. (Source: Geothermal Programme Review IX, DOE Report CONFE-913105, San Francisco, California, March 1991).
5. B.C. Lunis, Geopressured–Geothermal Direct Use Developments. *Geothermal Resource Council Bulletin*, 1990, p.531.
6. R. Di Pippo. International Development in Geothermal Power Production. *Geothermal Resources Council Bulletin*, May 1988.
7. *Survey of Energy Resources.* World Energy Conference, 1989.
8. *Survey of Energy Resources*, 16th edition, World Energy Council, 1992.
9. *Geothermal Energy Prospects for the Next 50 Years.* Electric Power Research Institute (EPRI), Palo Alto, California, 1978.
10. T.C. Shelling. Global Environmental Forces. *Energy: Production, Consumption and Consequences*, 1990.
11. *Characterization of US Energy Resources and Reserves.* US Department of Energy, 1991.
12. *Renewable Energy Excursion: Supporting Analysis for the National Energy Strategy – Geothermal Perspective.* US DOE, Washington 1991.
13. *Engineering News Record*, February 1992.

14. *Geothermal Technology Evolution: Rationale for the National Energy Strategy*. Geothermal Division, Office of Utility Technologies, US Department of Energy, 1980.
15. J.R. Frisch, K. Brendow, and R. Saunders. *World Energy Horizons 2000–2020*. Editions Technip, World Energy Conference, 14th Congress, Montreal, Canada, 1989.
16. R.L. Therkelsen. Valuing Emission Reduction. *Electricity Report 90*. State of California Energy Resources – Conservation and Development Commission, 21 November 1989.

BIBLIOGRAPHY

J. Barnea. *The Future of Small Energy Sources*. UNITAR, New York, USA, 1981.

S. Williams and K. Porter. *Power Plays – Profiles of America's Independent Renewable Electricity Developers*. Investor Responsibility Research Center, Washington, DC, USA, 1989.

R. Carella. Obstacles and Recommendations to Promote Geothermal Energy Development. *European Geothermal Update*, Florence, Italy, 1989.

R. Corsi. Technical Problems to Exploit Geothermal Energy (High and Low Enthalpy). *European Geothermal Update*, Florence, Italy, 1989.

Programmatic Objectives of the Geothermal Technology Division. Vol. 1, US Department of Energy, May 1989.

H.J. Olson. *Geology and Geothermal Development Potential of the Volcanic Pacific Island Nations*. Honolulu, Hawaii, 1987.

I.B. Fridleifsson. *Geothermal Resources – Present Status and Future Potential in the World Energy Supply*. 13th Congress of World Energy Conference, Cannes, France, 1986.

G. Allegrini et al. *Growth forecast in geothermoelectric capacity in the world to the year 2020*. 1991.

D.H. Freeston. Direct Uses of Geothermal Energy in 1990. *Geothermal Resources Council Bulletin*, July/August 1990.

C. Sommaruga. *Utilizzazione del Calore Geotermico Nel Mondo*. ENEL Internal Report, 31 Dic. 1988.

1990 International Symposium on Geothermal Energy, Parts I & II, Geothermal Resources Council, Davis, California, 1990.

Production of Electrical Energy From Low Enthalpy Geothermal Resources by Binary Power Plants. UNITAR/UNDP – Centre of Small Energy Resources, Rome, Italy, 1989.

G. Cuellar, W. Fangzhi and D. Rosing. *The Nagqu, Tibet, Binary Geothermal Power Plant at 4,500 Metres Above Sea Level*. 13th New Zealand Geothermal Workshop, 1991.

J.T. Lumb and M. Barr. *The Development of Geothermal Energy*. Regional Symposium of the World Energy Council, Perth, Australia, 1992.

CHAPTER 5

Biomass Energy

INTRODUCTION

Biomass is a term used in the context of energy to define a range of products derived from photosynthesis. Annually, through photosynthesis, solar energy equivalent to several times the world's annual use of energy is stored in the leaves, stems and branches of plants. Of the various renewable sources of energy, biomass is thus unique in that it represents stored solar energy. In addition, it is the only renewable source of carbon, and is able to be converted into convenient solid, liquid and gaseous fuels.

The purpose of this chapter is to assess the potential contribution of biomass to energy supply in the future. It will attempt to answer questions such as: which biomass resources are and will be available?; what are the consequences of large-scale production of biomass as fuel?; what technologies are available for converting the biomass to energy?; and which technologies should be further developed to ensure high efficiency of bioenergy utilization?

This first section provides a general introduction to biomass as a source of energy (bioenergy), looking at its current and historic uses. The second section surveys the biomass available from a number of sources – eg management of natural forests and woodlands for energy, waste products from the agricultural, forestry and timber industries, urban refuse and energy crops. The third section reviews the technologies currently available or under development for the conversion of biomass to energy or to more convenient biomass derived fuels. In the fourth through sixth sections, the constraints and opportunities for bioenergy are evaluated, and some visions for the long-term future are outlined. The final section contains conclusions and recommendations for future actions, followed by acknowledgements and references.

An Overview of Bioenergy Use

Biomass, principally in the form of wood, is mankind's oldest form of energy, and has been used to fuel both domestic and industrial activities. Traditional use has been primarily by direct combustion, a process still used extensively in many parts of the world. Biomass is characterized as a dispersed, labour-intensive, and land-intensive source of energy.

Historically, as industrial activity increased, the growing demand for energy depleted biomass natural reserves. The development of new, more concentrated and more convenient sources of energy has led to bioenergy being substituted by these. Biomass share in the primary energy supply for the industrialized countries is now not more than 3%, even though a number of developed countries still use an appreciable amount of bioenergy, such as Finland 15%, Sweden 9%, USA 4% and former USSR 3–4%.

Due to economic and social factors there is a serious problem with overuse and undersupply of biomass for energy purposes in a number of developing countries. Approximately 13–14% of the world's primary energy supply was derived from biomass in 1987[1]. Looking at individual countries, this percentage contribution goes even higher. Nepal obtains 95+% of its total energy from biomass sources, Malawi 94%, Kenya 75%, India 50%, China 33%, Brazil 25%, Egypt and Morocco 20%.

The use of biomass as a source of energy may be attractive not only for economic reasons (where the fuel is readily available at low cost), but also for economic development and environmental reasons. The systems that convert biomass into usable energy can be modular and efficient on a relatively small scale. Biomass is a renewable and indigenous resource that requires little or no foreign exchange. The agricultural and forestry industries that supply feedstocks also provide substantial economic development opportunities in rural areas.

The pollutant emissions from combustion of biomass are usually lower than those from fossil fuels. Furthermore, commercial use of biomass may avoid or reduce problems of waste disposal in other industries, such as forestry and wood products, food processing, and particularly MSW (municipal solid waste) in urban centres.

Examples of Major Bioenergy Programmes

The following three examples examine bioenergy technologies that have attracted much attention and been implemented on significant scales. Two of them (bio-ethanol and charcoal), have reached commercialization and produce bioenergy on an industrial scale. The third (anaerobic digestion), has been implemented on a large scale throughout rural areas in the People's Republic of China and in India and successfully demonstrated using a different approach (large biogas plants treating manure from several farms) in Denmark.

Bio-ethanol

The most impressive bioenergy programme is undoubtedly the Brazilian Proalcool project. Brazil's total primary energy supply in 1988 reached 6.1 billion EJ, of which 22% was supplied by wood and charcoal and 13% by sugar cane bagasse and ethanol[2]. The Proalcool programme now produces around 12 Gl/year of ethanol representing 62% of the country's automotive fuel[3]. A 22% blend of ethanol with petrol, requiring no engine modification and incurring no mileage penalty is used by 8 million cars. The use of hydrated ethanol in specially designed or modified Otto-cycle engines has also been adopted. More than 4 million cars are now using this fuel. They incur a 25% mileage penalty when compared with conventional or blend-powered vehicles.

Sugar cane cultivation in Brazil has always played an important role in the world sugar market. With the introduction of the Proalcool programme the area planted to sugar cane grew to 3.8 Mha in 1985.

One of the biggest problems related to the Proalcool programme has been the disposal of the stillage or vinasse produced during distillation. The volume ratio of stillage to ethanol can be as high as 13:1. In the past this was disposed of by dumping in rivers, but pollution levels resulting from this effluent have become unacceptable. Current methods of disposal include stabilization in lagoons, the recycling to the sugar cane fields as a liquid fertilizer, and increasingly, anaerobic digestion to produce biogas.

The Brazilian Proalcool programme was estimated to have cost US$ 6.5 billion (1986 prices) in the years 1976 to 1985. However, on the credit side, ethanol replaced petrol which was worth US$ 8.9 billion, thereby saving a substantial amount of foreign currency[4].

The USA, also driven by heavy dependence on foreign petroleum, embarked upon an alcohol fuel programme, and in 1989 was producing 3.2 Gl/yr. However, this is far from the intended levels of production envisaged (45 Gl/yr by 1990). The US alcohol fuels industry has been based on ethanol from maize, largely because of surplus production, and because of the political strength of the agricultural industry.

On the basis of the experience in Brazil and the USA, other countries have also developed ethanol programmes. In Zimbabwe, the Triangle Ethanol plant, which is adjacent to a sugar mill, saves the country several million US dollars per year in foreign exchange, by producing 40 Ml of ethanol from sugar cane. This provides about 12% of the country's petrol.

Malawi started producing fuel ethanol in 1982 with the commissioning of the Dwangwa Distillery. The annual production capacity is 10 Ml, with a distilling capacity of 60,000 l/day. The current petrol market of the country is approximately 50 Ml/yr, thus when running at full capacity, some 20% of the requirements are fulfilled by the distillery.

Charcoal

The main expansion in the use of charcoal in Europe came with the Industrial Revolution in England in the 17th and 18th centuries. In Sweden, charcoal consumption for iron making grew through most of the 19th century, and was the basis of the good quality tradition of Swedish steel.

Historically, the Brazilian iron and steel industry was born at the end of the 19th century in the state of Minas Gerais, a region rich in iron ore, fuelwood and hydropower resources. Throughout most of the first half of the 20th century, long after coke had substituted for charcoal for steel making in the industrialized countries, Brazilian industry decided to maintain the charcoal option. The charcoal-based steel industry, predominantly privately-owned and based on small and medium-sized mills, demonstrated a high potential of technological innovation and adaption, while maintaining a significant share in total supply.

Total Brazilian production of steel reached 25 Mt in 1989 and, for 19% of this, charcoal is used. Participation of charcoal in pig iron production was 38%. About 33 Mm^3 of charcoal were consumed by the iron and steel sector and 12 Mm^3 by other industrial sectors (cement, ferroalloys and others), for a total of 45 Mm^3 consumed in the industrial sector. Around 70% of this was produced from natural forests and woodlands and 30% from planted forests (basically eucalyptus). A total of 5 Mha of degraded land has been reforested in Brazil in the past 20 years, one half of it specifically for charcoal production.

Globally, this sector uses modern technology both for carbonization and for iron melting. Overall, from forest implantation to final product, about a three-fold increase in the efficiency in the use of land has been obtained in the past 20 years. As shown in Table 5.1, an additional five-fold increase could be expected in the future.

Future efficiency improvements are expected from better forestry with intensive use of biotechnology techniques, introduction of carbonization retorts (with recuperation of all the by-products of the pyrolysis), and widespread use of blast furnaces of improved design. The future of the industry is currently being discussed intensively in Brazil.

The integrated steel segment of the industry (30% of charcoal consumption) adopted the reforestation strategy and introduced modern blast furnace technologies to improve efficiency. Discussion on environmental effects of eucalyptus plantation is practically over, as environmentalists and public opinion now are convinced that the new forest technologies are ecologically compatible.

Table 5.1 *Charcoal use for the steel industry*

	Forest yield (t/ha-yr)	*Wood use (t/m³ of charcoal)*	*Charcoal use (m³/t of pig iron)*	*Land use (per t/yr of pig iron)*
Previous	10	1.2	3.5	0.42
Current	20	1.0	3.2	0.16
Short-term	30	0.9	2.9	0.09
Mid-term	60	0.85	2.45	0.03

Biogas

Very often organic wastes from plants, animals and humans are regarded as a nuisance. But such wastes potentially contain enough energy to contribute significantly to energy supply in many areas, particularly the rural regions of developing countries. By using a biogas digester or anaerobic fermentor, methane gas and a nitrogen-rich fertilizer can be produced, with improved sanitation as an added bonus.

For over 50 years, China has struggled to develop this technology. The early phases of biogas development emphasized the construction of large numbers of plants, with some seven million being built between 1973 and 1978. Recently, a co-ordinated effort has been made to disseminate the technology, and train personnel to maintain the digesters. Special biogas organizations were formed at the provincial and county level, in those areas thought to be most promising for the technology[5]. In total these employ some 12,000 people. In addition 716 so-called biogas service stations have been set up to undertake the construction of digesters, the purchase of materials, the provision of management and technical services, and the training of technical people. There are now five million family-sized digesters operating satisfactorily, mainly in the southern provinces of the country. A further 10,000 large and medium-sized digesters are operating at factories and large livestock farms.

India is another country with a long history of biogas use, with the first records dating back to the late 1800s. A co-ordinated programme has been in existence since the 1970s, and about 100,000 units were in existence by the end of 1981. With its 240 million head of cattle, India has tremendous potential for the use of biogas, and by the beginning of 1988, 940,000 digesters had been installed. On average these family-size digesters use the dung of four cattle. Of the total number, 85% are said to be operational, and each produces an average of four m^3/day of gas. This gas is estimated to be equivalent to 3.7 Mt of wood on an annual basis. India has an ambitious programme to increase the number of digesters to 12 million by the turn of the century.

In recent years the target of biogas technology has shifted from "energy recovery" to "environmental protection". This development is demonstrated in developed countries such as Denmark and the Netherlands, which have intensive agricultural production including agro-industries. During the past five years the technology of large biogas plants, treating manure from several farmers, has been developed in Denmark. The plants are now achieving a constant high production of gas. A total of 9 demonstration plants are in operation. The largest receives up to 500 m^3/day of biomass, mainly as manure, and produces more than 2 Mm^3/year of gas. The gas from the plants is used primarily for CHP (combined heat and power) production[6].

The immediate conclusions of the programme are:

- The technology has been significantly improved and, in general, the biogas plants are technically reliable.
- The gas production from the joint biogas plants is now high and stable, in part because of the addition of easily convertible organic waste.

- The costs of transportation weigh heavily in the budget of a joint biogas plant.
- Biogas plants are well suited for treating organic waste with a view to recycling it to farmland.

In addition to the energy and environmental advantages the farms profit from lesser need for commercial fertilizer and better utilization of the fertilizer produced by the farmers themselves. Economically, the demonstration plants have shown a very favourable line of development. The oldest of the demonstration plants have now been updated on the basis of experience gained and are operating extremely well today.

BIOMASS RESOURCES

A small portion of the solar radiation reaching the earth's atmosphere is captured in the photosynthetic process of plants. The photosynthate takes the form of a carbohydrate – eg starch, sugar, cellulose and hemicellulose. The amount of energy stored annually by photosynthesis is several times higher than the world's total current use of energy, and probably 200 times the world's current food energy consumption[7]. It is also informative to note that this biomass, which is 90% in trees, is equivalent to the current proven extractable fossil fuel reserves.

The maximum efficiency of photosynthetic conversion of solar energy is between 5 and 6%. However, in practice, taking into account the world's terrestrial areas, the overall average efficiency of photosynthetic conversion is about 0.3%. The average efficiency when improved agricultural techniques are implemented is around 0.5–1.0%[7].

Biomass resources suitable for energy production encompass a wide spectrum of materials. These range from fuelwood collected from farmlands and natural woodland, through agricultural and forestry crops grown specifically for energy purposes, agricultural and forestry residues, food and timber processing residues, municipal solid waste (MSW) and sewage, to aquatic flora.

Biomass tends to occur in a dilute form, unlike the currently used fossil fuels which are found in concentrated deposits. The cost of collecting large quantities of biomass for a commercial energy application can thus be significant since the material is dispersed, is often of low energy density, and frequently moist, if not wet. Consequently, most current commercial applications of biomass energy use material that has been collected for other reasons, such as timber and food processing residues and urban wastes.

Woodfuels

Woodfuel is a term used to describe all fuel types derived from forestry plantations, natural forests and natural woodlands – eg fuelwood and charcoal. This was the main source of energy used by mankind for centuries, not only for domestic purposes, but also for a wide range of industrial activities. Technological change and the discovery of new, more convenient sources of energy has led to displacement of woodfuels. This substitution is virtually complete in the industrialized world, but a number of factors, principally economic, have slowed the substitution in developing countries.

However, fuelwood is still the dominant source of domestic and small-scale industrial energy in the rural areas, while woodfuels remain important in the urban energy supply in developing countries. Since most domestic use of wood and other biomass fuels is outside commercial channels, it is extremely difficult to obtain reliable data on the amounts consumed. Sample surveys show that the quantities used by domestic consumers vary widely and depend heavily on local availability. Overall world consumption of wood for energy exceeds 1.2 Gt (450 Mtoe) annually.

If population-driven energy demands are to be met in the future, a massive programme of reforestation must be undertaken. The Food and Agriculture Office (FAO) of the UN estimated that more than 100 million people already face acute woodfuel scarcities[8], and that nearly 1.3 billion people live in woodfuel-deficit areas. The analysis suggested that for large numbers of people, woodfuels will become increasingly scarce and expensive[9].

Estimates of the rates of reforestation needed to meet future woodfuel demands are staggering. A World Bank study suggested, for example, that the rate of tree planting in Africa would have to be increased 15-fold if the projected demands for woodfuel in 2000 were to be met[10]. In Malawi, a 1986 review concluded that in the region of 800 000 ha of fast-growing trees would have to be planted to meet estimated 1990 "deficits" of 8 Mm^3, at a cost of well over US$360 million[11].

In the absence of any intervention, it is generally accepted that the "gap" between woodfuel supply and demand will widen considerably. In Kenya, there are serious woodfuel shortages in some areas, and responses to scarcity may involve very high economic costs. However, analyses approaching the woodfuel "crisis" on the basis of aggregate supply and demand figures (ie the "gap" approach), tend to greatly exaggerate and obscure the scope of the problem. Woodfuel supply/demand analyses are often quite misleading, because they fail to distinguish between demand, latent demand, consumption, and the extent to which current demands can be moderated without seriously affecting the access to basic energy needs.

The possible answers to the projected gap have been two-fold, and predictable: increase supply and reduce demand. Increasing supply means using professional forestry skills to develop woodfuel plantations, while reducing demand means the introduction of fuel-efficient woodfuel stoves.

More recent studies, critical of the past approaches, have turned the question around. They suggest that the best place to start looking for solutions to the woodfuel problem is in the day-to-day life and local environment of the people affected[12].

Deforestation of the rural areas of developing countries is another subject often discussed in connection with woodfuel use. These discussions are usually approached from two perspectives. First, woodfuel consumption is often identified as an underlying cause of deforestation. Secondly, in areas which have been deforested, woodfuels are thought to have become increasingly scarce. Neither of these observations clearly describes the norm, although they have been widely accepted and have formed the rationale for many tree planting interventions.

As the strain on existing resources becomes more intense, active strategies may be adopted such as rotational harvesting, the protection of naturally regenerating seedlings and the protection of valued tree types during land clearing. The most active of strategies involves nursery propagation of seedlings, tree planting and sustained-yield management[12].

There are numerous low-cost strategies for reducing fuel consumption when woodfuel is scarce, and these may involve no changes in food consumption patterns. Fire management is the most obvious of these, and studies have shown that the efficiency of a simple three-stone fire can be high if it is closely tended[13]. Changed methods of food preparation can also reduce fuel consumption effectively, at low cost, and without affecting food consumption.

The most high-profile, and highest cost, strategy to conserve woodfuels has been the introduction of "improved" cooking stoves for use by rural people. However, these have to a large extent failed to displace traditional modes of cooking.

It must be remembered that woodfuels are also an important source of energy for a large number of small, medium and large industries in the rural and urban areas of developing countries such as baking, brewing, fish smoking, sugar, tea, coffee, coconut, cocoa and tobacco processing, lime burning and brickmaking. Although conventional fuels are available in many rural areas, their cost per unit of useful energy are usually much higher than those of woodfuels. This may change as woodfuels become more scarce, but as yet, few industrial consumers pay the economic or replacement costs of woodfuels, and therefore have no incentive to switch.

In the developed world the prime woodfuel is residues from forest-processing industries. Utilization of this waste as a fuel at or near the source mill or processing plant has the advantage of incurring little or no additional transport costs because the wastes are a by-product. Surplus wood from industry processing is utilized almost 100% today, either as raw material in other industries or for energy purposes at the mills.

Garden waste for energy purposes includes logs, branches, and slashes. Hedges, isolated trees, and orchards represent a resource of still growing trees that can be utilized for energy purposes. Production of wood from roadside or isolated trees can be increased considerably if plantations along roads and railways, in the cities, or below power pylons are given a high priority.

Conventional wood resources consist of wood in excess of the needs of the traditional forest products industry. These resources are available from thinning of commercial stands or from clear-cutting to allow planting of improved stands. This is a very large resource if managed properly, especially if it includes the potential from unutilized lands that might be dedicated to future biofuels production.

Agricultural Residues

Vast amounts of agricultural residues are generated annually, and probably most of these are not fully utilized. One of the world's major crops, rice, consists of about 25% husk by mass, which represents 100 Mt per year. On a smaller scale, world production of groundnuts is about 10 Mt per year, of which about 45% is shell[14]. In general, although there are crops with both higher and lower residue yields, it is reasonable to assume that about 25% by mass of any dry agricultural product is residue.

The actual mass of residues which will be generated in any given place will depend on the climate, soil conditions and agricultural techniques being applied. The variations are so great that no generalizations are possible. However, even in the most underdeveloped agricultural areas, it is possible to obtain aggregate figures which are impressive. Barnard and Kristoferson carried out an extensive survey of agro-residues, and discovered that even in countries such as Chad, Mali and Peru, per capita cereal residue figures were of the order of 0.3 to 0.4 t/year[15]. On the basis of these having a heating value of 12–20 GJ/t, it means that it is possible in theory to envisage rural household fuel needs being satisfied from cereal residues alone.

Although these figures are interesting in giving a general perspective, they have to be looked at far more critically if any practical assessment of their potential for energy use is to be made. Agricultural residues can be divided into two broad categories – on-farm (or field) and off-farm (or process) residues. The former are those which are left behind in the field, while the latter are produced as a result of processing of the harvested crop at some central location.

The volume of agricultural residues in the industrial world available for the energy market in the future may be less than today if the present trend of producing grains with less straw continues and if the industry continues to buy larger quantities of straw in the future. However, a more constant surplus of straw for energy purposes may be expected if more efficient harvesting and transportation systems can be developed.

Although the use of agricultural residues for energy purposes may seem attractive in terms of its efficiency or cost-effectiveness, it must conform to the existing local agricultural dynamics. In developing countries it is likely that many of the trends in agriculture will help to make the use of residues for energy purposes more attractive. Among these are greater centralization of processing to serve urban or export markets, and the use of chemical fertilizers to raise yields and thus reduce dependence on organic fertilizers, themselves very often based on residues. An example is the shift to mechanized rice-milling at large centralized mills in most Asian countries. In the process of centralization, the rice-husk residue has become a waste product, which often cannot be used in the immediate environs of the mill in the same way that it once could be used around the rural homestead.

Another important factor in the use of residues is the economic value attached to them. It is very often assumed that they are wastes and therefore, almost by definition, a "free" good however costly to collect and transport. In practice they have an alternative use, which is also of some value. By way of illustration, it is difficult to find any commercial briquetting plants using residues for which they have paid nothing[14]. One of the alternative uses of field residues is as a fertilizer and soil conditioner, either through ploughing in or burning in the field. There seems to be little nutritional value in the direct returning of uncomposted residues to the soil. However, they may play a role in maintaining the soil quality by keeping up its organic content. Similarly, burning of residues may be important in supplying trace elements.

Energy Crops

The cultivation of crops specifically for energy purposes has been receiving much attention. One of the main driving forces behind this has been the crisis in which most industrialized countries' agricultural sectors find themselves. A removal of protection and support to agriculture would lead to the abandonment of an increasing amount of land in industrialized countries. The partial conversion of agriculture to the production of energy and industrial products could therefore be realistic.

However, in order to be successful, intensive agricultural production of energy and industrial products must not only be environmentally, but also economically sound. Modern agriculture has progressively increased its dependence on external inputs, in particular energy itself and energy-rich commodities such as fertilizers and pesticides. The use of liquid fuels occurs in crop establishment and harvesting, while the agricultural machinery itself represents a large indirect energy sink. The most important single energy use is in the provision of fertilizer, representing nearly 46% of the total energy input.

Farming as currently practised in industrialized countries is a net energy consumer. Its energy efficiency has been steadily decreasing, and the energy balance becomes increasingly unfavourable.

The types of crops likely to be considered for bioenergy are probably best represented by a product such as maize silage, which requires the following energy inputs for its production in the USA[16]:

	GJ/t (dry)
Labour	0.01
Machinery	0.75
Fuel	1.50
Fertilizer	1.28
Seeds	0.06
Herbicides	0.06
Electricity	0.29
Transport	0.13
Total	4.14

Farm machinery, fuel and fertilizer are the major components. The heating value of the silage itself is about 20 GJ/t (dry), but this is of course only available directly through combustion, and would more than likely be upgraded to some more useful fuel through further energy-intensive processes.

The following discussion gives brief descriptions of several types of potential energy crops.

Short-rotation Forestry

While tree plantations for the provision of industrial roundwood (ie for pulpwood and timber), are found all over the world, energy plantations are a relatively recent development, except in some Scandinavian and Far Eastern agricultural systems. In essence, an energy plantation does not differ much from a pulpwood plantation, the aim of both being to maximize bulk growth, and to operate on fairly short rotations. It is worth stressing that fuelwood production does not necessarily have to be separated from other wood production. The thinnings from commercial timber plantations can be used as fuelwood, and light timber for construction purposes can be extracted from energy plantations.

In Europe, it has been suggested that the cultivation of energy crops, among them short-rotation fast-growing trees, provide the only significant solution to the EU agricultural overproduction problem. To investigate the viability of this option, the LEBEN project has been set up within the EU.

In Brazil, eucalyptus plantations are used extensively for energy purposes (charcoal), and energy forests cover a total of 2 Mha. Another 2 Mha have been reforested with eucalyptus, pinus and other native or exotic species for other purposes, especially for pulp and paper as well as for fuelwood supply. In general, a eucalyptus plantation is ready for felling in its seventh year. After felling, there is a natural regeneration, and the normal practice is to repeat felling two more times before replanting. Production yields of the order of 30 to 50 t/ha-year have been reported[17].

Herbaceous Crops

Sorghum is receiving much attention as an energy crop. It is easily cultivated over a wide range of climates, has great genetic variability, and the expertise for its cultivation already exists in many parts of the world[18]. It is more drought-resistant than maize, has a very high water use efficiency, and unlike maize, has the ability to remain dormant during drought periods and restart activity quickly when humidity increases. It is also more adaptable to different soil conditions, and has low fertilization requirements. Some hybrid varieties have achieved yields of 10 t/ha of grain plus 100 t/ha of high-sugar-content green matter and stalks[19].

The sugar variety of sorghum has begun to be studied relatively recently, despite the fact that the syrup variety has been grown in the USA for more than a century, mainly for the production of sweet table syrups. It may be considered for ethanol production, for example as an off-season supply for the sugar mills.

Sugarcane is a crop grown in many parts of the world and which, besides its traditional use as a source of sugar and alcohol, has potential as a source of cellulose. Some high yielding "energy-cane" hybrids have been developed which have given experimental yields of 253 t/ha (76 dry t/ha)[18]. In Australia, estimates of ethanol yield from their currently grown high quality canes run at around 90 l/t of cane, fermenting all the sugar present[20]. In Brazil, the most efficient alcohol mills (12 units) have achieved an agricultural yield of 89 t/ha and an industrial yield of 79 l/t, giving a total yield slightly in excess of 7,000 l/ha year[21].

The Jerusalem artichoke is a tuberous plant, native to eastern North America. It is a strong plant able to withstand adverse conditions, especially with respect to cold, drought conditions, and relatively poor soils. This makes it a good crop for cultivation on marginal lands. The tuber is rich in the polymer, inulin, which can be subjected to acid or enzymatic hydrolysis to yield fructose, which is then fermented to ethanol. Experiments in France and the USA have produced about 85 litres of ethanol per tonne of tubers, which with yields of around 40 t/ha gives an ethanol yield of 300–400 l/ha[22]. In addition, the production of 7–8 t/ha of dried tops could contribute to the energy self-sufficiency of the process in terms of combustion fuel.

Cassava has attracted great attention as a substrate for the production of ethanol, especially in Brazil and Australia. In Brazil's drive to produce ethanol, cassava is seen as complementary to sugarcane, the cassava being cultivated in areas with acid, infertile soils and the cane in more amenable environments.

Vegetable Oil-bearing Plants

Vegetable oils can be used on their own or blended with diesel oil as a diesel engine fuel. Over 40 different plant-derived oils have been evaluated in short-term diesel engine tests, including soya, groundnut, cottonseed, sunflower, rapeseed, palm oil, and castor oil. Because of the experience already gained with the cultivation of sunflowers in many parts of the world, this has been the oil used in many experimental programmes.

There are important and attractive benefits to be derived from a plant oil fuel industry, such as:

- National and individual farm production of most plant oils could rapidly be increased if a demand developed, often requiring only one growing season's lead time.
- Most plant oil crops require the same agricultural techniques as those applied to common crops such as maize.
- One hectare of a good plant oil crop can yield sufficient fuel to supply the fuel for the cultivation of 8–20 ha of other crops.
- Because of the relatively simple technology required for pressing and filtering plant oils, they can be produced right on the farm, as well as on a co-operative or industrial scale.
- Most plant oil crops are very adaptable and hardy.
- Plant oils are safe to store and handle.
- The by-products from the oil press can often be used as a high-protein animal feed.

Hydrocarbon-bearing Plants

Proposals have been made to use plants directly for the production of gasoline and other hydrocarbons. For example, studies of the *Euphorbia lathyris,* for the extraction of hydrocarbons which have molecular weights very close to petroleum and produces sugar as a by-product, have been carried out in California. Under irrigation, yields of about 2 t/ha/yr of oil and 5 t/ha/yr of sugars have been obtained. There are at least five other trials in the world assessing the economic viability of this route, but conflicting results have been obtained by various researchers[18].

Urban Wastes

Municipal Solid Wastes

Municipal solid waste (MSW) is the solid waste generated by households, commercial and institutional operations, and some industries. It is somewhat misleading to assert that MSW is an energy resource in the traditional sense – ie that it is available for fuel use at the discretion of the consumer or power supplier.

MSW must be disposed of, and may through incineration or gasification provide electric power, process heat, or methane as a by-product. The market price of the energy will influence the decisions regarding what type of MSW disposal method to apply. Once plants are built, the energy product will be a permanent part of the local energy supply, irrespective of the value of the energy or the actual price paid to the MSW plant operator.

Vast quantities of MSW are still disposed of through landfilling operations. However, much of it could be used in some way as an input for energy production, recycled materials or compost, leaving only relatively small quantities to be disposed of. Gas from landfill sites can also be used for energy generation, and demonstrations have been conducted in several countries.

The calorific value of domestic refuse will differ significantly from area to area. This is illustrated by the ranges for the USA and West Germany; 7 to 14 MJ/kg and 4.2 to 10 MJ/kg, respectively.

Sophisticated methods of disposal include composting, salvaging and recycling, incineration, refuse-derived fuel (RDF), pyrolysis and hydrolysis. Any of the above methods will require separation and pre-processing to recover as much usable material as possible. Numerous such processes have been developed and are used in many industrialized countries.

Combustion of solid wastes may result in atmospheric pollution. Some of the pollutants generated require special attention. Because of the variety of materials contained in the waste stream, pollutants include derivatives of sulphur, chlorine, fluorine, nitrogen, chlorinated hydrocarbons and heavy metals.

One particularly important group of pollutants, polychlorinated dibenzo-p dioxins (PCDD) and the dibenzofurans (PCDF) are possibly among the most acutely carcinogenic compounds known. Understandably there is a great deal of concern internationally regarding the generation of PCDD/PCDFs by combustion processes. There is however also much controversy surrounding the dangers associated with these emissions. In the past decade, much research effort has been expended on the mechanism whereby PCDD/PCDFs are produced. The major objective of this research has been to identify possible process modifications that could result in the reduction or limitation of the production of PCDD/PCDFs. Thus far, methods of control have focused on: better control of combustion conditions, in particular combustion temperature; gas mixing and gas residence time in the combustion zone; control of boiler exit temperatures; and enhanced post-combustion gas cleaning.

Liquid Wastes

The sewage produced by human settlements has an appreciable energy potential. In the same way as animal manure, sewage can be anaerobically digested to produce methane gas. Anaerobic treatment of sewage has been practised for years, and in the past, much of the gas produced was used to power machinery and even to provide energy for street lighting. In many parts of the world, the gas is used for generating the energy required at the treatment plant for heating the digester. Alternatively, the gas can be used to run an engine for a generator, using the electricity generated in the plant.

It is appropriate under this section to consider a certain aquatic plant. The water hyacinth (*Eichornia crassipes*) is a native of the American tropics, but has spread to all the warmer regions of the world, and today is one of the major aquatic weeds. In many countries it has become so prolific that it has been declared a noxious weed and it is now an offence to encourage its growth. The largest concentrations are found in waters enriched by sewage and industrial effluent or by run-off from fertilized agricultural land. Its prolific growth and ease of harvesting make it a suitable carbon feedstock for anaerobic fermentation. Yields of up to 106 dry t/ha/yr have been reported[18].

Biomass Resource Volumes

The volume of resources from agriculture and forestry are very difficult to estimate due to the following factors:

- The statistics on harvest production and consumption of wood and straw are inadequate or totally missing.
- The production from agriculture and forestry depends on the climate, consequently giving great variations from year to year, and region to region.
- The market for agricultural and forestry products depends on market forces, and therefore the volumes to be used for energy purposes depend heavily on the price of the energy feedstocks.
- The volumes from the different sectors (agriculture, forestry, waste sector) interact in a complex way.

Consequently, biomass consumption is probably the most difficult to measure or even estimate. However, it currently contributes more than any other renewable resource to global energy consumption. Table 5.2 shows the most reliable up-to-date global data made available by the Biomass Users' Network (BUN). The table also shows the WEC data published in 1989[23]. The UN has published a report covering biomass resources prepared by the United Nations Solar Energy Group on Environment and Development (UNSEGED). This data is also included in Table 5.2[24].

Bioenergy can be divided into two categories loosely described as modern and traditional. The former category is understood to include all large-scale uses which seek to substitute conventional sources of energy, namely solid, liquid and gaseous fossil fuels. The bioenergy programmes in Brazil, Sweden and the USA are examples of this category. Modern biomass includes:

- wood residues (industrial);
- bagasse (industrial);
- urban wastes;
- biofuels (including biogas and energy crops).

The traditional category is mainly confined to the developing countries, and broadly speaking includes all small-scale uses, which are frequently, but not always beyond the market-place. The use of fuelwood for cooking in rural areas of the third world is a typical example. Traditional biomass includes:

- fuelwood and charcoal for domestic use;
- straw including rice husks;
- other vegetation residues;
- animal wastes.

Table 5.2 *Recent estimates of biomass use*

	Population in 1990 (millions)	*Land area (Mha)*	*Total energy Use (Mtoe)*	*Estimates of biomass use (Mtoe)*			
				*BUN**	*WEC (Trad*	*modern)*	*UN*
North America	276	1,833	2,277	96	38	19	
Western Europe	454	355	1,379	32	20	10	
E Europe + NIS	389	2,342	1,637	44	30	10	
Japan + Australia	144	827	503	1	4	7	
Industrial world	1,263	5,357	5,796	173	92	46	132
Latin America	448	2,016	417	88	125	46	
Mid East + N Africa	271	1,190	294	15	21	0	
Sub-Saharan Africa	501	2,363	291	180	141	5	
Pacific + SE Asia	1,663	1,281	1,091	331	347	16	
South Asia	1,146	752	498	296	204	8	
Developing world	4,029	7,602	2,591	910	838	75	748
World total	5,292	12,959	8,387	1,083	1,051	880	

Assumptions:
1 tonne of oil equivalent (Toe) = 42 GJ.
Biomass/fuelwood = 15 GJ/t (air dry, 20% moisture).

*Note:
Help in providing these estimates is gratefully acknowledged to the Biomass Users' Network.

Table 5.2 indicates that biomass contributes nearly 13% of the world's primary energy and is the most important (35%) energy source in developing countries where up to three-quarters of the world's people live. Biomass energy use is also significant in a number of industrialized countries. Although there are significant differences in the numbers for the individual regions, the world total compares reasonably well. As the basis for the development of the prognosis in the "Future of bioenergy" section below, the WEC data is used.

BIOMASS ENERGY CONVERSION TECHNOLOGIES

The technologies used to convert biomass into energy range from simple open fires used for cooking in the developing world, to sophisticated pyrolysis units producing solid, liquid and gaseous fuels.

Biomass conversion technologies can be separated into three basic categories: direct combustion processes, thermochemical processes and biochemical processes.

Direct Combustion Processes

Direct combustion is the principal process currently used to convert biomass into useful energy. The heat and/or steam produced is used to generate electricity or provide the heat requirements for uses such as industrial processes on all scales, space heating, cooking, or district heating in municipalities. Small-scale use, such as for domestic cooking and space heating, is usually very inefficient, and attempts have been made to introduce more efficient stoves both in developed and developing countries.

On the large industrial scale, furnaces and boilers have been developed for burning various types of biomass such as wood, wood wastes, black liquor from pulping operations, food industry wastes, and MSW. The larger units can be very efficient, nearly matching the performance of fossil fuel furnaces. The greater moisture content of most biomass, as well as the wide range of composition, makes it difficult to achieve comparable efficiencies at reasonable costs. The economic advantages of cogeneration, however, make it attractive for many industrial consumers with available biomass feedstock to install cogeneration facilities.

There are a variety of furnace designs suitable for burning chipped wood, wood residues and other forms of biomass. A few furnace designs, primarily for industrial use in developing countries, are described below:

- Dutch ovens are among the most common, but their basic technology is old and they have mostly been superseded by more efficient systems.
- Spreader-stoker furnaces are used in modern steam plants fueled by wood and bark. The fuel is fed into the furnace above the grate either by a pneumatic or mechanical spreader system. Some of the fuel is burnt while in suspension and the remainder falls on to a series of grates where the combustion is completed. This type of furnace is often employed with smaller furnaces, with steam capacities of typically 10 t/hr, but also up to capacities in excess of 200 t/hr.
- Fuel cell furnaces consist of two stages, the fuel being introduced from above on to a water-cooled grate in the primary furnace. Hot gases pass into a secondary combustion chamber, where final combustion occurs. These furnaces generally operate in the low-pressure range, with capacities ranging from 5 to 12 t/hr.
- In inclined grate furnaces, fuel is introduced at the top of the grate in a continuous cascade, and first passes over the upper drying section. From here it falls into the lower combustion section, and the ash is removed at the lowest part of the grate.
- Suspension furnaces are used to burn fine fuel particles quickly while suspended in a turbulent air stream. Configurations may be of the injection type, where the fuel and air are mixed inside the firebox, or of the cyclonic type, where the fuel and air are mixed in an external cyclone burner.

- Fluidized bed combustion systems use a heated bed of refractory sand kept in constant motion by an air stream. This bed essentially replaces the grate. The bed is preheated by an oil-, gas- or pulverized coal-fired burner, to a temperature capable of sustaining the combustion of the biomass fuel. At this point, the air flow through the bed is increased until the point at which the bed just begins to "boil" – ie it is fluidized.

  The method of introducing fuel to the bed depends mainly on its characteristics. Solid material with a mass greater than that of the bed material can be dropped on to the surface, where it will be engulfed. Whereas materials with low mass, such as sawdust or wood chips are introduced below the bed's surface. Liquids are introduced by means of water-cooled injectors.

Thermochemical Processes

The basic thermochemical process to convert biomass into more valuable or more convenient products (to be used as fuels or for other uses) is pyrolysis. The products formed are usually a gas mixture, an oil-like liquid, and a nearly pure carbon char. The distribution of these products is dependent on the feedstock, temperature and pressure of reaction, the time spent in the reaction zone and the heating rate.

High-temperature pyrolysis (1000°C) maximizes the production of gas (gasification). Low-temperature pyrolysis (less than 600°C) has been used for centuries for the production of charcoal. Another approach to produce liquid fuels and chemicals from biomass is direct catalytic liquefaction.

Gasification

Gasification is an established technology, the first commercial applications of which date back to 1830. It is essentially also a pyrolytic process, taking place at high temperature in order to maximize gas production. During World War II, biomass gasification systems appeared all over the world. Almost one million gasifier-powered vehicles helped keep basic transport systems running. Most of these were either wood- or charcoal-fueled gasifiers.

The energy "crises" of the 1970s rekindled interest in biomass gasification systems. The technology was perceived as a relatively cheap indigenous alternative, for small-scale industrial and utility power generation in developing countries, that were hard hit by high petroleum prices but had sufficient, sustainable biomass resources.

By the early 1980s, over 15 manufacturers in the industrialized world were offering wood and charcoal gasifier plants of capacities up to 250 kWe[2], but a real market never developed for this kind of equipment. A much more dynamic market developed in two developing countries, the Philippines and Brazil. In the Philippines extensive R&D programmes have been performed at local universities. The government created a public company to manufacture and sell small-size gasifiers for engine propulsion.

In Brazil more than 30 manufacturers offered equipment of different conceptions covering a large range of sizes, with no governmental subsidies or incentives. Industrial wood gasifiers of up to 3 MW (thermal) came into operation in semi-rural areas or for special applications. Over 100 plants have been installed throughout the country. Charcoal gasifiers were used in rural areas for small-scale power generation and to produce gas for engine-driven irrigation pumps or motor vehicles[26]. More than 1,000 of these were sold in 1980–82 by one manufacturer[27].

The best established of all gasifiers are the fixed bed type, also referred to as vertical shaft gasifiers. There are three basic types; counter-current, co-current and cross-current (also referred to as up-draft, down-draft and cross-flow). The exact choice of gasifier is more a function of the type of fuel to be gasified, and whether the gas is to be used for process heat (heat gasifier) or for powering an internal combustion engine (power gasifier).

As gasification showed so much promise as an economical source of energy for power and process heat in developing countries, extensive demonstration programmes were started by donor and national governments, besides the already described commercial developments in Brazil and the Philippines. The United Nations Development Programme (UNDP) and the World Bank have subsequently carried out a wide-ranging review of these programmes, and came to the conclusion that commercially proven power and heat gasifiers are available to run on biomass fuels such as charcoal, wood, coconut shells and rice husks[25]. Heat gasifiers are more tolerant of other biomass fuels, but only limited experience with biomass fuels other than the above is available for small-scale power gasifiers. The current economics of biomass gasification are marginal for most areas, with the viability of heat gasifiers being better than that of power gasifiers.

Investigations at Princeton University indicate that the biomass-integrated gasifiers, steam-injected gas turbines (BIG/STIG) combination has promise to compete with conventional coal, nuclear and hydroelectric power in both industrialized and developing countries[28]. (Although the use of the term "BIG/STIG" in this reference is specific, the term is used in this section to include various combinations of biomass gasification with high-efficiency power conversion cycles.)

Carbonization

The carbonization of biomass, or charcoal production has been practised for centuries. By carbonizing biomass, a higher energy density per unit mass can be achieved, thus making it more economical to transport. Charcoal is smokeless resulting in it being a sought-after fuel for use in domestic environments. In the industrial sector it is used in specialized sectors where specific fuel characteristics are required – eg high carbon and low sulphur content.

In the kiln process for making charcoal, part of the wood is burnt to sustain the temperatures required for the pyrolysis. Once the temperature reaches 280°C, the process becomes exothermic and air/oxygen supply to the kiln can be cut off. The simplest kilns, used in many parts of the developing world, consist of mounds of wood covered with earth, or pits in the ground. Carbonization in these kilns is a very slow, and often inefficient process, and the quality of the charcoal produced is poor. More sophisticated kilns, with high efficiencies and producing a high quality charcoal are in common use in many developing countries such as Brazil.

A number of large-scale processes for charcoal making are in use in the industrial world today, one of the best known being that of the Lambiotte company in Belgium. The method uses a vertical retort that works on a continuous basis. These units are designed for large-scale production (20,000 tonnes of charcoal a year), and are capital-intensive, which limits their applications in developing countries. Recently, Acesita Energetica, a subsidiary of a larger charcoal-using steel manufacturer in Brazil, produced a prototype retort with an annual capacity of 5,400 t of charcoal, 1,440 t of tar (for fuel) and 276 t of methanol, but commercial viability of this process is far from being demonstrated[29].

Fast Pyrolysis

Given the current interest in relatively simple biomass conversion technologies to produce low or high heating value gases and liquid fuels, pyrolysis has recently received a lot of attention. Extensive research into new processes of biomass pyrolysis with low residence time in the reaction zone and high heating rate (fast and flash pyrolysis) is being carried out world-wide. Some of these projects have reached demonstration stage, but none have yet been fully commercialized.

Catalytic liquefaction

Liquefaction is low-temperature, high-pressure thermochemical conversion in the liquid phase, usually with high hydrogen partial pressure and also a catalyst to enhance the rate of reaction and/or to improve the selectivity of the process. This approach gives a more physically and chemically stable liquid product compared to pyrolysis, thus requiring less upgrading to produce marketable hydrocarbon products[30].

Biochemical Processes

These processes make use of the biochemistry of the raw materials, and the metabolic action of microbial organisms, to produce gaseous and liquid fuels. Examples of bioethanol and biogas programmes were described earlier in this chapter.

Anaerobic Fermentation

This technology is versatile and relatively simple to use as a reliable and effective means of producing a gaseous fuel, known as biogas, from various organic wastes. The most common application has been the digestion of animal dung, agriculture waste, and domestic sludge. In the process of anaerobic fermentation, organic matter is completely degraded to the gaseous products, methane and carbon dioxide, with up to 90% of the energy content of the substrate being retained in the methane.

Distinctly different groups of methanogenic bacteria exist, depending on the temperature range in which the fermentation is taking place. Thermophilic microbial species are active in fermentation at temperatures of about 45–60°C, as compared to the mesophilic species found in the 30–45°C range and the cryophilic which are found at temperatures between 0 and 30°C.

The most publicized anaerobic digesters are those in rural China, of which there are reported to be five million in use, and India which has some 700,000 in operation. In China the digesters were initially conceived of as a means of stabilizing the sewage of rural communities, thereby reducing the incidence of disease caused by parasites and pathogens. However, they have subsequently been optimized for the production of biogas. The Indian gobar (cow dung) gas project, on the other hand, was initiated with the specific intention of producing gas to try to alleviate rural energy problems.

The Chinese digester has no moving parts, the digester itself acting as the gasometer, with the liquid level providing the gas compression. These digesters were originally designed to be operated on a family basis, or several families together, with the families and their animals providing the feed for the unit.

The conventional sludge digestion reactor consists essentially of a completely mixed one-step process operating on a continuous basis without solids recycle. Due to the low specific growth rate of the methanogenic bacteria, effective digestion of the waste in a conventional reactor can only be obtained at long retention times.

Increasingly stringent pollution control regulations coupled with the rising energy costs of aerobic treatment systems in the early 1970s greatly stimulated interest in anaerobic digestion as an energy-saving waste treatment technology. This interest led to the development of a range of digester designs suitable for the treatment of high and low strength soluble wastewaters of industrial origin. Retention of the active micro-organisms permits reduction of the liquid retention time from the 10–20 days characteristic of conventional and anaerobic contact digesters, to periods ranging from several hours to several days.

Sanitary landfilling is a method of controlled solid waste disposal. The basic operation is comprised of the processes of spreading, compacting, and covering the solid wastes. Estimating the potential energy available in a landfill is a difficult if not impossible task, the structural, physical, and chemical characteristics of landfills being infinitely variable.

The gaseous products of decomposition consist primarily of carbon dioxide and methane. However, trace gases are numerous and may cause significant problems. The landfill gases content of methane and carbon dioxide varies with the progressive stages of decomposition.

The major constituents of biogas generated by stable fermentations are given below:

Methane	50–65% (volume)
Carbon dioxide	35–50% (volume)
Moisture	30–160 g/m^3
Hydrogen sulphide	5 g/m^3

As shown, methane comprises about 50% to 65% of the biogas produced during anaerobic fermentation. It is a colourless, odourless, flammable gas which has an energy value of 37.3 MJ/m^3.

The degree of biogas cleaning needed depends on its eventual application, the least demanding in this respect being for direct burning. The minimum requirement in this case is moisture removal to prevent condensation in gas lines and excessive corrosion. The concentration of H_2S in the biogas is generally not high enough to be a health or environmental hazard when oxidized during burning.

Available methods to clean biogas able to meet very stringent standards have been evaluated[31]. This assessment showed that water scrubbing was the most economical process used commercially. Two other processes, the phosphate buffer and membrane separation processes, look promising and were comparable in cost to water scrubbing. However, neither of these processes have been extensively field tested.

The two predominant uses of biogas are direct burning and fueling of internal combustion engines. Examples of direct burning are for cooking, domestic lighting and for space heating of residential, commercial, or livestock facilities. Use in stationary internal combustion engines is for prime movers and for generating electricity. Biogas is unlikely to be used to fuel mobile vehicles on a large-scale because of its low pressure and high inert fraction.

Ethanol Fermentation

Ethanol fermentation from carbohydrates is probably one of the oldest processes known to man. Today, alcohol production from sugar, grain and other starches is widely regarded as an important potential alternative source of liquid fuels for the transport sector, an example of which was discussed early on in this chapter.

Raw materials for the production of ethanol from fermentation can be divided into three classes by type of carbohydrate: saccharine materials, starchy materials and cellulosic materials. The choice of raw material is critical, as feedstock costs typically make up 55–80% of the final alcohol selling price[32, 33].

Saccharine materials, with sugars available in fermentable form, require the least extensive preparation, but are generally the most expensive to obtain. Starch-bearing materials are often cheaper, but require processing to solubilize and convert the starch to fermentable sugars. Cellulosic materials are the most readily available raw material, cellulose being the most abundantly occurring organic compound in the world, but

Table 5.3a *Ethanol yield from carbohydrate-rich plants*

Raw material	*Carbohydrate* *(t/ha)*	%	*Ethanol* *(l/t)*	*(hl/ha)*
Beet	40–50	16	90–100	38–48
Sugar cane	50–100	13	60–80	35–70
Maize	4–8	60	360–400	15–30
Wheat	25	62	370–420	8–20
Barley	2–4	52	310–350	7–13
Grain sorghum	2–5	70	330–370	7–18
Potatoes	20–30	18	100–120	22–33
Sweet potato	10–20	26	140–170	16–31
Cassava	12–15	27	175–190	22–23
Jerusalem artichoke	30–60	17	80–100	27–54

Table 5.3b *Ethanol yield from ligno-cellulosic products*

Raw material	*(hydrolytic agent)*	*Dry matter* *(t/ha)*	*Ethanol* *(l/t)*	*(hl/ha)*
Softwood	(dilute acids)	9–15	190–220	18–31
	(concentrated acids)	9–15	230–270	22–38
Hardwood	(dilute acids)	9–15	160–180	15–25
	(concentrated acids)	9–15	190–220	18–30
Straw	(dilute acids)	1.5–3.5	140–160	2–5
	(concentrated acids)	1.5–3.5	160–180	3–6

they require the most extensive and costly preparation. Table 5.3 provides selected examples[22].

Fermentation is the heart of the ethanol production process, where micro-organisms convert fermentable sugars into ethanol. The only organisms currently used in large-scale industrial ethanol production are yeasts. Extensive research is being carried out on alternative fermentation organisms, and in particular, bacteria.

Until quite recently the technology used to produce fuel ethanol by fermentation was virtually identical to that used by the alcoholic beverages industry. At the start of a batch fermentation the rate of ethanol production is quite low, but as the number of yeast cells increases, the overall rate increases, and with rapid carbon dioxide evolution the beer appears to boil. After 20 hours a maximum in ethanol productivity is reached, and at this point the effects of reduced sugar concentration and ethanol inhibition become important. The fermentation continues at a decreasing rate until, at 36 hours, 94% of the sugar is utilized and a final ethanol content of 69 gm/l is achieved. The average volumetric ethanol productivity over the course of the fermentation is 1.9 gm/l/h.

For potable alcohol, great care has to be taken to minimize the evolution of undesirable co-products. However, with the production of fuel alcohol, where the requirements are not as stringent, the process can be run on a semi-continuous basis with recovery and recycling of the yeast. This permits the use of much higher yeast concentrations which enables a reduction in fermentation times.

The Mellé-Boinot process achieves a reduced fermentation time and increased yield by recycling yeast[33]. In this process, yeast cells from the previous fermentation are recovered by centrifugation and up to 80% are recycled.

Continuous flow fermentation processes have been used in industrial sulphite waste liquor fermentation since the 1930s[33]. The antiseptic qualities of this feed minimize the possibility of contamination and allow long continuous runs without shutdown for cleaning.

The Biostil process, developed by Alfa-Laval in Sweden, is a modification of the continuous fermentation process with yeast recycle, and involves close coupling of the fermentation and distillation processes and a high rate of stillage recycle[34]. The large liquid recycle provides an internal dilution so that very concentrated feeds such as undiluted molasses and grains can be processed. Liquid flows outside the recycle loop are greatly reduced, consuming less water, and resulting in very low or zero effluent flows.

A remarkable feature of the Biostil fermentation system is that it sidesteps the problems of bacterial infection. Because the bacteria are too light to be separated by centrifugation, they pass through this step and are destroyed by the high temperatures in the rectification column. Biostil plants have also been installed in Brazil, Australia, India, France and a number of other countries, where they are operating on molasses and sugarcane juice.

The distillation of an ethanol and water mixture is not straightforward, because of the constant-boiling ethanol/water azeotrope that forms under atmospheric pressure conditions. The maximum ethanol concentration obtainable by straightforward single stage atmospheric distillation is therefore about 95% (volume). If higher concentrations, up to anhydrous 99.7% (volume) ethanol, are required as product, more complex secondary distillation systems become necessary. This requires considerable energy input, which can often be from 40–60% of the total plant process energy requirement.

The traditional separation process consists of a preliminary distillation step, in a beer/rectifying column, to separate the ethanol from the fermentation broth and concentrate it to 95% (volume). The overhead product is then dehydrated to 99+% (volume) in an azeotropic distillation step, requiring two columns, an anhydrous and a recovery column. The total energy consumption of this configuration is 7.63 MJ/l, with the second azeotropic distillation step accounting for about 45% of this.

Ethanol fermentation produces considerable quantities of carbon dioxide, which can be easily recovered, compressed and used as an additive in the beverage and food industries. Alternatively it can be made into dry ice and used for refrigeration purposes, a useful product for tropical countries.

The fermentation yeast and other insoluble components of the fermented wort are rejected from the stills as stillage (also known as slops or vinasse). When starch is used as a feedstock, this has a high protein content, and can be sold as a livestock feed after evaporation and drying. In the USA, the economic viability of the industrial ethanol fermentation industry are determined to a large extent by the market value of this dried distillers grains and solids.

The stillage from sugar fermentation is of lower value, and also constitutes a major waste problem, since a large water load is transported through the system to emerge as an effluent stream of roughly 10–15 l/l ethanol. The polluting load varies widely depending on the substrate, process and efficiency, but levels of biological oxygen demand (BOD) between 10,000 and 60,000 ppm have been recorded. A molasses based distillery producing 60,000 l/day of ethanol could have the pollution load equal to a city of one million inhabitants.

Alcohol represents only one-third of total sugar cane energy. A similar amount is fibre residue (bagasse) and one-third is leaves and tops, normally burned or left on site. Part of the bagasse is normally burned to produce the heat and power needed in the process. The excess is used for the production of pulp and paper, as a component of construction materials and, after adequate treatment, for animal feed or fuel for cogeneration facilities.

Summary of Conversion Technologies

A variety of different conversion technologies are available and/or under development all over the world. Table 5.4 attempts to summarize the present development stage for the major conversion categories.

Among the most promising technologies seem to be ethanol, which is most versatile, and BIG/STIG (biomass-integrated gasifiers with steam-injected gas turbines). More detailed evaluations of degree of maturity are contained in references[24] and[35].

Table 5.4 *Development stage for primary conversion techniques*

Process	*Development stage*	*Next step*
Combustion	Fully commercial	Commercialization to power production
Gasification	Uneconomical but fully developed	Commercialization to power production
Carbonization	Fully commercial	To be determined
Pyrolysis	Developed to demonstration level	Commercialization
Anaerobic fermentation	Fully technically developed	Dissemination of technology
Ethanol fermentation	Fully technically developed	Identify cheaper feedstocks

PROSPECTS FOR BIOMASS USE

Bioenergy Use

Modern Bioenergy

Modern or commercial bioenergy was seminal for the Industrial Revolution in Europe during the 17th and 18th centuries, but had almost disappeared in the industrialized world in the first quarter of this century due to the rising use of more economical and transportable fossil fuels. After that time, bioenergy in the industrialized countries seemed to be limited to very specific applications, such as the use of agricultural and industrial residues and wastes (eg in the pulp and paper and in the timber industries). With no major opportunities apparent for increased use, technological development in this area stopped nearly completely. Technological obsolescence aggravated the lack of competitiveness of these techniques.

This situation started to change in the 1970s, when many industrialized countries became concerned with energy dependence. The change was accelerated at the end of the 1980s with the growing concern for global environmental effects of the intensive use of fossil fuels and for the accumulation of agricultural surplus. However, given the relative prices of production factors (labour, land) and the relative endowment of resources (capital, labour, energy resources) prevailing in the industrialized world, commercial bioenergy depends on the development of new and relatively sophisticated technologies. Great emphasis is being put on the technological development of the different bioenergy approaches, especially for the production of liquid fuels and electricity to facilitate integration with existing energy transport, distribution and utilization systems.

In some developing countries, under completely different economic environments, commercial bioenergy has never completely disappeared (eg the charcoal-based iron and steel industry in Brazil). The balance of payments of oil importing countries was strongly affected by the oil price shocks in the 1970s. Some of these countries were forced to develop domestic energy sources, including biomass. The alcohol programmes in Brazil and Zimbabwe and the gasifier industry in Brazil and the Philippines are the best known of this kind. The experience with these programmes raised a number of questions which are central to determine any increased role that bioenergy could play. One of the most important of these questions concerns competition for land use: should land be used to produce food or fuel, considering that both are crucial for sustainable development?

A second important question is raised by the fact that current agricultural practices demand high levels of energy inputs, which could result in little or no net gain in energy. Another major question to be answered revolves around the environmental and social effects of large-scale bioenergy production facilities. Social effects could be both positive and negative, in the sense that people could be displaced by large-scale cultivation of biomass for energy purposes, and positive in that more convenient sources of energy become available, and employment opportunities are created.

Finally, these programmes came under scrutiny in the late 1980s, when low oil prices threatened their economic viability. Under this point of view, it is very clear that their future depends on the future evolution of oil prices in the international market, environmental factors and of their success in improving efficiency and reducing costs.

Traditional Bioenergy

Despite a couple of decades of intensive investigation of the non-commercial bioenergy sector, it is still a relatively little understood area. Some work has been done on the determining of national biomass supply and demand in developing countries. However, these aggregate statistics very often mask real problem sites, especially in semi-arid areas and in and around large cities.

Various strategies have been adopted to deal both with supply and demand problems, but no simple solutions have been found. The complexity of this sector, and the fact that such a large proportion of the world's population is dependent on it, necessitates that it be given increased attention. In addition, the projected increase in the population of developing countries makes the need all the more urgent.

The Future of Bioenergy

It is extremely difficult to assess the technological prospects of bioenergy conversion processes. There is the split between the modern and traditional uses of bioenergy, the industrialized and non-industrialized, developed and developing economies. Certain processes such as direct combustion, anaerobic digestion and gasification could be regarded as mature, proven bioenergy technologies, in both the developed and developing economy contexts, although there is obvious room for improvement of all three.

Biomass contributes 13–14% of the world's primary energy and is the most important (35%) energy source in developing countries where up to three-quarters of the world's people live. The figures should be taken with caution, as the actual consumption of biomass, especially in traditional uses, is very difficult to measure or even estimate.

The Prospects to 2020

If current biomass use is difficult to estimate, future use is even more difficult to predict. At best, informed guesses can be offered.

Two contradictory trends have to be taken into account in estimating future biomass use. First, there is a growing transition in developing countries away from traditional biomass use to fossil fuels as biomass resources become more scarce and populations urbanize. On the other hand biomass consumption is likely to increase, for three main reasons:

- population growth in developing countries where the majority rely on biomass for their energy needs;
- environmental pressures in industrialized countries; and
- new biomass energy technologies that will increase efficiencies and reduce costs.

An approach for prognosticating the biomass use within each region was developed using expert consensus and was based on the population expansion and an anticipated development of the biomass use per capita. The expected population expansion is shown in Table 5.5.

Two scenarios have been employed. The "current policies" (CP) scenario postulates that the present pace of biomass development is maintained and even expanded in response to currently identified trends. For this scenario, in the industrialized countries, the traditional biomass use per capita is maintained while the use of modern biomass is expected to be expanded by a factor of 2–3 times within the next 30 years. In the developing world traditional use of biomass per capita is reduced due to increased limitations on availability. As the per capita level of new biomass is maintained at the current levels due to the limitation on available capital, the total biomass per capita may well be below what is needed for survival.

For the second, "ecologically driven" (ED) scenario it is assumed that a major effort will be made to enhance the use of modern biomass. In the industrialized world the traditional use per capita of biomass is reduced to approximately 70% of current levels in 30 years due to environmental concerns, and the utilization of modern biomass will be enhanced considerably. In the developing countries the increased availability of

Table 5.5 *Estimates for world population growth by region*

Population (millions)	*Year*			
	1990	*2000*	*2010*	*2020*
North America	276	292	309	326
Western Europe	454	465	477	489
E Europe + NIS	389	411	433	455
Japan + Australia	144	147	151	155
Industrial world	1,263	1,315	1,370	1,425
Latin America	448	537	626	716
Mid East + N Africa	271	362	453	544
Sub-Saharan Africa	501	732	963	1195
Pacific + SE Asia	1,663	1,866	2,069	2,273
South Asia	1,146	1,410	1,674	1,938
Developing world	4,029	4,907	5,785	6,666
World total	5,292	6,222	7,155	8,091

technology and capital coming from the developed countries allow for a major increase of new biomass per capita. This, in turn, allows for a substantial decrease, estimated at 60%, in the traditional biomass use and its accompanying negative side effects both for the persons involved and the environment.

The calculated prognoses for biomass are shown in Tables 5.6 and 5.7. The long-term development for the two scenarios is compared in Table 5.8.

Table 5.6 *Prognosis for current policies scenario*

	Traditional biomass				*Modern biomass*			
	1990	*2000*	*2010*	*2020*	*1990*	*2000*	*2010*	*2020*
North America	0.14 38	0.14 40	0.14 43	0.14 46	0.07 19	0.09 26	0.13 40	0.17 55
Western Europe	0.04 20	0.04 20	0.04 20	0.04 20	0.02 10	0.03 14	0.04 19	0.05 24
E Europe + NIS	0.08 30	0.08 32	0.08 34	0.08 36	0.02 10	0.03 12	0.04 18	0.05 23
Japan + Australia	0.03 4	0.03 4	0.03 5	0.03 5	0.05 7	0.07 10	0.10 15	0.13 20
Industrial world	0.07 92	0.07 96	0.07 102	0.07 107	0.04 46	0.05 62	0.07 92	0.09 122
Latin America	0.28 125	0.27 145	0.26 163	0.25 179	0.10 46	0.10 54	0.10 63	0.10 72
Mid East + N Africa	0.08 21	0.08 28	0.07 32	0.07 38	0.00 0	0.00 0	0.00 0	0.00 0
Sub-Saharan Africa	0.28 141	0.27 198	0.26 250	0.25 299	0.01 5	0.01 7	0.01 10	0.01 12
Pacific + SE Asia	0.21 347	0.20 373	0.19 393	0.18 409	0.01 16	0.01 19	0.01 21	0.01 23
South Asia	0.18 204	0.17 240	0.16 268	0.15 291	0.01 8	0.01 10	0.01 12	0.01 14
Developing world	0.21 838	0.20 985	0.19 1,106	0.18 1,216	0.02 75	0.02 90	0.02 106	0.02 123
World total	0.18 930	0.17 1,081	0.17 1,208	0.16 1,323	0.02 121	0.02 152	0.03 198	0.03 245

Note:
For each regions the top numbers are toe/capita and the bottom numbers are Mtoe.

Table 5.7 *Prognosis for ecologically driven scenario*

	Traditional biomass				*Modern biomass*			
	1990	*2000*	*2010*	*2020*	*1990*	*2000*	*2010*	*2020*
North America	0.14 38	0.13 38	0.02 37	0.11 36	0.07 19	0.10 29	0.15 46	0.21 68
Western Europe	0.04 20	0.04 19	0.04 17	0.03 15	0.02 10	0.03 14	0.05 24	0.07 34
E Europe + NIS	0.08 30	0.07 29	0.06 26	0.05 23	0.02 10	0.03 12	0.05 22	0.07 32
Japan + Australia	0.03 4	0.03 4	0.02 3	0.02 3	0.05 7	0.08 12	0.11 17	0.15 23
Industrial world	0.07 92	0.07 90	0.06 83	0.05 77	0.04 46	0.05 67	0.08 109	0.11 157
Latin America	0.28 125	0.26 140	0.23 144	0.20 144	0.10 46	0.14 75	0.20 125	0.26 186
Mid East + N Africa	0.08 21	0.07 25	0.06 27	0.05 27	0.00 0	0.01 4	0.02 7	0.02 11
Sub-Saharan Africa	0.28 141	0.26 190	0.23 222	0.20 239	0.01 5	0.02 15	0.03 29	0.04 48
Pacific + SE Asia	0.21 347	0.19 355	0.17 352	0.15 341	0.01 16	0.02 37	0.03 62	0.04 91
South Asia	0.18 204	0.16 226	0.14 234	0.12 232	0.01 8	0.01 20	0.02 40	0.04 68
Developing world	0.21 838	0.19 936	0.17 979	0.15 983	0.02 75	0.03 151	0.05 263	0.06 404
World total	0.18 930	0.16 1,026	0.15 1,062	0.13 1,060	0.02 121	0.04 218	0.05 372	0.07 561

Note:
For each region the top numbers are toe/capita and the bottom numbers are Mtoe.

Table 5.8 *Long-term comparison of two scenarios by region (in Mtoe)*

	1990		*2020 CP*		*2020 ED*	
	Trad.	*Modern*	*Trad.*	*Modern*	*Trad.*	*Modern*
North America	38	19	46	55	36	68
Western Europe	20	10	20	24	15	34
E Europe + NIS	30	10	36	23	23	32
Japan + Australia	4	7	5	20	3	23
Industrial world	92	46	107	122	77	157
Latin America	125	46	179	72	144	186
Mid East + N Africa	21	0	38	0	27	11
Sub-Saharan Africa	141	5	299	12	239	48
Pacific + SE Asia	347	16	409	23	341	91
South Asia	204	8	291	14	232	68
Developing world	838	75	1,216	123	983	404
World total	1,051		1,568		1,621	

Looking Beyond 2020

Naturally, no reliable prognosis for the time after 2020 is available. However, the USA Environmental Protection Agency estimates that given sufficient R&D and commitment of land, as much as 16,000 Mtoe/yr could eventually be produced from biomass. On a long-term basis, the EU is estimated to be able to contribute up to 300 Mtoe and the USA up to 700 Mtoe[36, 37]. Today, global recoverable residues are about 2,100 Mtoe.

In the short and medium term, waste and residues will dominate. In the longer run energy crops will dominate and constitute the renewable equivalent of fossil fuel. For long-rotation forests, decisions will have to be taken soon, since forests need 40–100 years to mature.

Hall et al have assumed that in 2050, one-third of biomass energy will come from residues and two-thirds will come from biomass crops. The land area available for energy crops by 2050 is assumed equivalent to 15% of the land now in forests and 40% (600 Mha) of the land now used as croplands[38]. Ample suitable land is available in Asia, Africa and Latin America (eg at present as much as 750 Mha of grassland, are burned off each year, which if used for energy crops could produce as much as 2,700 Mtoe annually)[39].

For modern biomass the technologies of preference seem to be ethanol, which is the most versatile, and gasification – eg BIG/STIG, which has a great potential, but must still be fully commercialized. It is estimated that if BIG/STIG technology was fully deployed in the 80 sugarcane growing countries, the generated electricity would be equivalent to 50% of the total electricity currently produced by utilities in those countries[40].

One published analysis indicates that carefully designed global afforestation programmes totalling 100 Mha, under a sustainable development scenario, would be physically capable of supplying more than 30% of current global electricity demand[41].

In the next 50 to 100 years, biomass has the potential to substitute for a significant fraction of the fossil fuels used in the transport and electricity generation sectors in many countries.

Feasibility of Expanded Biomass Use

It would be appropriate to assess the possibilities of achieving the 2020 numbers with respect to practical limitations such as productivity, availability of land and cost considerations. There are also many relevant examples of success today in making effective use of modern biomass.

Productivity

Increased yield is the key to providing larger feedstocks. Currently, the yield is 3–7 t/ha-yr. However, in some areas it has reached 35 t/ha-yr. Hall[42] cites that:

- In most forests trees grow at rates far below their natural potential.

- Nutrient availability is usually the most important limiting factor.
- Optimizing nutrient availability can result in a factor of 4 to 6 increase in yield.

Thus, doubling or even tripling the productivity does not seem too far-fetched, and a lot of research is aiming at just that.

Availability of Land

Hall also has made several different attempts to evaluate the availability of land for increased biomass production. Most instructive is the calculation for most of Africa indicated in Table 5.9, giving the land potential for biomass based on water availability and techno-economic criteria[42]. The biomass energy production shown is based on 10% of total land available for energy and the following assumptions:

2 t/ha-yr – low rainfall areas
5 t/ha-yr – problem areas
10 t/ha-yr – uncertain areas
10 t/ha-yr – flooded areas
20 t/ha-yr – good rainfall areas
(15 GJ/t, air dry, 20% moisture)

Table 5.9 *Analysis of potential biomass production in Africa*

Total land area for all purposes	*2572 Mha*
Low rainfall areas	73 Mha
Uncertain rainfall areas	97 Mha
Good rainfall areas	149 Mha
Naturally flooded	71 Mha
Problem land	358 Mha
Total land available for energy	748 Mha
Present energy consumption	$8,615 \times 10^6$ GJ
Potential biomass energy production	$9,919 \times 10^6$ GJ
Percent of present energy consumption	115%

An alternate methodology is to assume a yield of 5 t/ha-yr except for the arid regions of North Africa and the Middle East where 2 t/ha-yr is more appropriate. Table 5.10 gives the result when applied across the regions dominated by developing countries. Again the yield is in air dry tonnes equal to 15 GJ each. An area of 800 Mha would have to be cultivated with energy forests and energy crops to supply the bioenergy demand in the developing world in the year 2020, for the ecologically driven scenario.

Assuming an average direct cost of US$500/ha, total capital needs amount to US$400 billion, or nearly US$15 billion/year. It is doubtful that such an amount could be made available in these countries for this purpose, even if it seems relatively small compared with the total investment projected for the electricity sector.

Table 5.10 *Land area availability estimate for developing world*

Region	Land area (Mha)	Biomass use in 2020 (Mtoe)	Yield (t/ha-yr)	Land area for energy (Mha)	(%)
Latin America	2,016	330	5	184	9
Mid East + Africa	1,190	38	2	53	4
Sub-Saharan Africa	2,362	287	5	161	7
Pacific + SE Asia	1,281	432	5	242	19
South Asia	752	300	5	167	22
Developing world total	7,602	1,387	4.5	817	11

A number of similar calculations have been made with these kind of assumptions. For developing countries the calculations all show that there is ample room for providing the biomass energy assumed in the prognoses, but that investment capital is limited and must come from the industrialized countries[42].

For the industrialized world, however, the picture might be different mostly due to a much higher energy consumption normally set at 310 GJ/capita for North America and 140 GJ/capita for the rest of the developed world. Both the USA and the EU have at present plans to make as much as 90 Mha each idle due to the overproduction of agricultural products[36]. This area, if used for energy production, could yield:

$$90 \text{ Mha} \times 10 \text{ t/ha-yr} \times (15 \text{ GJ/t}) / (42 \text{ GJ/toe}) = 320 \text{ Mtoe}$$

This is more than twice the estimated modern biomass production in the industrialized countries for the EU scenario in 2020. Additional calculations for the EU have been made showing that the annual "economic potential" for the EU by 2000 is about 100 Mtoe (approximately 10% of total energy), while the "technical potential" is about 306 Mtoe[38].

Cost Considerations

Since biomass is generally widespread and used both in developing and developed countries, the price of biomass as a source of energy is quite variable and site-specific. Estimates of costs of delivered, air dry biomass chips in a number of countries have been developed[42]. The cost ranges from US$1.9 to 3.9/GJ (1987) and applies to both industrialized and developing countries. In the USA the cost of biomass from short-rotation plants ranges from US$2.7 to 3.9/GJ (1990). The cost targets for 2010 are US$1.9 to 2.7/GJ, which will make biomass competitive under a wide range of conditions.

The cost of electricity production based on newly constructed biomass technologies in the USA is reported as 5–6 US cents/kWh (1989)[39,43]. For developing countries the energy supply costs for electricity are stated to be of the order of 1.4 US cents/kWh, and for Europe (Netherlands) the price may be as high as 8–10 US cents/kWh.

For the Brazil Ethanol project different cost estimates have been made showing both profit and loss depending on the assumptions. Goldemberg has calculated the ethanol production costs in 1990 as US$0.20–0.25 /litre, depending on a number of assumptions and circumstances[44].

The true cost of the gasoline production is difficult to determine mainly due to lack of open information. The same applies to many other biomass projects all over the world.

The great variation of costs, some of which are competitive and others not, compared to the billions of dollars spent in subsidy to the agriculture industry in many countries shows that the cost issue is partly a political one. Present ideas on CO_2 tax on fossil fuels might suddenly make electricity from biomass very attractive in an increasing number of countries.

As to investments, the following ranges have been found from a variety of local sources[45, 46]:

Ethanol (fuel only):	30–300 million $/Mtoe
BIG/STIG (electricity/heat):	700–3,600 million $/Mtoe
Biogas plants (electricity/heat):	approx. 4,000 million $/Mtoe
Straw plants (heating only):	less than 10 million $/Mtoe
CHP on straw (electricity/heat):	1,400–4,200 million $/Mtoe

For the reasons cited below, using these values to predict the necessary investments in the two scenarios is not recommended. This applies of course to modern biomass. Traditional biomass is by definition usually beyond the market-place, and will have only a marginal impact on total investment over the next 30 years.

The necessary investment in modern biomass is a function of a number of conditions. Biomass as a renewable energy source is very labour-intensive, especially in the developing world. The expenses will therefore vary greatly according to the location of the biomass investment, and the local economic split between materials and labour.

The industrial world needs energy for electricity, heat and transport. The developing world needs energy primarily for electricity and transport. The investments for the different uses vary significantly, electricity being the most expensive. The extent of the different needs is not known and no assumptions in this regard seem justified.

The purpose of a facility might be totally different. In China a biogas plant may be constructed fairly cheaply and for the specific purpose of supplying energy. In northern Europe the prime purpose of a biogas plant is environmental and the energy produced is only an added benefit, and may be economical compared to other pollution control alternatives.

The same philosophy may be applied for MSW. In the industrial world MSW represents a "negative" value where the society is prepared to pay for incineration with energy recovery. In the developing world there is often room for disposal and any pre-treatment will hardly be feasible. In the developing world the trend is to have biomass energy primarily where the residues are readily available – eg from an industry. In the industrial world, not only residue projects but also schemes involving energy crops are likely to be developed.

Statements as to the competitiveness of biomass are subject to very high uncertainty. In some parts of the world it is a good idea and in other parts it is not. It all depends on the political climate and, as always, on the evaluated price of other alternatives.

Examples of Successful Programmes

Is it at all possible to expand rapidly the exploitation of biomass use for energy? Efforts aimed at this should begin with looking at existing success stories and potential advantageous programmes:

- The USA has over the past decade expanded electricity production based on biomass from 250 MWe to 9,000 MWe. That is an expansion of 36 times over 10 years.
- Brazilian alcohol production has expanded 20 times over 12 years and might have increased even more because more than 90% of the cars already used alcohol in 1983–87.
- In Ethiopia small eucalyptus plantations have been highly appreciated by urban and rural dwellers for almost 100 years. Approximately 20,000 ha are planted annually and the internal rate of return has been calculated to be 18–20%.
- Brazil produces yearly 7 Mt of charcoal to replace coal in pig-iron and steel production. The charcoal industry employs 267,000 people and generated about US$5 billion in 1989.
- The 80 sugarcane producing developing countries could produce electricity from cane residues and generate more than their electricity demand close to the rural areas that are most in need of new sources of electricity.
- In Mauritius bagasse currently provides 10% of the island's electricity requirement, and woody biomass supplies 63% of all the energy required for household cooking.
- Austria has during the past 10 years promoted wood chips use and increased the primary energy consumption from biomass from 0 to 10%. Today 11,000 district heating systems of 1 to 2 MWe average capacity use wood chips.
- In north-east Brazil existing sugarcane residues and future potential production of wood on dedicated plantations could be used to generate annually up to 41 TWe and 1,400 TWh of electricity, respectively, compared to present total annual generation of about 30 TWh. The cost of most of the biomass-derived electricity would be under 4.5 US cents/kWh, which compares favourably with marginal costs projected for hydroelectric projects in the Amazon region and would involve lower capital investment. Expansion of the system based on biomass rather than hydropower would also bring social benefits, including greater job creation.

  Currently a 25 MWe biomass gasifier/gas turbine combined cycle demonstration is being developed. Fuel will be biomass from a standing eucalyptus plantation. The design phase is scheduled to be completed in 1994. It is supported by a grant of US$7 million from the Global Environmental Facility which also has set aside an additional US$23 million for the construction phase[47].

- The Indian village of Hosahalli with about 200 people has demonstrated the feasibility of electricity production from the gasification of fuelwood for lighting, water pumping, milling and for future irrigation. The project includes a wood gasifier and a gas/diesel engine connected to a 5 kWe generator that supplies the specified needs of the village.

Other notable projects should also be mentioned, including[48]: an alcohol programme in Zimbabwe; straw, biogas and MSW in Denmark; landfill gas in the USA and UK; gasifiers in Finland; stoves in Kenya; fuelwood in Nepal; fuelwood and charcoal in Somalia; biogas in China; CHP from biomass in Sweden; and straw in the UK.

Many of the success stories could be copied and improved on in other parts of the world. Examples with widespread applicability include the generation of electricity from sugarcane bagasse, the production of alcohol fuels from sugarcane, and the production of electricity using advanced gas turbines fired by gasified biomass from various feedstocks.

With the development and dissemination of appropriate technology and experience, the land should be available and the productivity should be sufficient, if the sharing of financing and technical resources can be carried out on a broad scale involving both the developed and developing countries.

CONSTRAINTS TO WIDESPREAD USE

Bioenergy Economics

A number of economic constraints are evident, as summarized below:

- Feedstock may have competition from higher value application.
- Biomass projects suffer from not having "a level playing field" in comparison to conventional energy resources.
- Biomass residues from higher value products may become scarcer, and are subject to other market cycles.
- Substantial incentives may be necessary, especially for afforestation programmes.
- Available technologies may not be sufficiently mature to represent acceptable risk to private sector investors.
- Petroleum prices have been stable over the past decade, slowing development of bioenergy.

Numerous cost analyses for bioenergy have been published, with widely varying claims on the net benefits. It is impossible in a study of this nature to explore all the details of process cost estimates, but some general principles are worth stating. A careful reading of bioenergy cost analyses often reveals an unusually strong dependence on a feature such as tax subsidies or by-product benefits, thus shifting the viability of the process from the bioenergy itself to these credits.

It is generally accepted that, under current economic conditions, and in particular as regards conventional fuel prices, bioenergy is an expensive option in developed countries. This explains why its adoption has not occurred yet on a large-scale. Where bioenergy has been used on any significant scale, special circumstances are seen to apply. For example, governments may have intervened in the market through providing subsidies (eg the alcohol fuel programme in the USA), or the use of fossil fuels may have been penalized by high taxation. Alternatively, the biomass feedstock available in a particular instance may be a low-cost residue (eg wood waste in the timber industry), it may be a polluting waste (eg sewage and MSW), or it may be collected freely from surrounding countryside.

It is one of the vagaries of the market that the processed goods do not have a single value. Different prices are obtainable in different applications (eg charcoal or ethanol). The product is not merely valued for its heat content, but sometimes for aesthetic and other qualities. The demand for the high value products is, however, relatively easily satisfied, and is small compared with national or world energy demand.

In many developing countries a completely different picture could occur. Abundant natural resources and low labour costs could reduce bioenergy costs significantly, especially in decentralized facilities (reducing administration, management and transport costs). But high capital and overhead costs could penalize conventional energy production and imported energy could be very expensive.

On the other hand, available biomass technologies may not be sufficiently mature or technological information is not readily accessible to attract investor confidence. In many cases some initial support from government (credits, incentives) is necessary in the initial phase of a new programme or at the expansion of an existing programme. In many cases, both in the industrialized and in the developing countries, current trends to incorporate in energy costs its external social and environmental costs will significantly improve the competitiveness of bioenergy options.

The most favourable environment for the development of bioenergy is undoubtedly in an economy where it is cheaper than the alternatives. These conditions are currently unsatisfied in most of the industrialized countries, but do exist in a number of developing and industrializing countries. In the latter economies, both the population and per capita GDP may be rapidly growing but at the same time indigenous fossil resources may be lacking. However, land resources may be considerable, and agricultural labour may be available to be substituted for capital.

Environmental Effects of Bioenergy Use

The manufacture of fuels from biomass involves the processing of large quantities of complex organic compounds, and may result in the production of vast quantities of solid, liquid and gaseous wastes. Bioenergy processes may result in large amounts of pollutants, and the bioenergy process may convert one pollutant to another quite different pollution problem. Due to lack of large-scale operational experience the magnitude of the problem is yet to be determined.

However, the experiences in Brazil with the ethanol programme and its liquid effluent problem give some indication of the difficulties and how they may be solved. Even though some bioenergy processes are themselves waste treatment techniques, they also leave residues which will still have to be disposed of. The cost of treating these residues and the methods used will be highly site-dependent.

A number of important environmental issues, and potential constraints, surrounding the expansion of bioenergy are recognized:

- The need and/or desirability of maintaining bio-diversity. As the report of a workshop convened by the National Audubon Society and Princeton University in the United States in 1991 stated: "Bio-diversity has not yet received much attention as an environmental issue relating to biomass production". This generally remains the case. Modern biomass development should take place in a manner consistent with the Convention on Biological Diversity and the relevant chapters of Agenda 21 (see box). The most obvious threat is where natural forests – whether tropical or in more temperate zones – are replaced by monoculture energy plantations. Wetlands could also be threatened. Less obvious is the potential threat to land already ecologically impoverished but sustaining a diversity of agricultural uses and practices. Provided due care is taken with modern biomass development on degraded or impoverished land – such as through interplanting, the growth of standing timber, leaving areas to revert to wilderness and other measures to enhance local bio-diversity – modern biomass could have powerful net environmental benefits. Appropriate environmental/ecological criteria, effectively applied, will be needed to ensure that unfettered yield and revenue-maximizing practices are not allowed to dissipate potential environmental gains. One means of seeking to secure environmental gains could be to allow a minimum fraction of each biomass plantation to remain uncultivated (not less than 10%) or to be managed over a sufficiently long cycle in rotation with adjoining areas (ideally a much larger fraction than 10%). A network of such mature ecosystems would buttress bio-diversity by maintaining habitat connectivity. Modern farming practices have led to widespread loss of species, reduction in species population, habitat and ecosystem destruction and soil degradation. Modern biomass development could accelerate this process if pursued without due constraint and sound principles, or start to reverse this process if pursued with broad substainability criteria fully in mind. The plight of many species of birds highlights the scale of the problems and the potential for reversing the decline.
- There is therefore a need to maintain and further protect areas of great natural beauty, remarkable natural landscapes, ecologically sensitive and important sites, as well as species of flora and fauna.

The requirement here is usually to protect tropical forests and old growth forests elsewhere, wetlands, savannah, and other open habitats supporting rare resident and migrant species. It is also to seek to reverse past degradation within a framework of ecological restoration and reclamation. A further requirement is the need to constrain intensive forest and other biomass crop management in order to achieve balance with environmental/ecological goals.

- The need to control emissions and effluents arising from the provision and use of bioenergy and to ensure, through proper management of biomass resources, that sources and sinks of greenhouse gas emissions are balanced. There is too little awareness in some quarters of the emissions from, for instance, biofuels used in motor vehicles and the difficulty in calibrating carburettors accurately for the optimal use of such fuels. At the supply end, issues of land erosion, watershed protection, risks of diverting water channels, depletion of soil nutrients and water pollution are all real and will have to be carefully guarded against.

The social effects of large bioenergy projects could be severely negative. Competition of modern biomass with food production requirements in, for instance, South Asia with its rapidly growing population, could have serious consequences. In Latin America, major expansion of modern biomass development could accelerate depletion of tropical forests. In recent years, according to the World Bank, nearly two billion people have been displaced because of major projects – from irrigation to large hydropower schemes. Many have moved either to the squalor of urban environments unable to cope with the number of people involved, or to more marginal land. Various forces have been at work in recent years resulting in concentration of the ownership of land and growth of a more rootless population.

To set against these realities, it should be recognized that the Brazilian ethanol project today generates about 700,000 jobs which only demonstrates that not only economically, but also socially, biomass projects are very complex and tend to be site-specific.

THE UN CONFERENCE ON ENVIRONMENT AND DEVELOPMENT (UNCED) AND BIOMASS DEVELOPMENT

The Rio Earth Summit in June 1992 (UNCED) had the following main relevance to biomass development:

- **The Convention on Biological Diversity**: Although some objections were raised by individual countries relating to the terms on which biotechnology may be accessed, the treatment of genetic material and financial provisions, there was widespread agreement on:
  - Objectives: ie the conservation of biological diversity and the sustainable use of its components.

- Preamble: eg states are responsible for conserving their biological diversity and for using their biological resources in a sustainable manner; it is vital to anticipate, prevent and attack the causes of significant reduction or loss of biological diversity at source; the fundamental requirement of in-situ conservation of ecosystems and natural habitats and the maintenance and recovery of viable populations of species in their natural surroundings.
- Article 6: ie develop strategies, plans or programmes for the conservation and sustainable use of biological diversity; integrate as far as possible and, as appropriate, the conservation and sustainable use of biological diversity into relevant sectoral or cross-sectoral plans, programmes and policies.
- Article 10: ie integrate consideration of the conservation and sustainable use of biological resources into national decision making; adopt measures relating to the use of biological resources to avoid or minimize adverse impacts on biological diversity.

■ One of the greatest setbacks of Rio was failure to agree an international treaty to protect the world's forests – tropical, temperate and boreal. Nevertheless, the non-legally binding ''Statement of Principles for a Global Consensus on the Management, Conservation and Sustainable Development of All Types of Forests'' provided the basis for future negotiations, though widely regarded as flawed because it appeared to place trade considerations too far in front of environmental considerations.

■ **Agenda 21:** Several chapters of relevance, notably:

Chapter 10 – Integrated approach to the planning and management of land resources.

Chapter 11 – Combating deforestation.

Chapter 12 – Managing fragile ecosystems: Combating desertification and drought.

Chapter 13 – Managing fragile ecosystems: Sustainable mountain development.

Chapter 14 – Promoting sustainable agriculture and rural development.

Chapter 15 – Conservation of biological diversity.

Chapter 16 – Environmentally sound management of biotechnology.

Chapter 18 – Protection of the quality and supply of freshwater resources.

Chapter 32 – Strengthening the role of farmers.

Inevitably, perhaps, there is evidence of actual or potential incompatibility of objectives both between and within these chapters.

- **Rio Declaration,** notably:

  Principle 4 – In order to achieve sustainable development, environmental protection shall constitute an integral part of the development process and cannot be considered in isolation from it.

  Principle 7 – States shall co-operate in a spirit of global partnership to conserve, protect and restore the health and integrity of the Earth's ecosystem.

  Principle 16 – National authorities should endeavour to promote the internalization of environmental costs and the use of economic instruments, taking into account the approach that the polluter should, in principle, bear the cost of pollution.

Institutional Constraints

Despite the dramatic progress that has been made in the advancement of renewable energy sources, and reduction of their costs, as already seen, most are still not competitive with conventional sources on a direct-cost basis. Further research and development is needed, and better co-operation must exist between researchers, manufacturers and potential users.

Appropriate policies to stimulate such co-operation will vary from region to region, but these could include strategies such as favourable tax policies, and power-purchase agreements, which would create a flow of revenue to fund further developments. Such incentives often provoke questions of economic efficiency, but if these reflect the external social costs not represented in current energy market prices, then they offer a sound approach to the nurturing of new and renewable energy source development.

The following list illustrates some of the many constraints to the expanded use of biomass:

- Current energy policies are often biased against renewable energy sources.
- Taxes and subsidies often encourage fossil fuels, favouring operating costs over long-term investments.
- Energy prices do not reflect external social costs such as air pollution effects and nuclear risks.
- Co-operation between researchers, manufacturers and potential users is not well co-ordinated and needs to be improved.
- The electric utilities are often reluctant to purchase excess power or to offer back-up power.
- The notion of centralization (of power companies) is often counterproductive to development of bioenergy. Much of the biomass resources are beyond the established electricity grid in the developing countries.
- A substantial portion of the farmers' income in some countries is dependent upon subsidized food programmes.

- Biomass producers may not be willing to plant energy crops unless they are assured of markets for their output. Conversely, the power utilities may not be willing to build bioenergy facilities unless they are assured that feedstock will be available.
- There has never been any major co-operative marketing of biofuels.
- Biomass is still considered a "grassroots" fuel, belonging to the "green" ideology, and a small-scale industry.

OPPORTUNITIES

In the world as a whole, the resources of biomass energy are quite abundant. It has been estimated that in the whole marine and land ecological system on earth, the net annual production of dry organic materials is 164 Gt, 70% of which (110 Gt per year) are from land. This figure is several times higher than the equivalent of the current total world energy consumption.

In many regions, and especially in the tropical area of the world, production of biomass would be sufficient to cover the total energy demand or a great proportion of it. It should be stressed that very large amounts of biomass are already available as residues from agriculture and industries. The question here is how to make effective and rational use of the available biomass.

It should be reiterated that the use of bioenergy is highly site-specific, and it is therefore very difficult to arrive at any global assessment of bioenergy opportunities. What is of importance though, is that the opportunities that do exist on a country or regional level should be identified. This includes research and development aiming at increasing efficiency in producing and harvesting biomass as well as in converting it to energy. These programmes should concentrate on the types of biomass which are immediately available or are creating environmental problems and on processes where basic technology is dominating. From a long-range perspective, the R&D to be initiated now should aim at providing solutions that can be ready and finalized as the needs arise. The following technological options appear to be the most promising:

- Direct combustion of biomass to produce heat, steam or electricity.
- Production of liquid fuels through energy crops.
- Production of charcoal and char.
- Gasification of biomass (pyrolysis) for electricity generation, such as BIG/STIG.
- Anaerobic digestion of agricultural residues, manure, sewage or municipal solid wastes to produce biogas and eliminate environmentally hazardous disposal of these materials.

The most evident precondition for future world-wide, large-scale industrial use of biomass for energy is its cost-competitiveness with conventional sources of energy in a number of applications and socio-economic environments. Direct costs associated with production, transport and use of fossil fuels will be increasingly affected by governmental regulations on environmental protection, or direct taxation, which will favour clean fuels derived from biomass. It has been demonstrated that the use of modern technologies could increase conversion efficiencies, protect the environment and cut back costs of bioenergy production.

In many industrialized countries (Western Europe and the USA) governments are creating incentives or direct subsidies to bioenergy utilization to cut back food overproduction or to protect the environment. CO_2 taxes have already been introduced in Scandinavia.

It should be stressed that biomass is a dispersed form of energy, appropriate for small-scale utilization. Specific opportunities of use could only be perceived, analyzed and implemented locally. In many countries (the former USSR, Eastern Europe, some Western European countries and most of the developing countries), the energy industry has been centralized in huge state-owned corporations that developed centralized energy systems. In all these countries present trends to privatize these systems, on a local or regional basis, will open new opportunities for small-scale, local energy production on a competitive market. A number of advanced biomass technologies are under development which, if successful in terms of reliability, thermal efficiency, and specific investment, could be implemented by large energy companies as direct substitution for traditional fossil energy use.

Economics of bioenergy utilization will depend, in each case, on the specific situation regarding not only availability and costs but also social priorities of development in the three-dimensional balance of energy, society (economy) and environment. There is some evidence that, on a purely economic basis, competitiveness of bioenergy will tend to increase in the near future.

Reforestation and revegetation play important roles in reducing pollution. Biofuels make no net contribution to atmospheric CO_2 if used sustainably and can be regarded as a practical approach to environmental protection and global warming. Sequestering the carbon in the trees for 40–60 years would stabilize CO_2 levels in the atmosphere, and then using the biomass to replace fossil fuels would result in a decrease of CO_2 build-up. A better and more economically feasible strategy would be to substitute fossil fuel (mostly coal) with biomass fuel grown sustainably and converted into useful energy using modern conversion technologies.

Some published estimates show that 5.4 Gt/yr of carbon can be displaced by 2050 by replacing coal with biomass, about one-third of which comes from residues and two-thirds from biomass plantations[38]. The residues are prime candidates for the bioenergy systems. The required amount of biomass might be achieved with 600 Mha of plantations with an average productivity of 12 dry t/ha-yr. Yet estimates of the amount of tropical land, available for reforestation are of the order of 800 Mha. Moreover, some of the 1,500 Mha of tropical grassland in the world could be used for energy crops.

Besides, the removal of CO_2, forests provide many other ancillary benefits. Thus tree planting would improve the energy situation by providing fuelwood, may provide income generation, and would also have many important ecological effects associated with rehabilitating land such as soil erosion control, the maintenance of watersheds, improvement of local climates, and the prevention of the destruction of wetland and upland areas.

CONCLUSIONS AND RECOMMENDATIONS

Conclusions – The Future of Biomass

Biomass is the world's fourth largest energy source today. Biomass contributes nearly 14% of the world's primary energy and is the most important (35%) source of energy in developing countries where up to 75% of the world's people live.

It needs to be recognized that there is an enormous untapped biomass potential in improved utilization of existing forest and other land resources (including residues), and in higher crop productivity. It is essential that greater effort is put into producing and using biomass efficiently as a fuel since it is an available indigenous energy resource.

Biomass can be converted to electricity via gas turbines or liquid fuel via alcohols. Biomass can serve as a feedstock for direct combustion in modern devices and is easier to upgrade than coal because of its low sulphur content and high reactivity. Conversion devices for biomass range from small, domestic boilers, stoves and ovens up to larger-scale boilers and multi-megawatt size power plants. The most promising technologies seem to be ethanol production, which is most versatile, and gasification such as BIG/STIG.

Modern Biomass Outlook

Wider commercial exploitation on a sustainable basis awaits the development and application of modern technology to enable biomass to compete with conventional energy carriers. Much more useful energy could be extracted from biomass than at present, even without increasing primary bioenergy supplies.

In favourable circumstances, biomass power generation could be significant given the vast quantities of existing forestry and agricultural residues – over 2 Gt/yr world-wide. For example, studies of the sugarcane industry and the wood pulp industry indicate a combined power grid export capability in excess of 500 TWh per annum. Assuming that a third of the global residues resource could economically and sustainably be recovered by new energy technology, 10% of the current global electricity demand could be generated. In addition, a biomass energy crop programme on 100 Mha, could also supply more than 30% of current global electricity demand.

Efforts aimed at utilizing modern biomass energy could begin with looking at existing successes, such as described earlier in this chapter, and copying those in other parts of the world. But biomass energy still faces many barriers; economic, social, institutional and technical.

Biomass energy sources are very large and varied in nature, and the technologies for exploiting them span a very diverse range in terms of scale, stage of development and development requirements.

There are some very substantial environmental threats posed by modern biomass development if not handled in a sustainable manner, with full recognition of the environmental/ecological constraints that need to be observed. Careful interplanting, insistence in some areas on long cultivation cycles, use of interconnected uncultivated areas, and other measures could – if implemented judiciously and on a sufficiently large scale – enhance ecosystems and reverse the degradation which has occurred of far too many habitats.

It should be emphasized that the use of bioenergy is highly site-specific, and it is therefore very difficult to arrive at any global assessment of bioenergy opportunities. What is of importance though, is that the opportunities that do exist on a country or regional level should be identified. This includes research and development aiming at the need for better resource assessment and increasing efficiency in producing and harvesting biomass as well as in converting it to energy. These programmes should concentrate on the types of biomass which are immediately available, or are creating environmental problems now, and on processes where basic technology is dominating.

Traditional Biomass Outlook

The perspectives for traditional use of biomass are reflected in the prognosis summarized in Table 5.11. The complication in determining these future developments is due to the fact that traditional energy use is tied to a number of social and developmental factors outside of the energy sector. A large part of present and future use will be non-commercial and therefore outside of normal market forces.

The main factors determining the future use of traditional biomass are:

- General economic development in the countries with high traditional biomass reliance and the distribution of economic wealth.
- Population growth in general and the level of urbanization/urban growth, which may increase shift from non-commercial to commercial, but still in traditional biomass. This has importance for possible market-based policy responses.

Table 5.11 *Summary of prognosis for biomass development (in Mtoe)*

	1990		*2020 CP*		*2020 ED*	
	Trad.	*Modern*	*Trad.*	*Modern*	*Trad.*	*Modern*
Industrial world	92	46	107	122	77	157
Developing world	838	75	1,216	123	983	404
World total	930	121	1,323	245	1,060	561
Combined	1,051		1,568		1,621	

- Resource constraints, both for the rural non-commercial use, which is a highly complex issue, and for the urban commercial use where resource constraints are reflected in longer transport, higher prices and either shortages or fuel shifting.

Prognosis

Even though very limited information is available on the future development of bioenergy in the world, a current policies (CP) scenario and a ecologically driven (ED) scenario have been developed through a consensus of expert opinion.

In the CP scenario it is assumed that concern about local and global effects of the use of conventional energy sources on the environment, and population growth as well as lack of alternatives in the developing countries will maintain a steady growth in the use of biomass. Minor decreases in per capita use of traditional biomass are anticipated, as well as modest growth in the application and dissemination of modern biomass techniques. This may mean that biomass would retain, in the year 2020, approximately the same share in the total world energy supply as now.

In the ED scenario, the basic assumption is that the world society will focus on developing modern biomass resources, but at the same time also curb the utilization of traditional biomass. The outlook is for a similar total contribution from biomass, with a shift toward a more sustainable mixture of traditional and modern use.

Recommendations

Modern biomass offers a very significant potential for increased energy production in a more sustainable manner and substitution of conventional energy resources world-wide. Such a transition could have significant environmental and socio-economic benefits, if a number of constraints and possible risks are addressed.

In order to facilitate such a transition and realize the future potential a number of actions should be undertaken on local, national and international levels. These include information and data; research, development and demonstration; and economic and institutional.

Information and Data

- Better knowledge about the resources available should be promoted including future potential, with assessment of possible constraints due to other types of land use.
- An important need to assist the understanding of the possible role for biomass is the establishment of a bioenergy database. The Biomass Users' Network (BUN) could also assist with the task, in which the WEC could play a seminal role in conjunction with NGOs such as the UN and the World Bank. Data collected from various recent statistics, WEC reports and national data profiles, seminars and conferences, would provide a region and country-wide base, which

could then be updated with more precise data and assessments for each country on a uniform basis.

- Increased co-operation is also needed to bring together bioenergy research, development and demonstration work already in hand in different regions and countries. This information would greatly facilitate the identification and development of bioenergy opportunities where similar conditions exist, and avoid needless time being spent on duplication of learning in different countries. In addition, it would assist in promoting confidence in bioenergy and easier transfer of technology and knowledge.
- Investigations/research should be carried out into the area of the true costs of existing biomass technologies and if possible the probable costs of technologies currently under development. These data, combined with appropriate learning curve estimates, will provide policy makers with a necessary tool for allocating funds for future R&D.

Research, Development and Demonstration

- The traditional biomass sector is very complex and since a large proportion of the world's population depends on it, increased R&D attention should be placed on it.
- In the traditional biomass sector, R&D should be encouraged in conversion techniques – eg charcoal manufacturing, briquetting and stove experience.
- An increase in attention, possibly through WEC activities such as Regional Energy Forums, should be placed on the dissemination of knowledge of successful bioenergy projects to other parts of the world to promote reception of those projects and thereby increase the world-wide utilization of bioenergy.
- Emphasis should be placed on promising technologies such as the BIG/STIG concept to test its potentials – eg in utilizing sugarcane residues to produce electricity on a large scale.
- Governments should study, and if appropriate initiate, some of the large afforestation programmes (eg the ideas discussed by Elliot and Booth[41]), to provide electricity from sustainable development of large energy forests.
- Further work should be initiated both at the resource end to improve yield, in harvesting, transport and possible conversion techniques and obviously in some of the not fully developed energy production technologies. Focus should be on efficiency, environmental aspects, demonstration and transfer of know-how.
- The balance between environmental protection and enhancement on the one hand, and modern biomass development on the other, should be such that considerations of bio-diversity and species protection place the weight more in favour of the former than the latter. Costs, prices, yields, and conversion techniques should be adapted to encompass these environmental/ecological imperatives as part of the general criteria for energy provision and use. This will inevitably

impinge upon the potential for modern biomass development and may indeed indicate that in the long term the environmental and competitive advantage will lie with the various forms of more direct solar energy. Nevertheless, balanced development which expands biomass energy provision while enhancing the ecosystem will maximize satisfaction in the long run.

Economic and Institutional

- In many cases, both in the industrialized and in the developing countries, current trends to incorporate in energy costs its external social and environmental costs should be fully investigated. Incorporation through regulations, taxation and other economic measures is likely to significantly improve the competitiveness of bioenergy options.
- Financial mechanisms and funding arrangements for bioenergy projects should be supported. On an international level the institutions should increase their expertise and also ability to assist in biomass projects, and there should be emphasis on sustainable biomass development where environmental and socio-economic arguments validate this. With the employment intensity of most biomass activities in developing countries, the indirect effect of such efforts is expected to be significant and positive.
- In developing countries, one major incentive to the increased use of bioenergy should be to reduce the total investment in the energy sector and/or to reduce the burden in the balance of payments due to energy imports.
- Institutional constraints to the development of bioenergy uses should be removed. Access to the energy market on fair commercial conditions, and fair prices should be assured to the independent producer of energy.
- Alternative mechanisms for directing resources into technology development and demonstrations should be developed. The alternatives should be audited routinely and the successful ones identified for expanded use.
- Bioenergy programmes should be analyzed in the context of energy planning involving not only the individual regions supplies and needs, but also evaluating the advantages of biomass energy, both environmentally and agriculturally. Co-ordinated planning between government departments responsible for energy, agriculture, and environmental protection is essential.
- The governments should actively help linking biomass producers and power utilities in a long-term relationship assuring the biomass producers a market and the utilities a steady supply of raw material (biomass).
- Especially in the ED scenario, alternative (to traditional biomass) energy sources for the urban poor shall be promoted. The alternative fuel (eg kerosene or electricity), might very well be less expensive than woodfuel but investments in equipment and appliances can make it economically impossible without government assistance.

ACKNOWLEDGEMENTS

The biomass chapter has been prepared by the following members of the biomass subcommittee of the WEC Renewable Energy Resources Committee: J.J. Edens, Denmark (co-Chairman); P.R. Bapat, India (co-Chairman); S. de Salvo Brito, Brazil; John M. Christensen, UNEP; and A.A. Eberhard, South Africa.

We gratefully acknowledge all persons and organizations who provided information or comments on the work in progress, especially: J.R. Darnell, USA; J.R. Frisch, France; D.O. Hall, United Kingdom; M. Jefferson, United Kingdom; A. Oistrach, Spain; and A.T. Williams, South Africa.

REFERENCES

1. J.M.O. Scurlock and D.O. Hall. The Contribution of Biomass to Global Energy Use (1987); Short Communication. *Biomass* Vol. 21, No. 1, 1990.
2. *Energy in Brazil.* Brazilian National Committee of the World Energy Council, Rio de Janeiro, Brazil, September 1990.
3. C. Flavin. Slowing Global Warming. *State of the World 1990,* Worldwatch Institute, New York, USA, 1990.
4. D.O. Hall and F. Rosillo-Calle. Brazil Finds a Sweet Solution to Fuel Shortages. *New Scientist,* 19 May 1988.
5. D. Qiu, S. Gu, B. Liange and G. Wang (edited by A. Barnett). Diffusion and Innovation in the Chinese Biogas Programme. *World Development,* Vol. 18, No. 4, 1990.
6. *Large-scale Biogas Plants.* Danish Energy Agency, Denmark, 1991.
7. D.O. Hall. (1983); Biomass For Energy: Fuels Now and in the Future. *Biomass Utilization,* ed. W.A. Cote, Plenum Press, New York, USA, 1983.
8. *The Fuelwood Situation in the Developing Countries.* Map prepared by the Forestry Department, FAO, Rome, Italy, 1981.
9. E. Eckholm, G. Foley, G. Barnard and L. Timberlake. *Fuelwood: The Energy Crisis That Won't Go Away.* Earthscan, London, UK, 1984.
10. D. Anderson and R. Fishwick. *Fuelwood Consumption and Deforestation in African Countries.* Staff Working Paper No. 704, World Bank, Washington, USA, 1984.
11. D. French. Confronting an Unsolvable Problem: Deforestation in Malawi. *World Development,* Vol. 14, No. 4, 1986.
12. B. Munslow, Y. Katerere, A. Ferf and P. O'Keefe. *The Fuelwood Trap: A Study of the SADCC Region.* Earthscan, London, UK, 1988.
13. J. Gill. Improved Stoves in Developing Countries. A Critique. *Energy Policy,* April 1987.
14. S. Eriksson and M. Prior. *The Briquetting of Agricultural Wastes for Fuel.* Environment and Energy Paper 11, FAO, Rome, Italy, 1990.

15. G. Barnard and L. Kristoferson. *Agricultural Residues as Fuel in the Third World*. Earthscan Energy Information Programme, Technical Report No. 4, IIED, London, UK, 1985.
16. D.T. Boyles. *Bio-Energy: Technology, Thermodynamics and Costs*. Ellis Horwood, Chichester, UK, 1984.
17. F. Ackerman and P.E.F. Almeida. *Iron and Charcoal. The Industial Woodfuel Crisis in Minas Gerais*. Stockholm, SEI, Sweden, 1990.
18. D.O. Hall and P.J. de Groot. *Biomass for Fuel and Food: A Parallel Necessity*. Presented at World Resources Institute Symposium: Biomass Energy Systems, Building Blocks for Sustainable Agriculture, Virginia, USA, June 1985.
19. R. Landi. *Sorgo Breeding for High Biomass Production*. Proceedings of 5th EU Conference on Biomass for Energy and Industry, Lisbon, Portugal, 9–13 October 1989.
20. T.A. Bull and D.B. Batstone. *Potential for Multi-crop Processing Within the Sugar Industry*. Alcohol Fuels Conference, Sydney, Australia, 1978.
21. J.R. Moreira and D. Zilberstajn. *Ethanol Derived from Biomass and its Environmental Implication*. First World Renewable Energy Congress, Reading, UK, September 1990.
22. *Biomass for Energy*. OECD, Paris, France, 1984.
23. WEC Conservation and Studies Committee. *Global Energy Perspectives 2000–2020*. 14th Congress of the World Energy Council, Montreal, Canada, September 1989.
24. *Solar Energy: A Strategy in Support of Environment and Development*. UN Committee on the Development and Utilization of New and Renewable Sources of Energy (UNSEGED), 1992.
25. M. Mendis. Biomass Gasification: Past Experience and Future Prospects in Developing Countries. *Pyrolysis and Gasification: Proceedings of International Conference*, Luxembourg, May 1989.
26. C.A. Luengo and M.O. Cencig. Biomass Pyrolysis in Brazil: Status Report. *Biomass Pyrolysis Liquids Upgrading and Utilisation*, A.V. Bridgewater and G. Grassi (editors), Elsevier, London, UK, 1991.
27. W. Siquieroli. Uso de Gasogenio em Motores de Ciclo Otto. *Producao do Carvao Vegetal*, CETEC, Belo Horizonte, Brazil, 1982.
28. C.J. Weinberg and R.H. Williams. Energy from the Sun. *Scientific American*, Vol. 263, No. 3, September 1990.
29. F. Carazza et al. Carboquimica Vegetal: Aproveitamento do Licor Pirolenshoso. *ANAIS IV Congreso Brasileiro de Energia*, Rio de Janiero, Brazil, 1987.
30. A.A.C.M. Beenackers and A.V. Bridgwater. Gasification and Pyrolysis of Biomass in Europe. *Pyrolysis and Gasification: Proceedings of International Conference*, Luxembourg, May 1989.
31. E. Ashare et al. *Evaluation of Systems for Purification of Fuel Gas from Anaerobic Digestion*. Eng. Report COO-2991-19, Dynatech R/D Co., Cambridge, Mass., USA, 1978.

32. J.E. Marrow, J. Coombs and E.W. Lees. *An Assessment of Bio-Ethanol as a Transport Fuel in the UK*. Dept. of Energy, Report No. ETSU-R-44, 1987.
33. B.L. Maiorella. The Practice of Biotechnology – Current Commodity Products. Vol. 3 of *Comprehensive Biotechnology*, (editor-in-chief M. Moo-Young), Pergamon Press, 1985.
34. U. Ringblom. Novel Methods and New Feedstocks for Alcohol from Biomass. *Proceedings of 3rd EC Conference on Energy from Biomass*, (editors: W. Palz, J. Coombs and D.O. Hall) Venice, Italy, 1985.
35. *Evaluation of Biomass for Energy Supply* (in Danish). Sub-report No. 5 to Energy 2000, Danish Energy Agency, 1990.
36. R. Rabbinge. *Perspectives for Rural Areas in the European Community*. Agricultural University of Wageningen, Netherlands, 1991.
37. *Biomass – A New Future*. EU, Forward Studies Unit, 1991.
38. D.O. Hall, F. Rosillo-Calle, R.H. Williams and J. Woods. Biomass for Energy: Supply Prospects. *Renewable Energy: Sources for Fuels and Electricity,* Island Press, Washington DC, USA, 1992.
39. D.O. Hall, H.E. Mynick and R.H. Williams. *Alternative Roles for Biomass in Coping with Greenhouse Warming, Science and Global Security*. 1991.
40. S. Karekezi. *Energy from Biomass and Associated Residues: A Global Review*. ESETT'91, 1991.
41. P. Elliot and R. Booth. *Sustainable Biomass Energy*. Shell International, England, 1990.
42. D.O. Hall. *Analyse the Need for Fuelwood Plantations*. Shell International Petroleum Company and World Wide Fund for Nature, (draft in press 1992).
43. T.B. Johansson, H. Kelly, A.K.N. Reddy and R.H. Williams. *Renewable Energy, Sources for Fuels and Electricity*. Island Press, Washington DC, USA, 1992.
44. J. Goldemberg, J.R. Moreira, P.U.M. Dos Santos and G.E. Serra. Ethanol Fuel: A Use of Biomass Energy in Brazil. *Ambio*, Vol. 14, No. 4–5, 1985.
45. *Liquid Fuels from Biomass – Possibilities in Denmark* (in Danish). Danish Energy Agency, Denmark, 1992.
46. Renewable Energy Advisory Group. *Report to The President of the Board of Trade*. Energy Paper No. 60, The Department for Enterprises, HSMO Publications, London, UK, November 1992.
47. A.E. Carpentieri, E.D. Larson and J. Woods. *Future Biomass-based Electricity Supply in North East Brazil.* King's College, London, UK, 1992.
48. D.O. Hall, F. Rosillo-Calle and P. de Groot. *Biomass Energy: Lessons from Case Stories in Developing Countries*. 1991.

BIBLIOGRAPHY

BP Statistical Review of World Energy. London, UK, 1990.

A.V. Bridgwater. Economic and Market Opportunities for Biomass Derived Fuels. *Pyrolysis and Gasification: Proceedings of International Conference*, Luxembourg, May 1989.

U. Colombo. Energy and the Future of Biomass. *Proceedings of 5th EC Conference on Biomass for Energy and Industry,* Lisbon, Portugal, 9–13 October 1989.

Euroforum for Renewable Energies (EFNE). Proceedings of and International Congress, FR Germany, 1988.

FAO Forest Products Yearbook. FAO, Rome, Italy, 1989.

G. Foley. *Charcoal Making in Developing Countries*. Earthscan, London, UK, 1986.

J. Goldemberg, L.C. Monaco and I.C. Macedo (1992). The Brazilian Fuel-Alcohol Programme. *Renewable Energy: Sources for Fuels and Electricity*, Island Press, Washington DC, USA, 1992.

D.O. Hall. Carbon Flows in the Biosphere: Present and Future. *Journal of the Geological Society,* Vol. 146, 1989, pp. 175–181.

D.O. Hall and F. Rosillo-Calle. *CO_2 Cycling by Biomass: Global Bioproductivity and Problems of Devegetation and Afforestation.* Presented at WE-Hereaus Stiftung/European Physical Society Seminar on Balances in the Atmosphere and the Energy Problem, Bad Honnef, FRG, 4–7 February 1990.

D.O. Hall, H.E. Mynick and R.H. Williams. Cooling the Greenhouse with Bioenergy. *Nature*, Vol. 353, 1991.

D.O. Hall. Biomass Energy. *Energy Policy*, Vol. 19, No. 8, 1991.

D.O. Hall. Energy from Biomass: A Growing Trend. *The Economist*, 1991.

MIC/STI (Brazilian Ministry of Industry and Trade/Secretariat of Industrial Technology). *Oleos Vegetais: Experiencia de Uso Automotivo Desenvolvida pelo Programa.* OVEG 1. Brasília, Brazil, 1985.

World Energy Statistics and Balances 1985–1988. OECD, Paris, France.

R.E. Robinson. *Ethanol: Facts, Fables and Fiction.* SAIME Symposium on Alternative Transport Fuels, 1979.

F. Rosillo-Calle. Brazil: A Biomass Society. *Biomass – Regenerable Energy,* (editors: D.O. Hall and R.P. Overend), John Wiley and Sons, London, UK, 1987.

F. Rosillo-Calle and D.O. Hall. Biomass Energy Forests and Global Warming. *Energy Policy,* Vol. 20, No. 2, 1992.

Energy Statistics Yearbook 1987. United Nations, New York, USA.

UN Statistical Office (UNSO). *Energy Balances and Electricity Profiles*. New York, 1988.

E.A. Van Buren. *A Chinese Biogas Manual*. Intermediate Technology Publications, London, UK, 1979.

CHAPTER 6

Ocean Energy

INTRODUCTION

The oceans receive, store and dissipate energy through various physical processes. Energy exists in the form of waves, tides, temperature differences and salt gradients, each of which has been used or proposed for exploitation. Because of significant differences in the physical processes involved, exploitation techniques and state of development, these energy resources are best discussed separately, and parallel subsections are included for each one.

Wave energy is defined as the mechanical energy imparted by wind and retained in potential and kinetic form by waves of relatively short period. Energy transferred to the oceans from the rotational energy of the Earth through gravity of the sun and moon, and retained by long-period waves is classified as tidal energy. Energy stored in warm surface waters and available due to the lower temperature of the ocean depths is discussed under the heading of ocean thermal energy. Finally, the energy inherent in salinity differences between fresh water discharges and ocean water is discussed under the heading salt gradient energy.

Despite their diversity, these ocean energy resources share some common characteristics. In each case, the energy flux is large: about 2 TW for tidal and salt gradient energy; of the same order for waves; and at least two orders of magnitude larger for ocean thermal energy. However, these fluxes are spread over large geographical expanses so that energy densities are in fact quite low. One consequence is that much of the resource potential exists in areas remote from centres of consumption. For this reason alone, it appears inevitable that only a small fraction of the potential can be utilized in the foreseeable future.

Even after discounting the resource potential to reflect geographical limitations, it contrasts sharply with the negligible extent of present development. Except for tidal energy, this is at least partly a consequence of relatively recent technical interest, rather limited research support and immature technology. Low fuel prices for conventional generation have also had an effect. The extent of utilization over the next three decades will be greatly influenced by the pace and extent of technological improvement. Unfortunately, the present state of immaturity increases the importance of the question whether viable and competitive technologies will emerge through further research, and also makes it very difficult to reach dependable conclusions.

It has been claimed that: "Tidal energy, which entails the use of estuarine barrages at sites having high tidal ranges, offers the best prospects in the short to medium term. Not only are its components commercially available, but many of the best sites for implementation have been identified".[1] This claim, however, refers only to technical potential and overlooks adverse environmental impacts. The latter have (in the past at least) been given insufficient consideration in the Russian Federation – at the Kislaya Guba plant (near Murmansk) or at such potential sites as Lumborsk Bay (Barents Sea), Mezen Bay (White Sea), Tugur and Penzhin Bays (Sea of Okhotsk). Further development of tidal power in Canada's Bay of Fundy beyond the existing Annapolis plant (which has caused substantial fish mortality) in the Minas Basin, Shepody Bay or Cumberland Basin could, it is feared by some groups, have significant adverse impacts on the food and feeding space requirements of migratory bird populations. It has been said of the La Rance tidal power station in Brittany, France that: "The total closing of the estuary between 1963 and 1966 during the construction of the plant caused the almost complete disappearance of the original species".[2] Intensive environmental impact assessments of potential estuarial tidal barrage sites in the UK – Duddon, Wyre, Mersey, Conwy and Severn (of which only the last has significant installed capacity potential at around 8.5 GW) – have revealed similar seriously adverse environmental impacts. They include loss of invertebrate populations, ornithological impacts on sites of both national and international importance, siltation, interference with shipping movements, possible impact on hazardous chemical waste stored underground due to raising of the water table, and impacts on fish species (including extinction of a sub-species). Unless these problems can be satisfactorily resolved, it is doubtful whether this form of tidal power should have much of a future.

Near-shore and ocean tidal avoid most of the more obvious adverse environmental impacts. The technology, scale and economics of the former are relatively attractive and of interesting potential. Large-scale ocean tidal poses far greater challenges technically and economically, and given the significance of ocean fluxes in the carbon cycle, there are concerns that ocean tidal schemes on a substantial scale might significantly interfere with the natural rate of carbon absorption by the oceans.

INSTITUTIONAL CONSTRAINTS TO DEVELOPMENT

The economic acceptability of ocean energy, and indeed of most renewables, is affected by two factors of an institutional nature. One is the degree to which these resources must compete against conventional sources of energy priced on the basis of internalized costs which do not include, or only partially include, the social costs of environmental impacts. The other factor relates to the sensitivity of capital-intensive projects to interest rates and other market parameters. The future utilization of ocean energy will be influenced by both the treatment of social costs and the availability of financing under non-prejudicial terms and conditions. Both these factors must be considered in making projections.

The Treatment of Societal Costs

All energy obtained from combustion of geologically stored hydrocarbons results in release of carbon dioxide to the atmosphere. In addition, acid-forming oxides of sulphur and nitrogen are produced in quantities depending on engine or combustor design. Present emission standards in some countries result in minimizing such oxide emissions through use of scrubbers or fluidized bed combustion and the costs involved constitute internalization of a part of the social costs of pollution. However, at most it is only a part of the cost. No reduction in carbon dioxide, other than through improved efficiency, seems likely. Other pollutants are also emitted; for example, aldehydes from diesel engines and radionuclides from coal-fired boilers.

By contrast, ocean energy creates little or no atmospheric pollution. While other environmental impacts are anticipated, they are believed to be capable of minimization or avoidance by appropriate design, or by adoption of those forms which avoid the most serious and obvious environmental harm.

The wish to utilize renewable resources is at least partly founded on the perceived need to reduce the extent of atmospheric pollution from use of fossil fuels. But in order to achieve any significant degree of market penetration, ocean energy developments must be able to compete with the conventional alternatives. It is illogical under such circumstances to bias the contest in favour of conventional sources by failing to take external social costs into account. Yet in most cases this is the current decision framework.

The Financial Climate

The physical size and cost of machinery for extraction of wave, tidal and ocean thermal energy tends to be high for several reasons, including low energy density in the water, low efficiencies of low-temperature thermodynamic cycles and many wave devices, and intermittent operation in the case of tidal. In all cases, operating costs are low and fuel costs non-existent. This contrasts with the character of fossil-fired generation, where capital costs tend to be lower and life-cycle fuel costs tend to dominate.

In an economic comparison of two such dissimilar options, the results are influenced significantly by the interest rate. If the present value of life-cycle costs is compared, a high interest rate minimizes the worth of life-cycle fuel cost. If, on the other hand, the unit costs of product are compared, then a high interest rate increases the capital cost disadvantage of the non-fueled option. The result is the same in either case: high interest rates render capital-intensive projects less attractive.

Interest rates may affect the competitive status of capital-intensive projects in another way if, as is often the case, such projects require a long construction time. In this event, high interest rates increase the capital cost at completion, thus compounding their effectiveness in discouraging the displacement of fossil fuels.

The Effect of Perceived Risk

A further obstacle faced by capital-intensive projects is that a small margin of life-cycle benefits over costs may not be sufficient to establish financial viability, given the uncertainty in future events and fuel prices. The cost of capital is affected by risk, as perceived by lender or investor. The perception of risk tends to be high for all forms of ocean energy. It arises principally from immature technology in the case of wave and ocean thermal energy and from the site-specific nature of engineering and environmental problems relating to tidal power. Until ocean energy developments become commonplace, or at least acquire a fairly broad background of successful commercial operation, the risk factor will tend to raise the cost of commercial capital for such ventures.

TIDAL ENERGY

Resource Characteristics

Tides are caused by the gravitational attraction of moon and sun acting upon the rotating Earth. In the oceans, the moon's attraction increases the water height on both near and far sides of the Earth. These bulges are swept westward, due to the Earth's rotation, as deep ocean waves of period 12 hours 25 minutes and amplitude of less than 1 m. The sun's effect is similar but smaller, and with a period of 12 hours. It therefore appears as a modulation of the lunar wave, resulting in larger spring tides when sun, moon and Earth are co-linear, and smaller neap tides when they are in quadrature. In addition to these semidiurnal and lunar monthly cycles, the orbital motions of Earth and moon produce many other cycles with periods ranging from days to many years.

Where the ocean tidal waves impinge on continental shelves and coastlines, their ranges can be increased substantially through run-up, funnelling and resonance. Local tidal systems frequently exist, obtaining energy from the deep ocean tidal wave but exhibiting a unique regimen. Despite their complexity, tides at any location can be predicted with high accuracy.

Ocean tides obtain their energy from rotation of the Earth, slowing it down in the process. However, the slowing of the Earth is almost imperceptible over human time spans and would not be increased by tidal power developments. The energy is dissipated by friction in shallow seas and along coastlines at an estimated rate of 1.7 TW.

Extraction of energy from the tides is considered to be practical only where the energy is concentrated in the form of large tides and geography provides favourable sites for tidal plant construction. Such sites are not commonplace, but a considerable number have been identified. Recent estimates place the potentially economic energy at about 200 TWh per year[3], or roughly 10% of the technical potential given by Bernshtein's equation[2]. Locations of sites with high technical potential (disregarding environmental considerations) are indicated in Figure 6.1.

Tidal Technology

Tidal power is one of the older forms of energy used by man. Records indicate that tide mills were being worked on the coasts of Britain, France and Spain before 1100 AD. They remained in common use for many centuries, but were gradually displaced by the cheaper and more convenient fuels and machines made available by the Industrial Revolution.

Many methods of extracting potential or kinetic tidal energy have been tried in the past. Devices used have included waterwheels, lift platforms, air compressors, water pressurization and many others. Hundreds of patents have been registered during the past 150 years and the tides still entice inventors. None of the inventions appear to offer any advantage over the basic approach used for the old tide mills.

The tide mills typically were operated by filling a storage pond at high tide, later allowing the water to flow from pond to sea through a waterwheel. This is the simplest method of operation, now generally referred to as "single basin, single effect". In the modern version, the storage

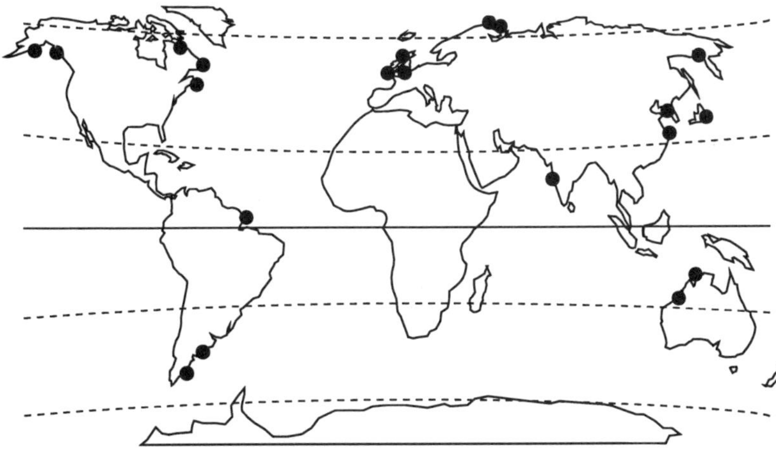

Figure 6.1 *Principal sites for tidal power development*

basin is equipped with gated sluices and the old waterwheel is replaced by a low-head turbine. The operating cycle consists of four steps:

1. basin filling;
2. holding the impounded water until the receding tide creates a suitable head;
3. releasing the water from basin to sea through a turbine until the rising tide reduces the head to the minimum operating point; and
4. holding until the tide rises sufficiently to repeat the first step.

This approach is termed "ebb generation". The cycle could be reversed, with generation from sea to basin (flood generation). However, the sloping nature of the basin shores usually makes ebb generation the more productive method.

Generation on both ebb and flood (termed single basin, double effect) is also feasible. In this case, the routine involves:

1. inward sluicing to fill the basin;
2. holding period;
3. ebb generation;
4. outward sluicing to empty the basin;
5. holding period; and
6. flood generation.

Whether greater output is obtained from single or double effect usually depends on basin bathometry, tidal range and plant design details.

Increased output can be obtained by pumping to increase the basin level and therefore the generating head. The energy required for pumping must be borrowed and repaid, but the pumping is done against a small head at high tide, whereas the same water is released through the turbine at a greater head, producing a net energy gain and limited ability to re-time energy.

In order to obtain continuity of supply, plant configurations and operating routines of greater complexity have been considered, including linked and paired basins. The linked basin scheme utilizes two basins, one topped up at high tide and the other emptied at low tide. In the simplest form, water is sent through a turbine from high basin to low basin whenever power is required. Layouts with other turbine locations have been proposed, including one in which the turbine discharge can be switched from basin to sea. Linked basins provide output with some firm capacity, typically only about 40% of the installed turbine capacity. Paired basins consist essentially of two single basin schemes. If one generates on the flood and the other on the ebb, output is almost, but not quite, continuous. These more complex schemes are only realizable where geography is favourable and have generally been found economically inferior to single basin schemes[4,5].

Tidal plants have usually been designed to produce electricity, although a pumping role is possible under certain circumstances. Single basin plants deliver one or two intermittent pulses of energy per tide, leading to low annual plant factors in the range 0.25 to 0.35 and providing no firm capacity. The output pulses recur at a period of 12 hours 25 minutes and thus move progressively in and out of step with the solar-dictated rhythm of human activity. Such an output can be absorbed without difficulty into electrical systems provided the tidal plant capacity is a small percentage of system capacity. In this case it has a small capacity value to the system in addition to its energy value.

Under other circumstances, including stand-alone operation, it would be necessary to modify the tidal output by use of some form of energy storage. Three commercially proven methods exist, of which the best requires the existence of river hydro plants with significant capacity and storage. If these conditions are satisfied, then the river hydro storage can be drawn down when the tidal plant is not operating and replenished when tidal output is available to displace normal river hydro generation. This method requires installing additional hydro generating capacity but has about 100% throughput efficiency.

Re-timing by pumped storage is feasible but incurs relatively high capital cost and attains a throughput efficiency of only about 75%. The remaining alternative, compressed air energy storage, incurs lower capital costs. Output is about 90% of the energy which could have been obtained if the inputs (tidal energy and gas turbine fuel) had been used separately. This approach incurs some atmospheric emissions: about a third as much as a combined cycle gas plant or a fifth as much as a coal-fired plant for the same output[6].

Relatively few tidal-electric plants have been built in the modern era. Of these, the first and largest is the 240 MWe single basin, double effect plant at La Rance, built for commercial production[7]. Others include the 20 MWe plant at Annapolis built to demonstrate a large diameter straight-flow turbine, the 400 kWe experimental plant at Kislaya Guba, the 3.2 MWe Jiangxia station, and a number of small or multi-purpose plants, all in China[4,8]. Further details are contained in Table 6.1.

Table 6.1 *Existing tidal plants*

Site	*Mean tidal range (m)*	*Basin area* (km^2)	*Installed capacity (MWe)*	*Approx. output (GWh/yr)*	*Date in service*
La Rance (France)	8.0	17	240.0	540	1966
Kislaya Guba (former USSR)	2.4	2	0.4	–	1968
Jiangxia (China)	7.1	2	3.2	11	1980[a]
Annapolis (Canada)	6.4	6	17.8	30[b]	1984
Various (China)	–	–	1.8[c]	–	–

Notes:

(a) First unit. Sixth (last) unit in 1986.
(b) At restricted headpond elevation.
(c) Excluding a plant only partially powered by tides.

Operating experience at La Rance, Kislaya Guba and Annapolis has generally been positive. The stators of La Rance generators and the field windings of the Annapolis generator required minor modification to overcome design weaknesses. Otherwise, they have proved reliable and maintenance requirements have been modest[4, 9]. On the basis of experience with these plants and at the present level of tidal technology, tidal power may be regarded as a technically proven, dependable and long-lived source of electric power.

Development Trends and Possibilities

Despite its long history, serious attempts to develop tidal power in the modern era span only the past three decades. Substantial improvements in machinery and techniques have been achieved during this period.

The bulb turbine was adapted to tidal use at La Rance. Since that time, other turbines for very low-head applications have been developed, including straight-flow, tube and pit types; the latter two for limited power ratings only. Also, there has been a steady increase in specific speeds as runner designs have improved.

Cofferdams were used in La Rance construction, but caissons, adapted for tidal applications at Kislaya Guba, have now become well established in a wide variety of marine applications. Ways to reduce construction time and secure early barrage closure through use of specialized caissons emerged from Bay of Fundy (Canada) studies. Severn Estuary (UK) studies have identified other possibilities for improving designs and reducing civil construction cost.

Design and feasibility studies in France, the former USSR, Canada and Great Britain have produced techniques for optimizing installed capacity; for maximizing plant output or the value thereof; for integrating tidal output into utility systems and computing its economic value. Hydrodynamic numerical models have been developed for prediction of tidal regime effects. In short, tidal technology has moved a long way towards maturity over the past three decades. This does not rule out further improvements, but it does mean that future cost reductions, improvements in performance and economic status can be expected to be relatively small for plants of this type.

The possibility of dispensing with dams, their associated costs and both actual and potential deleterious impacts on the environment, has encouraged consideration of extracting energy from tidal flows. Some alternative designs have been explored, including prime movers to extract energy from tidal currents (current generators), hydraulic compression and the use of air turbines, oscillating water columns and oscillating machinery.

For ultra-low head use, the Darrieus turbine has been examined in Canada and propeller machines in the United States. These approaches would avoid the cost of dams but suffer the handicap of energy densities at least an order of magnitude lower than those available from tidal heads. While any or all of these approaches may find economic application in the long term, it is considered unlikely that they will exert a significant effect on installed tidal power capacity over the study period.

In recent years, a number of small tidal power plants have been built in China as part of broader schemes of resource utilization involving, for example, aquaculture or navigational improvements. This approach has justified development where tidal power would have been too costly by itself.

Trends to be expected over the study period may be summarized as a marginal decrease in the real cost of large-scale developments due to improvements in machinery, materials and techniques, and some broadening of the scope for small installations, both as joint-use projects or as a result of developments in ultra-low head generation.

Economic Constraints

Tidal power incurs high capital cost per kWe of installed capacity which tends to vary inversely as the square of the tidal range. The capital cost per kWh is further affected by intermittent operation and consequently low capacity factor. Civil costs depend on the ratio of barrage length to basin area and on barrage height. All these considerations tend to make costs site-specific, but in any case tidal power is a very high capital cost energy source.

Whether it is economically viable depends both on the site and on the cost of alternatives in the market to be served. Viability does not require that tidal be cheapest, but only that it be cheapest for some segment of the market. Because hydro is often limited and nuclear is confined to base-load supply, the competition is mainly from coal, oil, or gas-fired thermal generation[10]. Apart from interest rates, which significantly influence tidal costs, and the non-internalized social costs of thermal generation, noted previously as institutional constraints, the main parameters of competitive status are the cost of hydrocarbon fuels, their rate of escalation, and the general inflation rate.

A further constraint exists in the case of large-scale projects. Their high capital costs, long lead times and other implications may place them beyond the capacity of all but governments.

Environmental Impacts and Benefits

Tidal power plants can alter tidal ranges and currents, water temperatures and quality to an extent depending on their size and location. Bay of Fundy projects are predicted to influence tidal ranges to a distance of several hundred kilometres. The benefits and impacts are likely to be site-specific and partly offsetting. Tidal barrages affect headpond ecology, such as providing conditions suitable for aquaculture, and also affect groundwater, erosion and drainage.

Legitimate concerns about the effects of tidal plants on bird habitats and populations have been raised. Extensive studies at various sites have been conducted and the general conclusion is that this impact is site-specific. Most of the sites so far investigated have been ornithological sites of national or international importance. There is no solid evidence to suggest that bird populations will remain unaffected by the elimination of the special ecological characteristics of that site, and the apparent availability of other sites further along a coastline. While the impacts of tidal power developments on migratory or indigenous birds should not be used as a general argument against this energy source, the effects must be studied at each individual site so that any potential problems can be avoided or mitigated. Most sites so far studied would not permit satisfactory avoidance or mitigation of these problems. The development of a major wetland site may be limited for such environmental reasons.

Potential impacts on fish may arise through turbine mortality and delayed passage of anadromous species[11]. Means of mitigating such impacts are available, but further research is necessary to find methods combining low cost and efficiency. Mortality problems are anticipated in connection with Severn and Fundy schemes. Indications of mortality at La Rance have only recently been found, but there is no evidence that any fish species have been significantly impacted by that plant.

The fundamental problem of tidal barrage schemes for estuaries remains the environmental consequences of building the barrage. There may be a few sites – harsh rocky terrain – where the local ecology is limited. In general, the development of tidal power in estuaries which do not have severe environmental impacts should await schemes which dispense with barrages. To avoid visual intrusion at low water, such schemes would involve the location of turbines at permanently submerged, near-shore locations. Effective steps to reduce or avoid fish mortality and other species loss would be required. These schemes would be non-polluting, and should have net environmental benefits at most sites if considered as displacing hydrocarbon fuels.

In short, this book distances itself from the findings of one recent study which claimed: "To date, no tidal energy plant has been subjected to extensive environmental monitoring. Nonetheless, most operational experiences have been positive. Although some environmental uncertainties remain and require future investigation, especially at the site-specific level, no major factors have so far been identified that would inhibit the wider implementation of tidal energy, assuming proper attention to scheme design".[1] Those sentences are misleading and insensitive concerning local environmental impacts. Tidal power development involving barrages should be regarded as conflicting with the objectives and sentiments of Agenda 21, Chapter 17 ("Protection of the Oceans, All Kinds of Seas, Including Enclosed and Semi-Enclosed Seas, and Coastal Areas ....") such as paragraphs 17.1, 17.4, 17.5, 17.6, 17.7, 17.19, 17.22, 17.29, 17.73 and 17.86.

Development Potential

In some areas, tidal power sites have been quite thoroughly investigated and there is continuing interest in development[12, 13, 14, 15]. The best estuaries in the UK might provide around 20 TWh/year of electricity at 3.5–4 p/kWh (5–6 US cents), with a further 8 TWh/year of potential under 5 p/kWh (7.5 US cents), if construction could be undertaken in the public sector with capital costs discounted at 5% per annum[16, 17]. However, the high capital costs and long construction times of large tidal barrages make the cost of electricity generation particularly sensitive to the method of financing. With construction financed in the private sector and capital costs discounted at, for example, 10% per annum, the cost of electricity from the best estuaries would double to, say, 7–8 p/kWh (10–12 US cents) (see Figure 6.2). This is another reason why the tidal *barrage* route remains unattractive.

Russia has been active in assessing its large potential both in the White Sea and in the Sea of Okhotsk. A pilot plant at Kolskaya is planned for testing the turbines proposed for Tugur, which would be the first major site developed. It is hoped to develop 15,000 MWe at Mezen by 2015–20.

Of the Canadian sites in the Bay of Fundy, Cumberland Basin is not considered competitive at present and Minas Basin, although economically more attractive, is thought to threaten unacceptable impacts on US coasts along the Gulf of Maine. Capital cost for the Cumberland Basin site is estimated at US$1,319 per kWe, plus US$513 per kWe for transmission and US$922 per kWe for re-timing by CAES. The resulting levelized cost of re-timed output would be in the range of 6.8 to 7.6 cents per kWh. For Minas Basin, costs are estimated to be US$995 per KWe for plant and US$214 per kWe for transmission, with levelized re-timed output cost of about 4.9 cents per kWh. All the above costs are in 1987 US dollars.

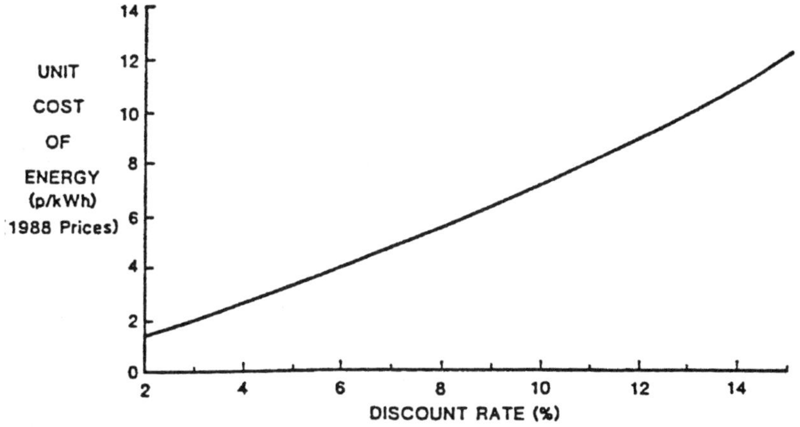

Figure 6.2 *Estimated cost of electricity from the Severn barrage (UK)*

Garolim in Korea is considered close to economic break-even. The Gulf of Kutch site in India is probably uneconomic at present fuel price levels. Mexico is investigating a site in the Colorado Estuary. China is conducting feasibility studies at two sites. The potential of sites in the Secure Bay area of Western Australia was recently re-examined.

Interest in some sites has lapsed because previous studies found costs too high. Among these are sites in Alaska, Passamaquoddy in the Bay of Fundy, San Jose in Argentina, and the north-east coast of Brazil. Other areas with large energetic potential but not the focus of present interest include Mont St Michel Bay in France and Ungava Bay in northern Canada. The Mont St Michel Bay site has been studied in depth, but financial and ecological constraints make development unlikely. The Ungava Bay site is too remote. The south-east coast of China is also considered to have large potential.

Site data, where known, is presented in Table 6.2. The data is merely representative, being subject to change with successive feasibility studies.

Table 6.2 *Sites considered for tidal development*

Site	*Mean tide range (m)*	*Basin area (km²)*	*Approx. installed capacity (MWe)*	*Annual output (TWh)*	*Annual capacity factor*
Argentina					
San Jose	5.9		6,800	20.0	
Australia					
Secure Bay 1	10.9			2.4	
Secure Bay 2	10.9			5.4	
Canada					
Cobequid	12.0	240	5,338	14.0	0.30
Cumberland	10.9	90	1,400	3.4	0.28
Shepody	10.0	115	1,800	4.8	0.30
India					
Gulf of Kutch	5.0	170	900	1.7	0.22
Gulf of Cambay	7.0	1,970	7,000	15.0	0.24
Korea					
Garolim	4.8	100	480	0.5	0.13
Cheonsu	4.5			1.2	
Mexico					
Rio Colorado	6–7			5.4	
United Kingdom					
Severn	9.0	520	8,640	17.0	0.22
Mersey	6.5	61	700	1.5	0.24
Conwy	5.2	6	34	0.1	0.20
Other, small			1,000	2.0	0.23
United States					
Passamaquoddy	5.5				
Knik Arm	7.5		2,900	7.4	0.29
Turnagain Arm	7.5		6,500	16.6	0.29
Russia					
Mezenskaya	6.0	2,640	15,000	42.0	0.32
Tugur	5.7	1,120	6,800	16.0	0.27
Penzhinsk	6.2	6,788	21,400	71.4	0.38
Kolskaya	2.3	6	32		

Development Estimates

Under a relatively unfavourable scenario, and extension of current policies involving only minor reductions in plant costs and real interest at 5%, it is estimated that within the study period sites in Britain, Canada and Russia could be developed, providing an annual output of 12 TWh (43.2PJ) from the tidal resource by 2020.

Under a more favourable (ecologically driven) scenario, with costs fully internalized for all sources of electricity and real interest at 4% it is projected that sites in Fundy, Korea, China and India could also be developed within the study period, raising total annual output to about 60 TWh (217 PJ).

Realization of either of these scenarios, equal to 2.7 Mtoe and 13.5 Mtoe, respectively, would probably require the involvement of national governments, because of the magnitude of some of the projects, the need for moderately priced capital, and the risks involved in trying to estimate the value of life-cycle output. Such involvement might take the form of ownership, or be limited to financial guarantees.

Estimates shown in Table 6.3 are based primarily on the criterion of economic viability, for which the real interest rate is the leading determinant. Site-specific factors, including tidal range, cost of competing energy resources, and the ratio of headpond storage to barrage cost, are also important. The table has not taken into account environmental considerations or the output potential of the various sites if tidal power schemes were to be developed at them without the building of dams (barrages). Economic status is not an infallible or even a reliable guide to the probability of site development, for reasons illustrated by the following example. The Cumberland Basin site would require a real interest rate of 4% or less to be competitive as a source of displacement energy. If existing river hydro storage capacities were available for re-timing, the site would be attractive at real interest of more than 5%. The site would probably be developed without regard to economic status if a restrictive cap were placed on CO_2 emissions. In the absence of such a cap, and

Table 6.3 *Tidal energy utilization estimates through 2020*

Scenario/ Year	*North America*	*Western Europe*	*CIS & E Europe*	*C & S Asia*	*Total TWh*	*Total (PJ)*
Less favourable scenario						
2000	–	0.5	–	–	0.5	(1.8)
2010	–	0.5	2.6	–	3.1	(11.2)
2020	3.4	2.1	6.5	–	12.0	(43.2)
More favourable scenario						
2000	–	0.5	–	–	0.5	(1.8)
2010	3.4	19.0	6.5	1.0	29.9	(108)
2020	17.4	19.2	20.4	3.2	60.2	(217)

even if economically attractive, it would not be developed until additional generation capacity was required in the region (2005 or later, according to present projections). Finally, development would not be permitted at any time unless means of minimizing environmental impacts were available.

This example shows that non-economic (and non-predictable) circumstances can be of pivotal importance in decisions to construct tidal plants. Where economic viability is likely under a scenario, it has been assumed that other enabling conditions will occur at some time during the study period and that the site will be developed.

The real interest rates assumed under the two scenarios are low compared to those usually available over the past two decades. However, they are not low in historical perspective: for example, they are 2% or 3% higher than rates on sovereign risks in the US, UK and Canada in the quarter century between 1950 and 1974.

WAVE ENERGY

Resource Characteristics

Waves are caused by the transfer of energy from wind to sea. The rate of transfer depends on wind speed and the distance over which it interacts with the water (the fetch). Potential energy is carried in waves by the mass of water displaced from mean sea level and kinetic energy by the velocity of water particles. The energy stored is dissipated through friction and turbulence at rates depending on wave character and water depth. Larger waves in deep water lose energy quite slowly, with the result that wave systems are complex, often deriving from both local winds and distant storms which have occurred days previously.

Waves can be characterized by their height, wavelength (distance between successive crests), and period (time between successive crests). Power is usually stated in kilowatts per metre, representing the rate at which energy is transferred across a line of 1 m length parallel to the wave front.

The strongest winds blow between 40 and 60 degrees latitude in both northern and southern hemispheres. Winds of lower velocity in the tradewind zones (within 30 degrees of the Equator) also produce potentially attractive wave climates because of their comparative regularity. Coasts with exposure to the prevailing wind and long fetches are likely to have the greatest wave energy density. The United Kingdom, the west coast of the United States and the south coast of New Zealand, for example, have excellent exposure and particularly good wave climates. Wave energy densities for selected areas are indicated in Figure 6.3[18].

Wave energy in the open oceans is likely to be inaccessible and the resource potential is therefore limited to the energy available near coastlines. Even with this restriction, the wave power dissipated on coastlines with favourable exposure is probably in excess of 2 TW.

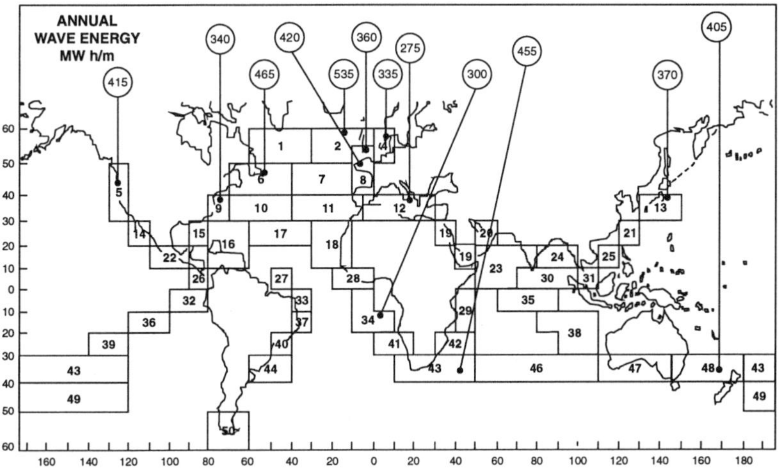

Figure 6.3 *Annual wave energy from specific areas*

Wave Technology

The prospect of extracting energy from ocean waves has attracted some attention from time to time over the past century, but serious efforts to establish an effective technology date from the mid-1970s. Since then, research has been undertaken in 13 countries and numerous devices have merged[19, 20, 21, 22]. These devices may be classified in several ways. For simplicity and clarity, a method (due to Hagerman and Heller[25]) based on actuating motion is used here. The motions are classified as heaving, heaving and pitching, pitching, oscillating water column and surge. A stylized representation of 12 wave energy devices is presented in Figure 6.4.

The simplest form of wave energy collector is the bell or whistle buoy where motion of the buoy is used to activate a striker arm or to push air through the whistle. In the latter case, a cylinder with the open end submerged has valves to draw in air from the atmosphere and to expel it through the whistle. Rising and falling water level in the cylinder does the pumping as the buoy heaves up and down.

Substitution of an air turbine for the whistle enables the same type of device to produce electricity. In 1978 and 1979, Japan put buoys with an output of 500 We in service and some still remain in use. A 3 kWe lighthouse generator was designed in 1983 and 1984. Also, a 120-ton buoy was tested, and reported capable of 300 kWe output. Other versions of buoy-mounted generators have been developed elsewhere. The former USSR tested a 3 kWe unit in the Caspian Sea. China has deployed over 50 wave-powered navigation buoys. Sweden tested a system in which the water tube was open at both ends and power was extracted by a hydraulic turbine or a piston.

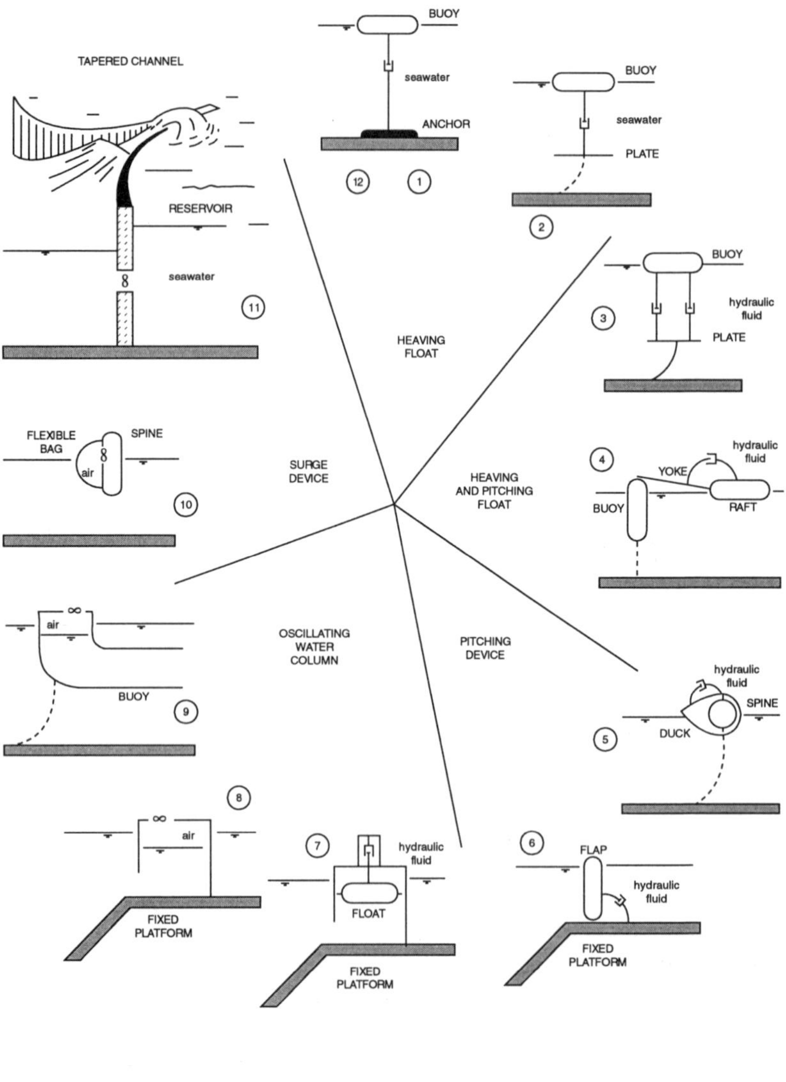

Figure 6.4 *Representation of wave energy devices*

Power may also be extracted from a buoy by means of an internally suspended weight, free to swing but with high inertia. Energy is extracted from the differential motion of buoy and pendulum. A Japanese version called the pendular has been analyzed and tested. The former USSR built a 50 kWe inertial device with generation from compressed air. Motion of a buoy with respect to a fixed point on the seabed has also been utilized. In the Danish version, a seabed-mounted piston pump is activated by motion of the attached buoy on the water surface. A design developed in the United States puts the pump on the float and uses the seabed structure only as a point of attachment. This unit was made from used truck tyres, is robust and relatively easy to deploy.

The Republic of China (Taiwan) has developed a wave power system actuated by a heaving float[21]. In one version, water is pumped to an onshore reservoir and passed through a turbogenerator. A demonstration project on Lan Yu Island is proposed.

Other methods of converting buoy motion into motion of a working fluid have been tried. Britain's "Sea Clam" uses the exchange of air between a series of flexible bags to drive an air turbine. A Swedish design features a submerged flexible tube as an air pump. Buoys are attached to it. Motion of the buoys stretches the tube, decreasing its volume to provide the pumping action.

Several designs have been based on motion of floats or flaps with respect to a fixed or floating reference point. Japan has tested prototypes of both versions. In either case, the machinery is housed in the terminating structure where linkages from the moving floats or flaps are used to pump oil at high pressure through a hydraulic motor which turns a turbine. The US-designed tandem flap system is closely related. The reference point is fixed to the seabed and the flaps, spaced at a fraction of the expected wavelength, extend up into the wave zone. Motion of the flaps activates a system similar to the Japanese device. A circular float with submerged damper plate is used as the reference point in another US design being considered for deployment off the Newfoundland coast.

An oscillating water column (OWC) is the energy collector in a number of designs. In contrast to buoys, where the motion of the buoy moves water in a tube, the OWC structure is fixed and the waves move the water column. Typically, such designs seek resonance to increase water column motion. The first system of this type was built in 1981 for an IEA experiment under the auspices of Canada, Ireland, Japan, the United Kingdom and the United States, with Japan as the operating agent. Barge-mounted for testing, the system had a collection efficiency of 4%. After some years of research, a Norwegian enterprise built a multi-resonant oscillating water column (MOWC) demonstration plant in a seaside cliff. It features a reception chamber for waves with modes of oscillation at different frequencies, permitting some resonant increase in amplitude over a range of wave periods. The vertical motion of the water column moves air past a unidirectional turbine (driven the same way regardless of the direction of air flow)[19].

Portugal is laboratory-testing an OWC model with the intention of installing a 300 kWe unit in the Azores. Ireland is running a field trial on an OWC provided by a natural rock chamber. The peak air power output is said to be 572 kWe.

A second type of onshore device developed in Norway is the TAPCHAN (for tapered channel). The waves taken in at the wide seaward end of the channel are funnelled by the tapered channel so that their crest height is increased. The wave crests then spill over the sides of the channel into a reservoir, from whence the water flows back to the sea through a turbine. Both TAPCHAN and MOWC are focusing devices. The TAPCHAN is not practical where there are significant tides.

In most cases, the output of wave energy devices is electrical, without firm capacity. The TAPCHAN is a possible exception because water stored in a sufficiently large holding pond could be sent through the turbine to meet demand. Also, some devices have been designed to produce high-pressure water, intended for desalination by reverse osmosis. In this case, the desalinated water could be stored, achieving a dependable supply. Only about 20% of the high-pressure seawater is actually passed through the reverse osmosis membrane and the remaining 80% could be discharged through a small impulse turbine. Heaving-buoy wave energy devices that use high pressure seawater as a working fluid thus have the capability to produce both electricity and potable water from a single unit. A commercial wave-powered desalination unit has been operating off the south coast of Puerto Rico in a 1.5 m wave climate.

Moreover, in tropical areas well-sited wave generators are likely to have some output at all times, and even where this is not the case, generators having capacity factors of 0.20 or greater have some capacity value when connected to large utility systems.

Development Trends and Possibilities

Wave power projects are planned or committed in various regions, mainly as demonstration plants. Two 1 MWe TAPCHAN plants have been contracted, one in Indonesia and one on an island off the coast of Tasmania. A 1 MWe MOWC unit is to be installed in Tonga and other oscillating water column plants have been built in the United Kingdom, India and Japan, with capacities of 75, 150 and 50 kWe respectively. The UK plant is located in a natural rock gully on the Scottish Island of Islay.

Installed capacity cannot be expected to increase very rapidly until more information on operations and more confidence in longevity are developed through demonstration projects. The emergence of wave power as a mature and viable technology will therefore require time. It will also require adequate support on a national or international basis in order to produce a climate conducive to progress. Areas where support would be appropriate and useful include:

- Eliminating gaps in wave climate data, including calibration of data accumulated by shipping reports to provide the basis for pre-feasibility studies of sites world-wide.
- Support of private research on a partial basis, particularly for materials testing under ocean conditions and for synthesis of experience to determine optimum directions for future research.
- At least partial support for demonstration projects. Potential host countries are not willing to bear the technological risks, and they may exceed the resources of developers.

Wave energy research has achieved an understanding of wave geometry and mechanics. Development has reached the point where a few small-scale devices are operating satisfactorily. While larger-scale systems have been designed, problems of cost, involving both equipment and its installation, remain to be resolved. Capacity to survive significant storms and to withstand marine fouling and corrosion need to be attained and demonstrated.

After 15 years of investigation, representing a large proportion of total world research on wave energy devices, the UK Department of Energy concluded that large-scale offshore devices were unlikely to become economic in the foreseeable future; that effort should be concentrated on small-scale modular shoreline devices; that although the largest part of the resource is offshore, much basic technology needed to be developed and proven before it could be exploited; and that such development would best be done at small-scale shore-based sites[23].

Cost per unit output is affected by all design considerations and particularly the attainment of long-term reliability at sea through the use of appropriate and proven hardware. Improvement of collection efficiency, which tends to be low for many of the devices described above, has been a design objective. For optimization of design, knowledge of the wave climate at the intended point of application is needed. While much data is available from the coastlines of industrial nations, there is little from less developed countries, where wave energy is likely to be most economically competitive.

Wave energy has been a fertile field for invention. Over 200 devices were tested by the United Kingdom alone in the period 1974–85 and it cannot be assumed that the invention phase has now come to an end. It would, however, help to focus inventive, research, and development effort if, through synthesis of experience or by parametric analysis, an understanding could be reached as to which of the several design categories are likely to be optimal and which inferior. This aspect is beginning to receive attention. Adequate periods of demonstration are required to reveal design weaknesses and maintenance problems. Attempts to commercialize larger-scale wave energy systems before such experience is gained could well be counter-productive.

Environmental Impacts

Wave power is essentially non-polluting and to the extent that it displaced hydrocarbon fuels, environmental benefits would be realized. No appreciable environmental effects are foreseen from isolated floating wave power devices. It would usually be both possible and necessary to avoid hazards to or from marine traffic by judicious siting and by provision of navigation aids.

Intensive development of wave power over a segment of coastline could extract enough energy to affect sediment and bed load transport. Depending on the character of the site, mixing, stratification and turbidity could also be affected. Such changes might be good or bad from an environmental standpoint, depending on site specifics. Aesthetic impacts could be important, particularly in populated or resort areas.

It is fair to assume that an installation of the type postulated would be placed in an energetic environment. If so, the probable result of removing some of the energy would be to make the environment habitable for a wider range of biota.

Economic Constraints

In the 1989 WEC Survey of Energy Resources, costs of 5–7 US cents/KWh are cited for low-capacity wave energy devices, with uncertain but somewhat higher costs for larger systems[24].

Table 6.4 *Wave power plant design summaries*

No.	*Plant location*	*Average annual incident wave power*	*Plant size (capacity factor)*	*Offshore capital cost*	*Annual O&M cost*
1	North Sea (Denmark)	9.1 kW/m	750 MWe (13%)	$875/kWe	0.8%
2	Baltic Sea (Sweden)	5.5 kW/m	25 MWe (37%)	$1090/kWe	6.8%
	NE Atlantic (Norway)	45 kW/m	64 MWe (32%)	$580/kWe	7.9%
3	NW Atlantic (Newfoundland)	21 kW/m	1.0 MWe (30% assumed)	$2080/kWe	4.0% (assumed)
4	Southern Ocean (Australia)	49 kW/m	0.5 MWe (60%)	$2280/kWe	3.0%
5	NE Atlantic (Scotland)	48 kW/m	2000 MWe (38%)	$1920/kWe	1.8%
6	Low energy (generic)	20 kW/m	4.4 MWe (24%)	$1640/kWe	4.0%
	Medium energy (generic)	35 kW/m	4.4 MWe (35%)	–	–
	High energy (generic)	50 kW/m	4.4 MWe (43%)	–	–
7	Central Pacific (Hawaii)	15 kW/m	1.0 MWe (30%)	$2320/kWe	5.0%
8	NE Atlantic (Norway)	25 kW/m	0.5 MWe (41%)	$1600/kWe	1.0%
	SW Pacific (Tonga)	21 kW/m	2.0 MWe (33%)	$4000/kWe	1.0%
9	Sea of Japan (Japan)	13 kW/m	0.5 MWe (16%)	$4830/kWe	4.0% (assumed)
	Central Pacific (Hawaii)	23 kW/m	1.0 MWe (20%)	$3210/kWe	–
	NE Atlantic (England)	51 kW/m	3.5 MWe (20%)	$1320/kWe	–
10	NE Atlantic (Scotland)	48 kW/m	2.0 MWe (31%)	$1040/kWe	5.0%
11	NE Atlantic (Norway)	25 kW/m	0.35 MWe (57%)	$3870/kWe	0.4%

In a recent study by Hagerman and Heller, cost estimates for a number of devices were compared on a uniform basis. Those included in the study are illustrated in stylized form in Figure 6.4 and the estimates are shown, with other data, in Table 6.4. They are at various stages of development, ranging from scale model to demonstration unit. For present purposes, the Hagerman and Heller cost estimates have been converted to lifetime levelized constant dollar costs. The resulting energy costs are heavily dependent on wave climate, and the relationship indicated by regression analysis being:

$$\text{Cost (US cents/KWh)} = 26.49 \,/\, (\text{Wave power})^{0.357}$$

where wave power is stated in kW/m.

Energy costs are further influenced by life expectancy of the various devices, which in the absence of any dependable data have been assumed in the above calculation to be 10 years for devices using flexible membranes, 15 years for other tethered devices, 20 years for fixed platform devices and 50 years for devices using existing hydro technology. A further economic constraint is imposed by the variability of wave climate which may result in little or no firm capacity value for standalone installations producing electricity.

Due to immature technology and incomplete cost data from various technology vendors, the results indicated above must be regarded as tentative. However, it appears that under existing conditions, the potential for increased wave power is limited. In many areas with the most favourable wave climates, ample energy is available at moderate cost. For example, Australia has cheap coal, New Zealand has considerable hydro and geothermal power potential, and Britain has coal and nuclear power.

Competitive Status

Market areas in proximity to wave energy resources cover the full spectrum from those served by a well-developed, low-cost electric grid to isolated loads served by diesel generation. Comparisons are made on the basis of lifetime levelized constant dollar costs in 1989 US dollars. Four levels of cost from conventional sources are assumed, as defined in "Basis for economic comparisons" section below. Three levels relate to coal-fired thermal generation with various degrees of emission control and the fourth to diesel generation. Costs displaced are considered to be energy costs, plus one-third of capacity costs for coal, on the assumption that where the competition is coal a well-developed utility system exists, in which case even an intermittent source has some capacity value.

Wave power costs are based on three wave climates and on two levels of capital cost; the level defined above by Hagerman and Heller data and a 25% reduction therefrom. Results are shown in Table 6.5.

Table 6.5 *Wave energy market evaluation*

Wave power (kW/m)	*Cost level %*	*Competing source*			
		Diesel	*Coal Level 3*	*Coal Level 2*	*Coal Level 1*
10	100	N	N	N	N
	75	N	N	N	N
30	100	N	N	N	N
	75	C	C	N	N
50	100	C	N	N	N
	75	C	C	N	N

Notes:
C = wave power competitive.
N = wave power not competitive.
Level 1: no air pollution control.
Level 2: acid gas emission control.
Level 3: enhanced emission control.

The competitive balance depends heavily on the degree to which costs of atmospheric pollution are internalized to conventional sources of electricity. However, even under present conditions, wave energy promises to be capable of displacing diesel fuel in isolated areas and in developing countries where even modest wave climates exist. In that role it would have to compete with other renewables, particularly wind and solar photovoltaic power.

Projected Utilization

In estimating the possible market penetration of wave energy under a favourable scenario, it is assumed that a moderate level of support is maintained for research, development and demonstration over the next decade, by which time the technology will have matured to a reasonable extent and equipment of demonstrated dependability and longevity will be available.

In that case, wave energy would constitute a modular source well suited for the displacement of electric energy from diesel or other sources in isolated areas. It is assumed the islands of the Pacific and Caribbean would constitute a primary market area; that electric use is about 2 kWh per day per capita; that half this energy can be displaced; and that wave energy can capture one-third of the displacement market by 2020.

It is assumed that a second market niche, water desalination, is also penetrable by wave energy; that the market area for desalination includes windward coasts in arid regions as well as islands, and that the market area is therefore twice as large as for displacement energy; that water requirements are about 40 litres per day per capita; that by 2020, increasing populations will result in a need for 50% of the demand to be desalinated and that wave energy can capture one-third of the market.

If continuing research and development result in wave energy becoming cheap enough to displace thermal energy in utility systems (an improbable scenario, not further considered), utilization would be at least an order of magnitude higher than indicated above.

Under an unfavourable scenario, it is assumed that RD&D support is minimal, no further internalization of environmental costs occurs, and no further gains in cost-effectiveness are made. Wave energy would probably still be able to displace diesel in isolated areas with very good wave climate, but market penetration would be comparatively low, both for electricity and desalination.

These assumptions result in the estimates presented in Table 6.6.

Table 6.6 *Wave energy utilization estimates through 2020 (in PJ per year)*

	Less favourable			*More favourable*		
	2000	*2010*	*2020*	*2000*	*2010*	*2020*
1 North America	–	–	–	–	0.2	0.6
2 Latin America	–	0.5	1.7	1.0	6.0	12.8
3 Western Europe	–	–	–	–	1.1	2.7
4 CIS & E Europe	–	–	–	–	0.1	0.3
5 N Africa & Mid East	–	–	–	–	0.9	1.9
6 Sub-Saharan Africa	–	–	0.1	0.1	2.2	4.6
7 Pacific/China	–	0.5	1.6	0.9	5.0	10.7
8 Central & S Asia	–	–	0.2	0.1	3.5	8.2
Totals	–	1.0	3.6	2.1	19.0	41.8

Conclusions

- Wave energy technology is immature from several points of view. It is not possible to judge with any degree of certainty whether the types of devices already developed include the best technical approach or whether something better remains to be found. There is insufficient experience to predict the longevity of present designs in actual service. There is insufficient experience to predict operating and maintenance requirements or to minimize them through appropriate design.
- Under the circumstances, significant improvements in cost and overall performance can be expected as the technology matures.
- The time to maturity will be affected by the availability and consistency of appropriate support for research and development. The risks at the present stage of development are high and the potential rewards too far in the future to attract the necessary investment only from private forces.
- The time to maturity will also be affected by the availability of support for demonstration projects, as well as incentives for early commercial projects.

- Estimates of future market penetration are now necessarily tentative, but it appears likely that wave power could, under a moderately favourable scenario, supply over 12 TWh (42 PJ) per year by 2020. The corresponding figures for a worst case scenario would be 1 TWh (4 PJ) and for a best case (but improbable) scenario, 100 TWh (360 PJ). These values correspond to 2.5, 0.2 and 22.3 Mtoe, respectively.

OCEAN THERMAL ENERGY

Resource Characteristics

Ocean thermal energy exists in the form of a temperature difference between the warm water at the ocean surface and the cold water of the deep ocean. In most tropical and sub-tropical areas, the temperature difference between the surface and water at 1,000 m depth exceeds 20°C, regarded as the minimum difference for practical energy conversion. Thus the energy resource has an extent of nearly 60 million km^2 and a sustained power output capability of many terawatts.

The exploitable output is much lower because many areas are too remote and also because the extraction process is limited to very low efficiency by the laws of thermodynamics. But even after due allowance for these factors, the potential output is very large. Moreover, the areas of greatest temperature gradient coincide with the location of many developing countries and constitute for them an indigenous energy resource. The distribution of thermal gradients is indicated in Figure 6.5.

Ocean Thermal Technology

Ocean thermal energy conversion (OTEC) plants can operate on open, closed or hybrid cycles. In the closed cycle, warm surface water is used

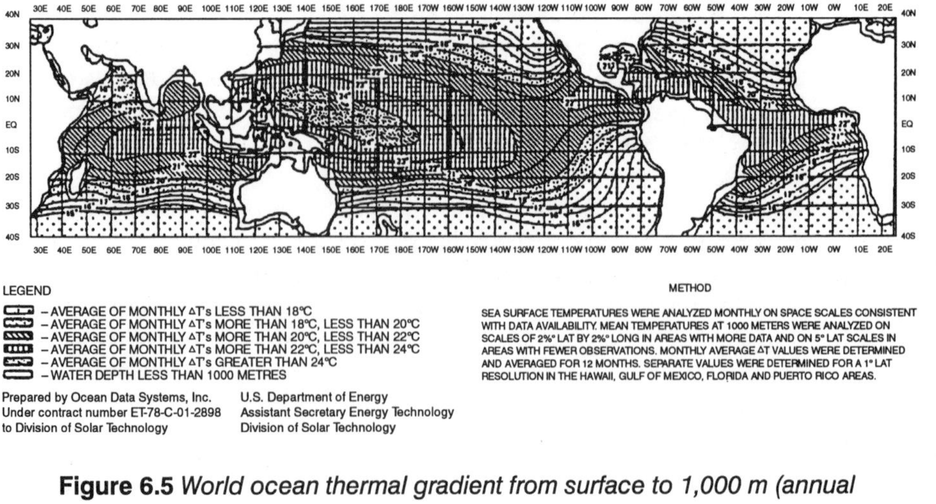

LEGEND
– AVERAGE OF MONTHLY ΔT's LESS THAN 18°C
– AVERAGE OF MONTHLY ΔT's MORE THAN 18°C, LESS THAN 20°C
– AVERAGE OF MONTHLY ΔT's MORE THAN 20°C, LESS THAN 22°C
– AVERAGE OF MONTHLY ΔT's MORE THAN 22°C, LESS THAN 24°C
– AVERAGE OF MONTHLY ΔT's GREATER THAN 24°C
– WATER DEPTH LESS THAN 1000 METRES
METHOD
SEA SURFACE TEMPERATURES WERE ANALYZED MONTHLY ON SPACE SCALES CONSISTENT WITH DATA AVAILABILITY. MEAN TEMPERATURES AT 1000 METERS WERE ANALYZED ON SCALES OF 2½° LAT BY 2½° LONG IN AREAS WITH MORE DATA AND ON 5° LAT SCALES IN AREAS WITH FEWER OBSERVATIONS. MONTHLY AVERAGE ΔT VALUES WERE DETERMINED AND AVERAGED FOR 12 MONTHS. SEPARATE VALUES WERE DETERMINED FOR A 1° LAT RESOLUTION IN THE HAWAII, GULF OF MEXICO, FLORIDA AND PUERTO RICO AREAS.
Prepared by Ocean Data Systems, Inc.
Under contract number ET-78-C-01-2898
to Division of Solar Technology
U.S. Department of Energy
Assistant Secretary Energy Technology
Division of Solar Technology

Figure 6.5 *World ocean thermal gradient from surface to 1,000 m (annual mean)*

to evaporate a working fluid such as ammonia or Freon. The vapour flows through a turbine and is then cooled and condensed by cold water pumped up from the ocean depths. In the open cycle, the surface water is itself the working fluid. It is evaporated at a pressure less than its vapour pressure, passes through the turbine and is then cooled and condensed in the same way as Freon in the closed cycle. In both cases, condensation creates the vacuum which impels the vapour, turning the turbine and its connected generator.

The operating cycle, whether open or closed, is similar to that of a conventional thermal generating station, with the difference that the operating temperature is lower and there are no fuel bills. The heat of the surface water is used instead of the heat of combustion. An unfortunate, but inevitable, consequence is that the theoretical maximum efficiency is very low: 6.8% with a 20°C temperature difference and 9.0% at a 27°C temperature difference. Attained efficiencies after providing auxiliary loads such as pumping are in the 2.5% to 4% range.

Both open or closed cycle plants can be mounted on a ship or built on shore. The ship option requires an underwater power cable or manufacture of a transportable product, while the shore option requires a longer cold water pipe, which may have to traverse steep underwater slopes.

Unlike wave and tidal energy, OTEC is a source of firm power. Apart from maintenance requirements, such plants can operate indefinitely and are well suited to base-load generation. The process has other important advantages in its ability to create by-products. The open cycle is an inherent producer of fresh water: the condensed vapour is nearly salt-free and can easily be kept separate from the cooling water. In both open and closed cycles the cooling water, taken from the ocean depths, is nutrient-rich and can be used for mariculture.

Use of ocean thermal energy was first proposed by the French physicist d'Arsonval in 1881, and in the 1930s one of his students, Claude, built a test-scale plant in Cuba. The process did not attract any further attention until the 1970s, when rising oil prices and perceptions of resource scarcity triggered serious development efforts. Since that time, and partly due to advances in other fields of engineering, progress has been made to the point that prototype plants are in operation and there is confidence that OTEC will become a viable and competitive energy source in certain situations.

Research and development at first concentrated on the feasibility of various design options, on engineering aspects of floating and submerged platforms and on components. Attention tended to focus on floating, closed cycle plants and first net power (15 kWe) was produced by a 50 kWe test plant of this type off Hawaii in 1979. The Tokyo Electric Power Co (TEPCO) built and operated a 100 kWe gross shore-based closed cycle plant in the Republic of Nauru. It had a net output of 15 kWe.

Interest in open cycle plants has developed because of their inherent ability to produce desalinated water as a by-product. USA research on open cycle OTEC has led to the definition and experimental evaluation of spout evaporators and both direct and contact surface condensers.

Significant technical progress has been achieved on heat exchangers, power transmission, cold water intakes, corrosion control and anti-fouling measures.

Variations of the basic operating cycles have also been researched. One open cycle variation is the "mist-lift" cycle. Warm water is flash-evaporated at the bottom of a large vertical tube. The warm vapour rises; is trapped and condensed at the top of the tube; and is thence discharged back to sea level through a hydraulic turbine. A variation combining aspects of both open and closed systems is called "latent-heat-transfer-closed-cycle-OTEC" or hybrid cycle. Warm water is first flash-evaporated as in the open cycle. However, the vapour is then condensed against ammonia, which evaporates and is used to drive a turbine[26].

Development Trends and Possibilities

Present or planned projects by the eight countries active in OTEC research and development include the following:

- In view of favourable results from its 100 kWe plant at Nauru, TEPCO is planning a 20 MWe plant on the same island. The Japanese government is designing a 10 MWe floating plant and is also considering a land-based plant. The government is also researching a combination of OTEC and mariculture using a 5 kWe plant.
- Feasibility studies for a 10 MWe floating plant in the Dutch Antilles and a 100 kWe land-based plant for Bali were completed by the Netherlands, but funding on OTEC has now been terminated.
- After an R&D programme and feasibility studies, France decided to build a 5 MWe land-based pilot plant in Tahiti. Problems incurred in laying the cold water pipe postponed the intended 1988 project completion date.
- Taiwan has plans for a 9 MWe unit to utilize discharged cooling water from a nuclear plant at Hon Tsai. This would provide a high temperature difference for the OTEC plant and reduce environmental impacts of the cooling water discharge. No commitment has been made to date. Preliminary study of a multiple product ocean project has been funded. The project would involve a 5 MWe OTEC plant combined with mariculture and recreation facilities at Ho Ping on the east coast of Taiwan.
- A United Kingdom firm has completed designs for a 10 MWe closed cycle plant to be installed in the Caribbean or Pacific. Also, another UK firm has designed a 500 kWe closed cycle land-based plant combining fresh water production and mariculture activities, to be installed in Hawaii.
- A United States firm is proceeding with conceptual design of a 100 MWe closed cycle plant for Puerto Rico and a government-supported design for a 40 MWe plant for Oahu, Hawaii has been carried beyond the conceptual stage.

Commercialization of both closed and open cycle OTEC plants will require demonstration of technical performance, reliability and cost effectiveness, and this is likely to be the main thrust of future development activities.

Environmental Considerations

Studies on the environmental impacts of OTEC have identified a number of potential concerns. Of these, the majority relate to the necessary intake and expulsion of large volumes of water. It is anticipated that biota including eggs, larvae and fish species could be entrained and destroyed. The entrainment of biota may be an operational as well as an environmental problem, perhaps controllable by appropriate siting of intakes. Changes in local temperature and salinity might also affect the local ecosystem, impact coral, and influence larger-scale processes such as ocean currents, temperature and climate.

Release of carbon dioxide from warm surface water to the atmosphere is a possibility, particularly from open cycle systems, but under worst-case conditions would be only 1/15 that of oil or 1/25 that of coal. It could be reinjected into the warm water discharge. Use of biocides to prevent fouling is cited as a source of pollution, but releases (of chlorine) can be held to about 1/10 the limit allowed by the USA Environmental Protection Agency.

Economic Constraints

Based on present designs, OTEC capital costs are in the vicinity of $10,000/kWe. Because of the general nature of the processes and particularly the absence of high temperatures, it is anticipated that plants can easily be designed for unattended operation and that operating costs will be low. Constant dollar unit costs would now be in the vicinity of 12 US cents/kWh. At this level, OTEC cannot compete with oil, gas or coal-fired thermal generation, but can compete in isolated areas where electricity is supplied by diesel generators. The cost of electricity from open cycle plants can be lowered considerably where there is a market for fresh water.

Significant capital cost reductions have been forecast. A 1987 OECD report estimates that capital costs can be reduced 25% in 5 years and 67% in the long term[22]. The 1989 WEC *Summary of Energy Resources* predicts a reduction of 40% by 1995[22]. Realization of such figures will require refinement in materials, plant and process design, implying the need for continuing research and development support at adequate levels.

Even if projections of capital cost are fully realized, a time lag can be expected between the identification of cost-reducing changes and their incorporation in OTEC plants. It therefore appears likely that OTEC's ability to compete will remain limited, at least until the end of the century. If in the long term a reduction of 67% were achieved, OTEC would be competitive with oil-fired thermal generation even where no revenue was available from by-products.

In summary, it is concluded that the ability of OTEC to compete with conventional sources of electricity is likely to increase with time, and that the rate of increase will be affected by the level of support accorded to research, development and demonstration. At the present level of costs, it appears that government support will be necessary if prototype and demonstration plants are to be built on an appropriate scale.

Competitive Status

Market conditions for electricity in areas in proximity to OTEC resources can be classified according to the type of generation used to meet marginal system loads. The conventional sources postulated for comparison purposes are the same as those used in the case of wave energy and defined in the "Basis for economic comparisons" section below. The basis of comparison is also the same, except that OTEC is assumed to have an 85% capacity factor and earns full capacity credit. A related consideration is whether a market for fresh water exists. Table 6.7 indicates the cases in which OTEC would be competitive under various assumptions relating to capital cost reduction and interest rates. Comparisons are based on constant-dollar costs and real interest rates. No escalation of oil prices is assumed.

Table 6.7 *OTEC market potential*

OTEC cost reduction (%)	*Water value ($/m³)*	*Competing Source*							
		Diesel		*Coal*[3]		*Coal*[2]		*Coal*[1]	
		W	*N*	*W*	*N*	*W*	*N*	*W*	*N*
0	0.21	C	–	–	–	–	–	–	–
20	0.21	C	C	C	C	–	–	–	–
40	0.21	C	C	C	C	C	–	–	–
60	0.21	C	C	C	C	C	C	C	–
0	0.42	C	–	C	–	–	–	–	–
20	0.42	C	C	C	C	–	–	–	–
40	0.42	C	C	C	C	C	–	C	–
60	0.42	C	C	C	C	C	C	C	C

Notes:
C = OTEC is competitive.
W = Water marketable.
N = No market for water.
[1] No air pollution controlled.
[2] Acid gas emission controlled.
[3] All air pollution controlled.

The results indicate that even at the present state of development, OTEC is capable of displacing diesel-generated electricity, but that considerable reductions in capital cost (or gains in process efficiency) would be necessary to enable OTEC to compete with conventional base-load generation.

Projected Utilization

At present and for much of the study period, the ability of OTEC to compete with conventional sources of energy is projected to depend heavily on the marketability and value of its by-products. Of these, air conditioning, mariculture and desalinated water (in that order) are estimated to exert the greatest leverage on the unit cost of electric output.

Cold water discharge from a 1 MWe OTEC plant can displace 10 MWe of air-conditioning loads. After demonstration of dependability OTEC would be economic at present capital cost levels where substantial air-conditioning loads exist. However, this application would only be practical where deep cold water is available for shore-based plants. Against this background, Pacific islands, the Caribbean and coastal areas of the Gulf of Mexico appear to be the regions of greatest market potential. OTEC can be sized to meet limited electrical demands, in situations where large, efficient conventional base-load plants are not feasible[27, 28, 29].

For this primary market area, recent estimates of potential OTEC penetration are 4 GWe for islands in the displacement of diesel or in multi-purpose applications and an additional 36 GWe if OTEC becomes competitive with utility base-load generation. No similar information is available for other regions, but the potential is considered comparatively small because of the lack of deep water near shore. The above estimates are increased to 5 GWe and 40 GWe respectively to include all regions.

Large additional potential would exist if OTEC costs could be reduced to the lower end of the normal range of conventional electricity costs, for in that case in-situ electrochemical processing based on offshore OTEC installations would be attractive[30]. However, this possibility is remote within the study period and is not further considered.

For purposes of this report, the most favourable scenario assumes that adequate support for OTEC RD&D is provided and that environmental costs of fossil-fueled thermal generation are further internalized, with the result that OTEC becomes a competitive source for grid supply. Under this assumption, OTEC would displace existing diesel, would replace existing coal- or oil-fired thermal generation only at the end of existing plant life, and would serve new loads. Therefore, OTEC market penetration would be time-dependent and only part of the estimated potential could be realized by 2020. Much of the OTEC potential exists in regions where capital is scarce and expensive. The best case scenario assumes that capital necessary to realize economically justified OTEC applications would be made available on reasonable terms.

A less favourable scenario, including minimal R&D support and no further internalization of fossil-fueled generation costs, is also examined. It has many elements of a business-as-usual scenario. Without any supportive policies, by 2020 OTEC may reach the cost reduction levels now predicted to be reached by 2000, in which case market penetration will be limited to multi-purpose applications at the most favourable sites.

The estimated OTEC contribution to energy supplies under each of these scenarios is summarized in Table 6.8. Electric outputs are calculated on the basis of 85% capacity factor. This would only be realizable where other sources of peaking capacity were available. The total contribution by 2020 is 37.4 Mtoe (168 TWh) for the more favourable scenario, and 7.8 Mtoe (35 TWh) for the less favourable scenario. Indirect contributions to energy supplies are also shown in the table. It is considered that most OTEC plants will produce desalinated water as a

Table 6.8 *OTEC energy utilization estimates through 2020 (TWh/yr)*

		North America	*Latin America*	*Mid East/ N Africa*	*Sub-Saharan Africa*	*Pacific/ China*	*S/Central Asia*
Less favourable scenario							
2000	Output	–	–	–	–	–	–
	DSW savings	–	–	–	–	–	–
	A/C savings	–	–	–	–	–	–
	Total	0	0	0	0	0	0
2010	Output	–	2	–	–	4	–
	DSW savings	–	–	–	–	1	–
	A/C savings	–	1	–	–	2	–
	Total	0	3	0	0	7	0
2020	Output	–	8	–	–	14	–
	DSW savings	–	2	–	–	3	–
	A/C savings	–	4	–	–	4	–
	Total	0	14	0	0	21	0
More favourable scenario							
2000	Output	–	1	–	–	1	–
	DSW savings	–	–	–	–	–	–
	A/C savings	–	–	–	–	–	–
	Total	0	1	0	0	1	0
2010	Output	–	10	–	4	10	4
	DSW savings	–	2	–	1	2	1
	A/C savings	–	4	–	1	5	1
	Total	0	16	0	6	17	6
2020	Output	21	26	1	17	32	17
	DSW savings	–	6	–	1	8	3
	A/C savings	–	12	–	8	10	6
	Total	21	44	1	26	50	26

Notes:
1. Western Europe, Eastern Europe and CIS are estimated to have no OTEC potential.
2. DSW savings = energetic equivalent of desalinated water produced.
3. A/C savings = energy saved due to displacement of electricity for air-conditioning.

by-product and that a minority will deliver cold water for air-conditioning. Energy otherwise required to produce the desalinated water is estimated to average 25% of net electric output. Energy savings for air-conditioning are estimated at five times the rated output of plants supplying this by-product.

SALT GRADIENT ENERGY

A great difference in osmotic pressure (equal to a 240 m head) exists between fresh water and seawater. In theory, if this pressure could be utilized, every cubic metre of fresh water flowing into the ocean from a river could generate 0.65 kWh of electricity. A stream with a 1 m^3/sec flow would translate into a power output of 2,340 kWe.

Conceptually, the theoretical head could be developed by allowing the fresh water to flow through a semi-permeable membrane into a reservoir of salt water. The pressure would be sufficient to raise the level of the reservoir 240 m, assuming that the salinity was not reduced in the process. The water could then be released through a turbine to recover the energy. In theory, and supposing all the rivers of the world to be harnessed with devices of perfect efficiency, an output of 2.6 TW would be achieved.

Even more dramatic is the potential of salt deposits. The osmotic pressure between fresh water and brine in the Dead Sea corresponds to a head of 5,000 m, as compared to 240 m for ocean water. The very large deposits of dry salt in salt domes contain energy in an even more compact form. A 1978 paper pointed out that the salt in salt domes represents more energy than the oil contained in even the most productive domes.

A number of investigations were undertaken in the 1970s to find practical ways of reclaiming salt gradient energy. Some of the practical difficulties can be visualized by referring to the simplistic example cited above. The fresh water would in practice dilute the salt water and in order to maintain the salinity gradient, more salt water would have to be introduced into the reservoir. If the process were continuous, the surface level of the reservoir would be 240 m above sea level and it would require a great deal of power to pump the salt water against that head.

Unfortunately, the best practical systems identified by research were very expensive. The reverse electrodialysis process, in effect a salt battery, was proposed to extract energy from brine. Capital cost was reported in a 1978 paper to be US$50,000 per kWe. Reverse osmosis, used to raise water to supply a turbine, was reported to have a prospective cost of 10 to 14 US cents per kWh.

A third process considered technically feasible was based on the different vapour pressures of water and brine. Water was to be evaporated and condensed in salt water, the flow of vapour being used to turn a turbine. This process would produce turbine conditions similar to open cycle OTEC, and would entail machinery of roughly equal cost. However, it would be strategically inferior, consuming fresh water, whereas the OTEC cycle produces desalinated water.

Such results tended to discourage further work and salt gradients have not attracted research efforts in recent years. It is concluded that other forms of ocean energy are likely to be more rewarding targets for development efforts.

BASIS FOR ECONOMIC COMPARISONS

In this section, the economic viability of ocean energy developments is tested by cost comparison with conventional energy sources. Costs are estimated in constant United States dollars of 1989 purchasing power and levelized over plant life. Comparisons are made in terms of unit (per kWh) costs. This is the method used by the OECD Nuclear Energy Agency for comparison of coal and nuclear generation and it is appropriate where the options being compared have different exposures to inflation or fuel cost escalation. It should, however, be emphasized that the costs so computed are not actual costs; essentially they are measures of comparative merit.

Coal-fired thermal generation is selected as the typical conventional source in those regions where well-developed utility systems exist. This choice reflects both the expectation that coal will continue in widespread use through the study period and the fact that it is readily available in all coastal zones served by ports.

Three cost levels, corresponding to varying levels of environmental impact, are considered:

- pulverized coal, no flue gas desulfurization;
- fluidized bed with limitation of acid gas emissions;
- limitation of both acid gas and CO_2 emissions.

In areas not served by well-developed utility systems, the existing source is assumed to be diesel.

Both capital and operating costs of coal-fired generation vary from country to country. For present purposes United States costs, about average for the 16 countries included in OECD estimates, are used[31]. Based on a real interest rate of 5%, 150 MWe unit size and a constant dollar coal cost of $2.40 per MJ, coal-fired electricity costs are estimated to be:

Degree of environmental protection	*Cost in US cents per kWh*		
	Variable	*Fixed*	*Total*
Level 1	2.7	1.5	4.2
Level 2	3.2	1.8	5.0
Level 3	5.1	3.2	8.3

At a diesel fuel cost of US$4.77/MJ (US$5.03/MBtu), diesel-produced electricity is estimated to have a variable cost of 7.0 US cents per kWh and a total cost of 9 cents per kWh.

It is appropriate to compare wave and unre-timed tidal energy with the variable costs of production from conventional sources. However, studies have indicated that while such sources have no firm capacity, they nevertheless have some capacity value in an integrated system. The capacity value would be considerable for wave devices located in a good wave climate. Thus the basis of comparison is conservative.

SUMMARY AND CONCLUSIONS

The oceans of the world contain vast quantities of energy in the form of tides, waves and thermal gradients. The technical feasibility of extracting this energy has been firmly established and demonstrated.

A salt gradient exists between the oceans and inflowing fresh water from the world's rivers. This gradient represents an additional source of energy. While extraction of salt gradient energy is possible in theory, the practicality of doing so has not been established. The prospects for salt gradient energy do not warrant further consideration for the near future.

Extraction methods for wave, tide and ocean thermal energy are highly capital-intensive. This is a consequence of low energy densities typical of ocean energy, and while there is scope for reduction of capital costs through technical improvements, the capital-intensive character of extraction technologies is inherent. In consequence, the developable resource potential is severely restricted.

Wave, tidal and ocean thermal energy resources are for all practical purposes non-depletable, non-polluting and, while not devoid of adverse impacts, environmentally benign in so far as they do not burn hydrocarbons. Tidal power schemes involving the building of barrages in estuaries do, however, have such wide-ranging adverse environmental impacts on natural habitats that they are generally unattractive on this score. Schemes which avoid the building of barrages and creation of headponds may demonstrate in future that estuarial tidal power still has considerable potential. The potential social benefits justify continued development support for these ocean energy resources.

The total estimated contribution of ocean energy resources is summarized in Table 6.9. The estimated cumulative investment necessary to support this contribution by 2020 is given in Table 6.10.

Conclusions – Tidal Energy

The following conclusions were reached concerning tidal energy:

- The availability of tidal energy is limited to areas where large tidal ranges occur.
- Tidal technology is relatively mature, but there is still scope for improvement of cost-effectiveness in the conventional approach to energy extraction. Investigation of unconventional approaches warrants limited support.
- The competitive status of tidal power is site-specific in many respects, but depends generally on the level of interest rates and the

extent to which the social costs of power generation from hydrocarbon fuels are internalized.

- Projections of development indicate an output of 12 TWh (43.2 PJ) per year from tidal plants by 2020 AD under a scenario based on present economic conditions and treatment of social costs, or 60 TWh (217 PJ) per year under a favourable scenario. These equate to 2.7 Mtoe and 13.5 Mtoe, respectively.
- Such schemes, where they involve the building of barrages and significant destruction of the natural habitat, would have unacceptable impacts for those who take a consistent and comprehensive environmental view. If such views prevail then tidal barrage schemes will not play a significant role in reaching the potential indicated here.

Conclusions – Wave Energy

The following conclusions were reached concerning wave energy:

- Wave energy is available on coasts with good exposure to prevailing winds through most of the tropic and temperate zones.
- Wave energy technology is immature and there is considerable scope for improvement in cost-effectiveness. Realization of such improvements will require a level of continuing support for research, development and demonstration.
- The contribution of wave energy to world energy supplies will depend on the availability of adequate support thorugh the next decade as well as on economic conditions and treatment of social costs.
- Projections of development indicate annual output of wave energy at 1 TWh (4 PJ) under a worst case scenario, 12 TWh (42 PJ) under a favourable scenario, and 100 TWh (360 PJ) under a best case (but improbable) scenario. These equate to 0.2, 2.5 and 22.3 Mtoe, respectively.
- Care will be required to conserve local biota and minimize fish mortality.

Conclusions – Ocean Thermal Energy

The following conclusions were reached concerning ocean thermal energy:

- Ocean thermal energy is potentially available through most tropic and sub-tropic regions.
- Technology is immature and there is considered to be significant scope for improvement in cost-effectiveness. Realization of such improvements will require a level of continuing support for research, development and demonstration.
- Commercialization of OTEC will require demonstration of technical performance, reliability, cost-effectiveness and avoidance of unacceptable environmental impacts.

Table 6.9 *Potential utilization of ocean energy through 2020 (Mtoe/yr)*

		Less favourable case			*More favourable case*		
Region	*1990*	*2000*	*2010*	*2020*	*2000*	*2010*	*2020*
Tidal energy							
North America				0.8		0.8	3.8
Latin America							
Western Europe	0.1	0.1	0.1	0.5	0.1	4.2	4.3
CIS & E Europe			0.6	1.4		1.4	4.7
Mid-East & N Africa							
Sub-Saharan Africa							
Pacific/China							
Central & S Asia						0.2	0.7
Sub-total	0.1	0.1	0.7	2.7	0.1	6.6	13.5
Wave energy							
North America							
Latin America				0.1	0.1	0.4	0.8
Western Europe						0.1	0.2
CIS & E Europe							
Mid-East & N Africa						0.1	0.1
Sub-Saharan Africa						0.1	0.3
Pacific/China				0.1	0.1	0.3	0.7
Central & S Asia						0.2	0.5
Sub-total	0	0	0	0.2	0.2	1.2	2.6
Ocean thermal energy							
North America							4.7
Latin America			0.7	3.1	0.2	3.6	9.8
Western Europe							
CIS & E Europe							0.2
Mid-East & N Africa							0.2
Sub-Saharan Africa						1.3	5.8
Pacific/China			1.6	4.7	0.2	3.8	11.1
Central & S Asia						1.3	5.8
Sub-total	0	0	2.3	7.8	0.4	10	37.6
Total ocean energy							
North America	0	0	0	0.8	0	0.8	8.5
Latin America	0	0	0.7	3.2	0.3	4	10.6
Western Europe	0.1	0.1	0.1	0.5	0.1	4.3	4.5
CIS & E Europe	0	0	0.6	1.4	0	1.4	4.9
Mid-East & N Africa	0	0	0	0	0	0.1	0.3
Sub-Saharan Africa	0	0	0	0	0	1.4	6.1
Pacific/China	0	0	1.6	4.8	0.3	4.1	11.8
Central & S Asia	0	0	0	0	0	1.7	7
Total	0.1	0.1	3	10.7	0.7	17.8	53.7

Note:
Sources of less than 0.1 Mtoe/year are not included in this table. Such sources existed in 1990 in North America and Central/South Asia for tidal, and Western Europe for wave.
1 Mtoe = 4.487 TWh.

Table 6.10 *Estimated cumulative investment (US$bn) for ocean energy projections*

	Less favourable scenario			*More favourable scenario*		
Region	*2000*	*2010*	*2020*	*2000*	*2010*	*2020*
1. North America	–	–	2.5	–	2.5	21.4
2. Latin America	–	2.4	6.8	1.3	9.3	16.7
3. W Europe	–	–	1.6	–	14.9	15.6
4. Former USSR & E Europe	–	–	2.7	–	3.8	12.6
5. Mid-E & N Africa	–	–	–	–	0.2	0.9
6. Sub-Sahara	–	–	–	–	3.7	10.1
7. Pacific	–	4.4	11.6	1.3	9.1	19.4
8. Central & S Asia	–	–	–	–	4.7	13.3
Total	0	6.8	25.2	2.6	48.2	110.0

- At the present state of development and subject to satisfactory demonstration, OTEC is considered capable of displacing diesel generation in isolated areas and of providing the cheapest method of water desalination.
- Significant capital cost reductions and/or improvements in operating efficiency would be required to attain competitiveness with conventional base-load generation.
- Projections of development indicate annual output of OTEC energy will reach 35 TWh (126 PJ) per year by 2020 under a less favourable scenario, and 168 TWh (605 PJ) under a more favourable scenario. These totals include energy savings due to by-products (desalinated and chilled water), and equate to 7.8 and 37.4 Mtoe, respectively.
- Potential interference with the natural CO_2 absorption rates of the oceans would need to be kept under review if OTEC developments were to proceed on a large scale.

Long-term (Post 2020) Potential for Ocean Energy

Some favourable tidal sites exist in remote areas; for example the Secure Bay in Western Australia, Ungava Bay and Baffin Island in Northern Canada. Growth of population, electrical demand and transmission systems could be expected to make such sites more attractive at some time during the 21st century.

The projections indicate that only a small part of the wave potential in coastal areas will be developed by 2020. Given cost-effectiveness and demonstrated longevity, wave devices could steadily increase market share up to five or ten times the projected 2020 levels.

The favourable scenario shows OTEC at about one-third of its estimated potential for coastal deployment by 2020. Under continuing favourable circumstances, it would be reasonable to expect a gradual increase to 50% of potential over the following 30 years. The potential would be much larger should OTEC ever become competitive for electrochemical processing.

ACKNOWLEDGEMENTS

This chapter was prepared by the Ocean Energy Subcommittee of the WEC Study Committee on Renewable Energy Resources, under the leadership of Dr. D.L.P. Strange of Canada. Major authors included T. Tung and G.C. Baker, also of Canada. Contributions and reviews were also provided by G. Hagerman and L.F. Lewis of the USA and R.H. Clark of Canada, but many others from a variety of countries provided comments and helpful suggestions. The help of all these people is gratefully recognized.

REFERENCES

1. Thomas B. Johansson et al. *Renewable Energy: Sources for Fuels and Electricity*, Island Press, Washington DC 1993, p. 513.
2. M. Rodier (Electricité, de France) in R. Clare (Ed). Tidal Power: Trends and Developments. *Proceedings of the 4th Conference on Tidal Power*, Institution of Civil Engineers, London, 19–20 March 1992, Thomas Telford, London, 1992.
3. *1986 Survey of Energy Resources.* London, UK, World Energy Conference, 1986.
4. L.B. Bernshtein. *Tidal Power Plants.* Energoatom Publishing House, 1987.
5. *Feasibility of Tidal Power Development in the Bay of Fundy.* Ottawa, Canada, Atlantic Tidal Power Programming Board, 1969.
6. *Assessment of Retiming Tidal Power from the Bay of Fundy Using Compressed Air Energy Storage.* Prepared by Canadian Atlantic Power Group Limited for Energy, Mines & Resources Canada, Ottawa, Canada, 1988.
7. R. Gibrat. L'Usine Maremotrice de La Rance. *Revue Française de l'energie.* No. 74, 1957, pp. 234–341.
8. X. Cheng. *Tidal Power in China.* Beijing, China, Ministry of Water Resources and Electric Power, 1986.
9. M. Banal and A. Bichon. *Tidal Energy in France: The Rance Estuary Tidal Power Station – Some Results After 15 years of Operation.* Cambridge, England, Paper K3, Second Symposium on Wave and Tidal Energy, September 1981.
10. G.C. Baker. *The Competitive Status of Fundy Tidal Power.* ECE Symposium on the Status and Prospects of New and Renewable Sources of Energy, Sophia Antipolis, France, 1987.
11. D.J. Solomon. *Fish Passage Through Tidal Energy Barrages.* Harwell, Oxfordshire, UK, ETSU TID 4056, Energy Technology Support Unit, 1988.
12. *Tidal Power from the Severn Estuary.* (2 vols.) London, UK, Energy Paper No. 46, Her Majesty's Stationery Office, 1981.
13. *Fundy Tidal Power Update.* Halifax, Canada, Tidal Power Corporation, 1982.

14. *Fundy Tidal Power Stage 1.* Halifax, Canada, Tidal Power Corporation, 1985.
15. *The Severn Barrage Project: General Report.* London, UK, Her Majesty's Stationery Office, 1989.
16. *Tidal Power from the River Mersey: A Feasibility Study.* Harwell, UK, ETSU TID 4047, Energy Technology Support Unit, 1988.
17. *Annual Report of the Tidal Energy R&D Programme 1988/89.* Harwell, Oxfordshire, UK, Energy Technology Support Unit, (Confidential), 1989.
18. J.M. Leishman and G. Scobie. *The Developments of Wave Power.* NEL Report No. EAU M25, 1976.
19. M.E. McCormick and C. Young Kim (Eds.). *Utilization of Ocean Waves – Wave to Energy Conversion.* Proceedings of an international symposium, New York, NY, USA, American Society of Civil Engineers, 1986.
20. T. Lewis. *Wave Energy – Evaluation for CEC.* London, UK, published by Graham & Trotman Ltd, 1983.
21. F.H.Y. Wu and T.T.L. Liao. *Wave Power Development in Taiwan.* ICOER '89, Hawaii, USA, 1989.
22. *Renewable Sources of Energy*, Paris, France, OECD/IEA, 1987.
23. P.G. Davies (Ed.) et al. *Wave Energy.* Harwell, UK, Report No. ETSU R26, Energy Technology Support Unit, Her Majesty's Stationery Office, 1985.
24. *1989 Survey of Energy Resources.* London, UK, World Energy Conference, 1989.
25. G. Hagerman and T. Heller. *Wave Energy: A Survey of Twelve Near-term Technologies.* Proceedings of the International Renewable Energy Conference, Hawaii, September 1988.
26. *Ocean Thermal Energy Conversion – An Overview.* Golden, CO, USA, Solar Energy Research Institute, 1989.
27. *The Potential of Renewable Energy.* An Interlaboratory White Paper for the USA Department of Energy, 1990.
28. *Comprehensive Ocean Thermal Technology Application and Market Development (TAMD) Plan.* (6th annual update) Washington, DC, USA, US Department of Energy, 1989.
29. *US Department of Energy Ocean Energy Technology Programme: Economic Viability of Ocean Thermal Energy Conversion.* Washington, DC, USA, US Department of Energy, 1990.
30. D.E. Lennard, *Prospects and Potential for Ocean Thermal Energy Conversion.* London, UK, World Energy Conference, 1986.
31. *Projected Costs of Generating Electricity From Nuclear and Coal-fired Power Stations for Commissioning in 1995.* Paris, France, Nuclear Energy Agency, OECD, 1986.

CHAPTER

7

Small Hydropower

INTRODUCTION

Mankind has used the energy of falling water for many centuries, at first in mechanical form and since the late 19th century by further conversion to electrical energy. Historically, hydropower was developed on a small scale to serve localities in the vicinity of the plants. With the expansion and increasing load transfer capability of transmission networks, power generation was concentrated in increasingly larger units and to benefit from the economies resulting from development on a larger scale.

Sites selected for development tended to be the most economically attractive; in this regard, higher heads and proximity to load centres were significant factors. For this reason, development was not restricted to large sites, and hydro stations today range from less than 1 MWe capacity to more than 10,000 MWe. The efficiency of hydroelectric generation is more than twice that of competing thermal power stations.

While definitions in common use vary, small hydro is defined for purposes of this section as including all plants of capacity 10 MWe or less[1]. Small hydro is thus taken to include the categories mini- and micro-hydro, which are usually confined to strictly local use but are statistically indistinguishable from the balance of the small hydro capacity range.

River courses suitable for accommodating hydro plants of up to 10 MWe capacity are common. There is also a considerable demand throughout the world for increased electricity supplies in rural and outlying areas; only about 5% of the population living in such areas in the developing countries are thought to have access to an electricity supply. Progress in introducing small hydro plants into rural and outlying areas has been slow. But concern over pollution from thermal plants plus shortened construction time and lower development cost by independent power producers have led to renewed interest in small hydro in the industrialized countries. This chapter examines the future potential of small hydro as an essential component in the exploitation of renewable energy resources.

TECHNOLOGY AND DEVELOPMENT STATUS

The Status of Hydropower

It is appropriate to review first the present situation in the hydropower sector as a whole, bearing in mind that small hydro occupies the lower end of the capacity scale. In 1990, hydro plants accounted for some 22.9% of the world's total installed electric generating capacity, but the combined output was only 18.4% of the world electricity supply total.

Most hydro plant designs include considerably higher generating capacity than that which would be needed to handle average water flows. An appreciable proportion of plants connected to transmission grids are designed for peaking service, and in all cases extra generation capacity is necessary to accommodate river flow variations and reduce spillage. This has the effect of reducing the average annual hydro generation plant factor or capacity factor to 39.3%, compared with an average of 51.6% for other conventional power plants. Statistics for regional hydro capacity and production in 1991, published in *Water Power and Dam Construction* (August 1992), are given in Table 7.1[2].

Table 7.1 *Capacity and production of global hydropower*

Region	*Capacity (GWe)*	*Production (TWh)*
North America	133.7	579.8
Latin America	94.0	390.0
Western Europe	136.7	405.3
E Europe and CIS	82.3	260.2
Mid East and N Africa	13.1	40.2
Sub-Saharan Africa	16.5	45.1
Pacific	12.1	38.7
China	37.9	124.8
Asia	100.7	397.4
Total	627.0	2,281.2

Industrialized countries account broadly for two-thirds and developing countries for one-third of present hydro production. Based on statistics provided in the 1992 *Handbook of Water Power and Dam Construction*[2], and supplemented where possible by the survey information obtained by Russian analysts for this WEC committee, the statistics for regional small hydro (under 10 MWe) capacity and production are given in Table 7.2. It is estimated that the installed capacity of small hydro is about 19.5 GWe in 1990, or 3.1% of world hydro capacity. Annual output for small hydro is 81.7 TWh. These estimates indicate that small hydro accounts for 3.8% of hydro electricity production.

Table 7.2 *Capacity and production of small hydro in 1990*

Region	*Capacity (MWe)*	*Production (GWh)*
North America	4,302	19,738
Latin America	1,113	4,607
Western Europe	7,231	30,239
E Europe and CIS	2,296	9,438
Mid East and N Africa	45	118
Sub-Saharan Africa	181	476
Pacific	102	407
China	3,890	15,334
Asia	343	1,353
Total	19,503	81,709

Characteristics and Costs

Small hydro technology is mature and proven. Construction is straightforward and involves simple processes which offer opportunities for a large degree of local participation, both in terms of labour and materials. Construction lead times are short. Already established and proven design concepts offer considerable scope for adaptation to local circumstances, both in construction and operation (the latter may range from simple manual attention to fully automatic and computerized systems).

The main civil works of a small hydro development are the dam, spillway or diversion weir, and the water passages to the powerhouse. The dam directs the water into the powerhouse through water passages. The powerhouse contains the turbine with the mechanical and electrical equipment required to transform the potential and kinetic energy of the water into electrical energy.

New small hydro developments are usually run-of-river developments where water is used only as it is available, and with no water storage reservoir. The cost of large dams can rarely be justified for small projects. Therefore, a low dam or diversion weir of the simplest construction is usually built.

Small projects lack the advantage of scale and their cost per installed kWe can therefore be quite high. Present investment costs for projects in the range of 500 to 10,000 kWe can be expected to lie in a range of about US$1,500 to 4,000 per kWe. Higher costs are likely to be justifiable only in special cases. Lower costs might be experienced at particularly favourable sites or where there is an appreciable amount of lower cost local input. Project costs per kWe usually decrease with increasing capacity and increasing head but design parameters are normally defined by local conditions and allow little latitude for change. As it is often quite inexpensive to add a power component to an existing water supply or irrigation scheme, multi-purpose projects could well provide one of the main platforms for the expansion of small hydro in the future.

For smaller plants, there is greater need to concentrate on capital cost reduction, even at the expense of efficiency of operation. Non-essential features are eliminated and local materials are used whenever possible.

Selection of the appropriate control equipment requires a compromise between complexity and cost, and the choice may be constrained by available skills. Achieving satisfactory performance with simple manual control requires relatively high skill by the operator. A high degree of automation, on the other hand, puts the emphasis on adequate maintenance and the procurement of spares, which in turn could entail greater reliance on remote/foreign support and imports. However, it is important to recognize that small hydro capacity spans three categories: micro (less than 100 kWe), mini (100 kWe–1,000 kWe) and small (1,000 kWe–10,000 kWe). At the low end of the range, simplicity of design and control is usually essential for economic viability, while at the high end of the range, the magnitude of the investment is likely to warrant fairly sophisticated protection and control devices.

Seasonal water flow variations can be overcome to some extent by incorporating a storage reservoir in the hydro scheme and adapting water release through the power station to the requirements of the power market. The installed capacity can then be more closely tailored to the electricity demand. For run-of-river projects where no significant water storage exists, the firm capacity will be limited by low-flow water conditions and will only be a small fraction of the installed capacity. In such cases the value of electricity generated is equivalent to the displacement energy value only. Nevertheless, it is usually not economically justifiable to create a storage reservoir solely to serve a small hydro plant.

The economic feasibility of small hydro projects is enhanced if the project is developed to supply a local load. If not, then transmission plays a vital part in hydro projects, especially where it has to be developed specifically for a new scheme. Transmission costs, which then form an integral part of plant investment, will help to inflate the costs of the project.

Environmental Aspects

Hydro schemes may create environmental impacts of various kinds. Such impacts tend to be highly site-specific and to be influenced by the civil design adopted.

Some loss of habitat may occur due to the need for access roads, transmission lines, and the construction of the dam structure. Impacts of this type may be counterbalanced by improvements in resource utilization, recreational or other social benefits. Small hydro plants generally do not displace large numbers of people as large projects sometimes do. At extra cost, overhead transmission lines can be avoided. This is part of the general problem caused by the visual intrusion of overhead transmission lines and towers, which can be relatively acute in remote areas where small hydro has significant technical potential.

Where construction of reservoirs or enhancement of natural storage is involved, water quality may be impacted, siltation may occur and the character of river flow may be altered. Water quality may be affected by leaching of pollutants from newly inundated land, increased bacterial oxygen demand and stratification.

Small hydro does not usually involve the construction of significant storage and in general avoids significant impacts of the types noted above. However, the diversion of water from natural channels and its passage through a turbine is inevitable and may create impacts due to fish mortality or impediments to the movement of migratory fish species. The technological status of fish diversion design and methods for the prediction and determination of mortality leaves much to be desired. Recent experience has shown an order of magnitude variation in the results of mortality tests, depending on techniques used. The effectiveness of fish passages at operating hydro plants and devices to divert fish to them is known to vary widely. This is an area where improvements would be of value in increasing both the environmental and the economic status of hydro developments, both large and small.

In general, the environmental impacts noted above are not significant with small hydro projects, certainly relative to the benefits of the projects or to alternative means of supplying electricity. The impacts of small hydro are usually much smaller than those of large hydro development, but sensitivity to avoid siting near areas of great natural beauty and delicate ecological settings is necessary.

DEVELOPMENT TRENDS

The price escalation of fossil fuels experienced in the early 1970s focused attention on new and renewable sources of power production, among which hydropower is the most prominent. A number of large and medium-sized hydro schemes were developed in the industrialized countries (and to a lesser extent in the developing countries) but even so, only about 18% of the world's exploitable hydro potential has been developed to date.

Hydropower development has been given a new impetus with the realization that fossil fuels, which still supply the bulk of the primary energy used for electricity generation, are not only subject to unpredictable price movements but contribute materially to environmental pollution[4]. Large hydro schemes are commonly associated with large reservoirs which give rise to much environmental concern. There is a growing trend of development toward medium or smaller-sized run-of-river plants with only a small upstream pondage.

There is also a strong global trend towards privatization. The high indebtedness of developing countries leads to private power generation as an effective solution to the need for additional electricity supplies. Medium-sized and small schemes will be more attractive for a private sector developer[5].

Small-scale hydropower has continued to attract attention, and indeed now some prominence, because:

- The long asset life, operational robustness, price stability and its renewable nature makes hydropower an attractive option for an electricity supply to small and isolated load centres.
- Transmission networks connecting power plants and load centres are of restricted extent and there are many areas, particularly in the developing countries, where electrification has to depend on local generation on a small scale.
- Small rivers which can be exploited for power production are fairly widespread.
- The environmental impact of small hydro is usually very slight.
- Other uses of water, irrigation and water supply for example, can often be made more attractive by the incorporation of a small hydro component.
- In the industrialized world, small hydro frequently serves as a power source for local industry, but it can also contribute to the public supply in suitable locations.
- Rehabilitation of old small plants at existing dams or structures is often justified by the favourable cost of the energy produced.

China, where many areas are not served by transmission grids, has the greatest number of small hydro plants, aggregating to 4 GWe for projects in the range of 10 MWe or less[6]. The USA is in second place with some 3.5 GWe. In that country, there is legislation requiring utilities to purchase the output of non-utility-owned plants at avoided cost (as mandated by the Public Utilities Regulatory Policies Act (PURPA)). This triggered a large number of private developments, many of which involved small hydro. Over 733 MWe of small hydro have been installed since 1983[7].

In some European countries, incentives are being used to promote the development of new and renewable sources of energy. Britain uses a stick-and-carrot approach involving a levy on utility thermal generation, which is used to increase the price paid for energy from renewable sources.

In Canada, provincial utilities are establishing site rehabilitation and non-utilities generation (NUG) programmes for encouraging private independent power producers (IPP)[8]. Some progress has been made for displacement of diesel generation in remote communities, including native, provincial and federal parks communities. Native communities are interested in small hydro as an economic development tool.

In terms of technological development, a number of governments and international organizations such as the European Commission have actively encouraged R&D through programmes of demonstration projects. Canada has a co-operative research programme with industry for the development of low-cost manufacturing processes, design, and standardization. As a result of these incentives, the number of private developers has increased, although there is still a great disparity between institutional arrangements within the regions mentioned above. As a consequence, there are strong incentives for private developers in some countries or regions but bureaucratic barriers in others.

In Asia, Africa, and Latin America, where the needs of rural communities can often be met more appropriately by small or micro-hydro schemes, the power generating stations are usually owned by national or local government utilities. Of these hydro schemes, some are operated in isolation and others are connected to local transmission grids. In these regions, international co-operation is playing a major role, as those with long-term experience of small hydro development set up co-operative projects to transfer technology.

In broad terms, it appears that where national policy is favourable (eg provides aid and encouragement to financially viable projects) towards renewables, the pace of small hydro development has been significantly increased.

DEVELOPMENT POTENTIAL

Inventories of small hydro sites are not complete for many areas of the world, nor is the capacity range of individual sites assessed in any reliable way. Various definitions of "small" exist and this exacerbates the problem of determining the present capacity and future prospects. Figures can be quoted to only a very broad degree of approximation.

The total hydro potential is better defined and is here used as a basis for estimates relating to small hydro. The extent to which the currently estimated economically exploitable hydro potential has been developed is shown in Table 7.3. The table points in particular to the very large resources remaining available for future development.

Table 7.3 *Development extent of world hydropower resources*

Region	*Net* exploitable (TWh/yr)*	*Exploited (% of exploitable)*
North America	801.3	72.4
Latin America	3,281.7	11.9
Western Europe	640.9	63.2
E Europe and CIS	1,264.8	20.6
Mid East and N Africa	257.1	15.6
Sub-Saharan Africa	711.2	6.3
Pacific	172.0	22.5
Asia and China	1,165.6	44.8
World total	8,295.0	27.5

Notes:
Based on statistics given in *Water Power and Dam Construction*, August 1992[2], with adjustment of North America figure (Canada). The figures include all hydropower but exclude pumped-storage.
* This is the economically feasible hydropower capability. The technically feasible capability is approximately double the figures quoted here.

Some 59% of the net exploitable hydro resources lie in the developing world. Hydro sources contribute at the present time around 5% of the world's primary energy supply. The share of hydro energy supply in the industrialized countries is not expected to change significantly in the next 25–30 years but is thought to become more prominent in the developing countries, as illustrated in Table 7.4.

Table 7.4 *Estimates of share of hydro sources in primary energy supply (% of primary energy)*

	Industrialized countries	*Developing countries*	*World total*
Year 2005			
Moderate growth	6.0	6.8	6.3
Low growth	5.8	6.2	6.0
Year 2020			
Moderate growth	6.5	9.7	7.7
Low growth	6.2	8.5	7.3

Note:
Based on UN 1989 projection and World Bank sources in 1991.

To improve the position of hydropower in comparison with other energy sources, very considerable investments will be needed and it must also be assumed that environmental constraints can be largely overcome. A recent statistical review by British Petroleum pointed out that, at present, the annual increase in hydro reserves, due to revised inventories, exceeds the annual increase in hydro production.

The current supply of small and large hydro energy dwarfs that of all other forms of renewable energy except biomass. The amount of energy derived from traditional biomass, which includes firewood, is estimated to be about twice that coming from hydro, but hydro contributes about four times the current amount supplied by modern biomass. For the next decade, the growth rate of hydro electricity is anticipated to be around 2.5% per year, and hydropower will remain by far the largest source of renewable energy well into the 21st century. Economic and environmental factors could, however, seriously restrict the realistically exploitable hydro potential.

Figures for the exploitable small hydro potential can be no more than speculative. Small water courses are more common than large ones but the exploitable potential is restricted for technical and economic reasons, as well as by the limited transmission availability between the power source and the load. Many countries which should have significant small hydro resources have yet to develop realistic estimates[9]. A comprehensive estimate of small hydro potential for sites less than 10 MWe is still not available from surveys provided by the periodical *Water Power and Dam Construction* and by the World Energy Council with its *Survey of Energy Resources*[10]. An estimate of 5% of the total hydro potential currently thought to be exploitable is perhaps of the right order of magnitude. This gives a net exploitable (technical) potential of 630 TWh of which just under 10% has been developed. Approximately 3,000 MWe of small hydro capacity is reported to be under development, two-thirds of this in China. Canada has recently completed an inventory of small hydro sites, which identified over 3,600 sites having a total technical potential of about 9,000 MWe. Approximately 15% of this potential was found to be economically viable. An estimate of total global small hydro development in the range of 1,000–2,000 MWe per year would be considered realistic.

CONSTRAINTS TO DEVELOPMENT

Economic and Technical Factors

The high initial capital cost of hydro schemes acts as an impediment to the expansion of hydropower, especially in the developing countries where the funding problems are most acute. Generally, a small hydro scheme is built to meet a small increment of demand from a water resource available within a feasible transmission distance. Apart from the preconditions of technical and environmental feasibility, the scheme has to conform with acceptable economic selection criteria. This limits the scope for small hydro in many places.

Siting and cost considerations tend to discourage the building of substantial reservoirs in association with small schemes. Small hydro is therefore usually a run-of-river development incorporating a small pondage capacity where this is feasible. As a consequence, the output from small schemes is critically dependent on the temporary water run-off and hence subject to the prevailing hydrological cycles. This can cause problems if an isolated small hydro plant is required to meet the year-round demand for electricity. Such a situation may require backup support, leading to substantially increased operating costs and pollution associated with diesel or thermal generation. Such problems will not arise if small hydro serves as a supporting plant in an interconnected network.

The financial attractiveness of small hydro is heavily influenced by the real interest rate at which financing can be obtained. The high real rates which have predominated in some countries over the past two decades have had an inhibiting effect on small hydro development.

Small hydro is still often considered to be a derivation from large hydro and the pre-investment effort (eg, site survey and feasibility study) is dimensioned according to large hydro experience. The pre-investment work, and its cost, can then become excessive and out-of-scale relative to the size and importance of the scheme. This limitation is particularly serious if a first feasibility study is not followed by a firm decision for or against the proposed development, but rather leads to further studies at a later stage, or indeed repeated feasibility investigations, as is quite often the case. As a rule of thumb, the pre-investment work should not cost more than about 10–15% of the ultimate investment and a great effort is needed to contain these costs, even if they are supported by government or foreign aid.

Economic constraints would be minimized and the economically developable potential of small hydro would be increased if a reduction in a project's capital cost could be achieved. Potential avenues to the realization of such reductions include: development of design tools to provide site-specific estimates of water flow characteristics, automation of some aspects of feasibility studies; development of cheap but adequate controls for very small units; and a decrease of costs of prime movers, generators and civil works through design improvements and standardization. Such objectives have been pursued in recent years with some success, but considerable potential for further improvement appears to exist.

Socio-economic Factors

Much attention has been given to the training and technology transfer which would enable developing countries to play a more significant and autonomous role in small-scale hydro development but the results have not always been very positive. Developing countries still tend to rely to a great extent on technical support from abroad; the technology gap is not being bridged as effectively as may have been hoped. This fact, added to the costs of importing foreign expertise, materials and equipment, even if supported by foreign aid, has greatly impeded the more widespread introduction of small plants.

The socio-economic merits of electrification and of local resource exploitation are well established but their quantification is still in its infancy. As a consequence, they do not usually enter into the evaluation of economic merit, and projects which could bring considerable advantages to the local population are in danger of being discarded by conventional economic analysis. Much more thought needs to be given to secondary and less tangible features which could nevertheless revitalize local living conditions.

Although small hydro is of considerable socio-economic importance for the lesser-developed countries, the installed capacities so far are not significant and the outlook is very uncertain. Of 150 schemes judged to be technically and economically feasible in pre-investment studies funded by the United Nations, only two have been built to date. Progress has been impeded not only by infrastructural and developmental considerations but also by funding problems. Funds available for power sector expansion tended to be concentrated on the central urban and industrial areas where they are judged to make the most immediate and measurable impact on national economic progress.

The concentration of national development on the central areas in developing countries and the resulting difficulties of funding electrification programmes in rural and outlying areas have drawn attention to the possibility of greater involvement of the private sector in the promotion of small hydro schemes, not necessarily by commercial developers but local interest groups or co-operatives. Some pilot schemes have been very successful and it is quite possible that small hydro development will find much encouragement from more widespread and direct participation of the potential beneficiaries in an electrification campaign.

Water supply and irrigation will be the key challenges in the future. Massive investment is earmarked for this sector, with a substantial increase in the number of dams and reservoirs to be built. There is thus considerable scope for the inclusion of a power generating component in water supply schemes and hence in multi-purpose development which has been rather stagnant in recent years. Even in the USA, only 5% of the dams over 25 m in height have a facility for power generation. The potential contribution and benefits of combined development can be quite large.

Environmental and Regulatory Factors

In some countries (especially in North America and Western Europe), the environmental regulation and control process has become so cumbersome, onerous and costly for a proponent that it is now an important factor inhibiting small hydro development. Approvals must sometimes be obtained from a number of different agencies with different and potentially conflicting requirements. The cost of dealing with the regulating agencies and the cost of the necessary environmental investigations are formidable for a small project. In addition to such direct costs, the indirect cost of delay must not be overlooked.

Constraints arising from the environmental assessment process could be minimized in two ways:

- Improvement through research of the reliability of environmental measurements and impact mitigation measures. The estimation of fish mortality and the design of fish diversion devices, discussed in the "Environmental aspects" section above, are prime candidates for improvement.
- Simplification of the environmental approval process.

OTHER ISSUES AFFECTING DEVELOPMENT

Experience with Incentive Programmes

The merits of small hydro as a source of cheap and reliable local power are well known and well established. Its renewable and benign nature, especially in comparison with fossil fuels, is recognized and has rekindled interest in the industrialized countries in its potential for supplementing supplies from other sources under environmentally beneficial conditions. Financial support is offered in some European countries for use of renewables in replacing local generation from fossil fuels and greatly enhances its economic advantage for the private developers. If a similar approach should find its way to the developing countries, the prospects for small hydro may be greatly improved.

The thrust of inventive programmes is to ensure that the cost of all energy sources reflect their full cost of production, including clearly identifiable environmental impact costs, so that they can compete on an even basis. The following are some of the latest developments.

United Kingdom

Incentives offered in England and Wales for electricity supply from renewable sources raised the price offered to producers from 2–3 p/kWh to 5–6 p/kWh (3–9 US cents/kWh) for a fixed period of time[11]. Few hydro plants had been built to supply grid power in the years prior to the introduction of incentives. Following the introduction, 26 small hydro schemes with total capacity of 11.9 MWe were accepted in 1990, and a further 12 schemes with total capacity of 10.4 MWe were accepted in 1991. Of these, about one-third were new developments or redevelopments of disused sites. The rest were existing schemes.

The secure market and guaranteed price offered under the incentive programme has increased the economically developable potential and spurred entrepreneurial interest. However, the pace of development is limited by the number of proposals accepted by utilities.

The British House of Commons Energy Select Committee[12] has called on the government to increase its renewables goal for the year 2000 from the present 1,000 MWe to 3,000–4,000 MWe. It also urges the extension of the Non-Fossil Fuel Obligation, which expires in 1998.

European Union

A European Union proposal[13] for a tax on carbon dioxide emissions and energy was submitted to the European Council. Renewable energy sources including small hydro less than 10 MWe would be exempt from tax.

In the meantime, the European Union is developing a programme, known as ALTENER. It calls for adopting the EU's objectives for renewable energy under the following targets by the year 2005:

- Doubling the contribution of renewables to energy supply, to 8% of the total from 4% today.
- Tripling the amount of electricity produced from renewable energy sources (excluding large hydroelectric power stations).
- Increasing the market share of biofuels to 5% of total fuel consumption by motor vehicles.

Co-ordination of Aid Programmes

A number of bilateral and multilateral agencies offer support to developing countries. These range from the identification and exploration of promising sites, to institutional, managerial and technical strengthening of utilities and other developers (including, to an increasing extent, non-governmental organizations). Transfer of technology and skills is intended to improve the capacity of the counterpart organizations in the developing countries to take an active part in establishing small-scale hydropower programmes, to lead these to a successful conclusion and to avoid the pitfalls of faulty conception and operation[14].

This activity has grown considerably in scope and importance during the past 10 years and has led to the formation of well-trained cadres in several countries. It has also resulted in establishing focal points for mutual support and co-operation between developing countries.

Co-ordination of aid programmes in a given country is needed, to avoid duplication of effort, and also the introduction of different designs and machine specifications which make training programmes, operation, maintenance and spare parts inventories difficult to manage. Costs and management problems might be minimized if small hydro programmes are co-ordinated and executed under common management and supervision, even if supported by different sources of aid. Any foreign party providing expertise, equipment and capital on acceptable contractual and financial terms should retain responsibility for the satisfactory functioning of the scheme over a sufficiently long period to ensure stable operation and reasonable cost recovery.

Local manufacture of hydro plant components is advocated by a number of agencies as a means of reducing foreign currency expenditure and to facilitate financing. There are various simple items that can be fabricated in local workshops. In general, smaller plants will lend themselves to a large share of locally made components, because sizes and weights will be more manageable and components will probably be less stressed when in sevice. Unless there is a substantial local market and promise of an expanding hydro programme, the local manufacture of more complex items will not be worthwhile.

Technological Development and R&D

In an era of growing environmental sensitivity, the development of hydropower requires new strategies and improved design concepts to bring out the beneficial features of hydropower and avoid environmental pitfalls. The need for better management of water resources will require more attention being paid to the siting of powerplants and the study of their effect on river systems. Optimum utilization of water resources will call for more extensive investigations before a development plan can be set out.

For mitigation of environmental impact, planning and design need to be less conventional and more readily adapted to the circumstances at particular sites. Rigid adherence to guidelines established elsewhere should be avoided.

The need for greater efficiency in the use of capital, for coping with the increasing scarcity of investment funds and for paying greater attention to environmental and socio-economic issues, has created a new stimulation for a more appropriate technology. Principal advances will be in design, construction methods and materials so that the changing requirements can be adequately addressed. Progress in plant and equipment design will also be made, but perhaps at a slower rate. Schemes of all sizes will benefit from whatever advances are made.

Surveys of hydrological resources, especially small-scale resources, will provide more reliable information on optimum siting of new plants. Computerized data acquisition and handling will facilitate classification and evaluation of the data needed for site selection and environmental impact assessment. Techniques for monitoring water resources and their use will offer more reliable information on hydrology, which is the backbone for hydro generation. Such knowledge is also vital for multipurpose applications. More attention will be paid to comprehensive resource utilization studies covering all purposes, rather than sectoral studies for one particular application only. Scarcity of water will dictate such an approach.

Much progress will be made in the use of dam materials which can make the structure less sensitive to ground conditions and simpler to build. A move away from large reservoirs may cause fewer large dams to be built. Their place will be taken by medium-sized or even small schemes, backed by small reservoirs which have limited environmental impact. Flexibility in power system operation will overcome the disadvantages of only limited blocks of power being available from each scheme. Costs and construction lead times will be significantly reduced.

Attention will focus on improvements in efficiency and reliability, on reduction in cost and on standardization. Other features to be addressed include greater flexibility to accept changes in throughput of turbines, without undue loss in efficiency and simplification of maintenance and repair. More is likely to be done to develop more appropriate machines for small schemes, introduce greater uniformity into their design and improve robustness. As an example, a major change in hydro control technology has facilitated the production of low-cost generating units with capacities of up to about 150 kWe. This has been achieved by a change in philosophy, in which the load is controlled instead of the usual method of controlling flow. The result is a load management system which maintains a constant electrical load on the turbine, that is allowed to operate at full load.

The concept is used in situations where the minimum stream flow is more than the full-flow capacity of the turbine. When there are several months of low-flow stream periods, two turbines are installed, one large and one small. During low-flow periods, only the small turbine is operated. This simplifies the mechanical components, eliminating wicket gates, governors and oil pressure systems at a major reduction in cost.

The output of a turbine is a function of the water flow times the head. As the head decreases, more flow is required to obtain the same output.

More flow means larger water passages and turbine runners, which in turn means higher cost. Horizontal axis, axial flow turbines can be used down to heads of about 3 m. Below this level, the costs rise sharply.

The production of hydropower in a reliable and low-cost manner can be enhanced through R&D activities. Contrary to statements that hydropower is an established technology and that research is not necessary, there are numerous areas where R&D is expected to result in significant improvement. Research is an effective way to deal with environmental problems, to incorporate technological developments, and to examine operational changes that would improve efficiency. A recent North American Hydro Research and Development Forum[15] identified more than 150 specific projects in 6 broad topic areas (operations, planning/analysis, environmental assessment and mitigation, hydromechanical issues, forecasting and structures/hydraulics). The high unit cost of small hydro projects will gain many benefits from the results of R&D activities. A Canadian government small hydro market analysis indicates that project cost reduction of 16–19% will double the economically viable potential capacity in Canada. Capital cost reduction of 10–15% can be achieved through innovation and simplification of site assessment, design and construction work, innovative and standardized turbine/generating equipment and controls, and new engineering concepts.

ESTIMATES OF REALIZABLE POTENTIAL

The estimates provided here for the realizable potential of small hydro includes sites with capacity of 10 MWe or less. Estimates have been provided of capacity, electrical energy and millions of tonnes of oil equivalent (Mtoe) for two scenarios: "current policies" and "favourable".

The current policies scenario is based on a projection of existing trends. Some of the circumstances leading to the "favourable" scenario are national commitments to support the development of renewables, minimization of development constraints, reasonable real interest rates, and adequate financing for worthy projects in developing countries. Estimates of generation capacity and electrical output by region through 2020 are presented in Table 7.5 for the two scenarios. Table 7.6 presents these same estimates in terms of Mtoe.

Basis of the Estimates

The information for 1990 is taken primarily from *Water Power and Dam Construction*[3]. That information is supplemented where possible by information obtained by Russian analysts for this WEC study by means of a questionnaire to WEC member committees.

Current Policies Scenario

The forecast capacity for Canada is based on the following approach: for 2000, the 1990 capacity plus the capacity under construction in 1990

Table 7.5 *Small hydro capacity and generation potential by world regions to 2020*

Region	*Current policies case capacity (MWe)*					*Favourable case capacity (MWe)*				
	1990	*2000*	*2005*	*2010*	*2020*	*1990*	*2000*	*2005*	*2010*	*2020*
North America	4,302	4,861	5,154	5,464	6,152	4,302	6,829	8,604	9,849	12,906
Latin America	1,113	1,992	2,607	3,444	5,751	1,113	2,125	2,937	3,839	6,557
Western Europe	7,231	8,822	9,704	10,577	12,587	7,231	11,478	14,462	16,555	21,692
E Europe and CIS	2,296	2,801	3,082	3,359	3,997	2,296	3,645	4,592	5,257	6,889
Mid East and N Africa	45	81	108	140	233	45	86	119	156	266
Sub-Saharan Africa	181	324	434	560	935	181	345	477	624	1,065
Pacific	102	124	137	149	177	102	162	204	234	306
China	3,890	6,963	9,331	12,036	20,101	3,890	7,428	10,264	13,415	22,915
Asia	343	614	823	1,061	1,772	343	655	905	1,183	2,021
Total	19,503	26,583	31,441	36,790	51,706	19,503	32,753	42,564	51,112	74,616

Region	*Current policies case generation (GWh)*					*Favourable case generation (GWh)*				
	1990	*2000*	*2005*	*2010*	*2020*	*1990*	*2000*	*2005*	*2010*	*2020*
North America	19,738	22,304	23,645	25,067	28,225	19,738	31,332	39,476	45,188	59,214
Latin America	4,607	8,246	11,050	14,255	23,805	4,607	8,795	12,155	15,890	27,138
Western Europe	30,239	36,891	40,580	44,232	52,636	30,239	48,000	60,477	69,231	90,715
E Europe and CIS	9,438	11,514	12,665	13,805	16,428	9,438	14,981	18,875	21,606	28,313
Mid East and N Africa	118	212	285	367	613	118	226	313	410	699
Sub-Saharan Africa	476	850	1,139	1,469	2,454	476	906	1,253	1,638	2,797
Pacific	407	511	562	613	730	407	666	838	962	1,257
China	15,334	27,448	36,780	47,446	79,235	15,334	29,280	40,458	52,881	90,328
Asia	1,353	2,420	3,243	4,184	6,987	1,353	2,582	3,567	4,663	7,965
Total	81,789	110,396	129,949	151,438	211,113	81,709	136,768	177,413	212,470	308,426

Table 7.6 Estimates of realizable small hydro potential by world region (in Mtoe)

	Current policies case				*Favourable case*		
Regions	*1990*	*2000*	*2010*	*2020*	*2000*	*2010*	*2020*
North America	4.4	5	5.6	6.3	7.0	10.1	13.2
Latin America	1.0	1.8	3.2	5.3	2.0	3.5	6
Western Europe	6.7	8.2	9.9	11.7	10.7	15.4	20.2
E Europe and CIS	2.1	2.6	3.1	3.7	3.3	4.8	6.3
Mid East and N Africa	0	0	0.1	0.1	0.1	0.1	0.2
Sub-Saharan Africa	0.1	0.2	0.3	0.5	0.2	0.4	0.6
Pacific	0.1	0.1	0.1	0.2	0.1	0.2	0.3
China	3.4	6.1	10.6	17.7	6.5	11.8	20.1
Asia	0.3	0.5	0.9	1.6	0.6	1.0	1.8
Total	18.2	24.6	33.8	47.1	30.5	47.4	68.7

plus one-half the additional capacity planned in 1990; for 2010, the 2000 capacity plus the remaining half of capacity planned in 1990; for 2020, the 2010 capacity plus the same amount of capacity added in the period 2000–2010. The electrical energy derived from the capacity is based on historical capacity factors in Canada for small hydro.

The forecast capacity for the USA is based on information obtained from a renewable energy white paper[7]. The electrical energy derived from the capacity is based on historical capacity factors in the USA for small hydro.

The forecast capacity and electrical energy for other regions of the world was based on the following estimate of the rate of annual increase of electricity production in the world:

	To year 2005	*2005–2020*
North America, Pacific, Europe (East and West) (industrialized regions)	2.0%	1.75%
All other regions (developing regions)	6.0%	5.25%

These electric capacity annual growth figures are derived from UN/World Bank Estimates. Electrical energy has been converted to Mtoe by dividing gigawatt-hours (GWh) of electricity by 4487.

Favourable Scenario

For the favourable scenario, the current policies scenario production for the years 2005 and 2020 for developing countries were increased based on estimates set out in Table 7.4. The "moderate growth" figures were applied to the favourable scenario for developing countries while "low growth" were applied to the current policies scenario. The estimates for industrialized countries were based on the assumption that their programmes for renewable energy can be achieved.

Assuming that the primary energy demand remains the same under each scenario while the portion supplied by hydro increases as shown in Table 7.4, small hydro production would increase (relative to production estimated for the current policies case) as follows:

	Industrial countries	*Developing countries*
Year 2005	Double 1990	10% increase
Year 2020	Triple 1990	14% increase

The above process provided estimates of electricity production for each area of the world in the years 2005 and 2020. An annual rate of increase in small hydro production between these years was then determined, and estimates were prepared from that for small hydro production for the years 2000 and 2010.

Estimates of the Total Investment

The total investment for the two scenarios including both R&D and plant investment were estimated based on Table 7.5. The site-specific nature of small hydro and unit price variation in different regions lead to a range of required investment unit costs. A range of about US$1,500–4,000 per kWe installed can be expected as stated in the "Characteristics and costs" section above. An assumption of US$2,000 per kWe average for the total investment is assumed to be appropriate. The following are cumulative total investment in billions of US dollars:

	2000	*2010*	*2020*
Current policies	54	76	100
Favourable	66	100	150

Post-2020 Perspectives

Based on the above projections, by year 2020, approximately 40% of the net exploitable technical potential of small hydro will be developed. Economic and environmental factors could restrict the development of the remaining large hydro potential after this time. It would be reasonable to expect that small and medium-sized hydro sites would be developed with steadily increasing market share.

CONCLUSIONS AND RECOMMENDATIONS

- Under a current policies scenario, it is estimated that small hydro can contribute 211,000 GWh (47 Mtoe) by 2020. The annual global average installation rate is about 1,000 MWe per year.
- Under a favourable scenario featuring government incentive programmes, minimization of development constraints, availability of reasonable real interest rates, and adequate financing for worthy projects in developing countries, it is estimated that small hydro can contribute 308,000 GWh (69 Mtoe) by 2020. The annual global average installation rate is approximately 2,000 MWe per year.
- In spite of the widespread availability of water courses which could support small hydro schemes, progress has not been as extensive as may have been hoped. The constraints to the development of small hydro schemes are numerous and challenging.
- Electricity supply is now becoming more decentralized, with greater involvement of local and private organizations. The resulting move away from centralized supply organizations places greater emphasis on local power production. This could help to put small hydro in a more favourable light and improve its economic attractiveness.
- Rehabilitation and refurbishment of older hydro plants and dams can form a very advantageous way of securing better utilization of an existing resource at low cost. The matter should be given high priority in development plans
- Environmental sensitivity places more emphasis on renewable and benign forms of power production. The outlook for small hydro would be improved if suitable precautions are taken in its conception.
- Subsidies are now becoming available for electricity generation from non-fossil fuel sources. Small-scale hydropower could benefit from such subsidies, particularly if they should become available in developing countries, where the scope for small hydro is greatest.
- Governments and aid agencies recognize that social and economic progress in rural and outlying areas of developing countries depends critically on an adequate supply of energy. They are encouraging development which could contribute to enhanced local electrification. This would offer greater scope to small hydropower and introduce a more accommodating attitude into the assessment of its merit.
- Better co-ordination of technical assistance and aid programmes would facilitate the introduction of country-wide small hydro programmes. This could result in savings in investment costs and spare parts inventories, and lead to substantial easing of operational and maintenance tasks as well as overall project management.
- On balance, small-scale hydropower has an important role to play in future energy supply scenarios provided the challenges specific to this form of electricity generation are recognized and effectively dealt with. Principal among these problems are the high investment

requirements for the power plants and the need for effective institutional arrangements, and an adequate supply of suitably trained and skilled local personnel. Technological development and research, especially on mitigation methodologies of environmental impacts, should be pursued further. Technology transfer and applications should be appropriately matched with regards to the types of local physical conditions.

- A single international organization should be established or designed to give focus and leadership to the increased use of small hydropower. This organization should co-ordinate with other agencies and collect resources data, make market projections and give guidance on technology transfer issues.

ACKNOWLEDGEMENTS

During the preparation of this chapter, many people contributed to the final product. At the beginning of the work of the WEC Study Committee on Renewable Energy Resources, the effort was directed by A.A. Zolotov and L.P. Mikhailov of Russia. Additional work was performed by F. Jenkin of the UK, and further editing was carried out by D.L.P. Strange and T. Tung of Canada. Contributions are also noted from K. Goldsmith and A. Bartle of the UK, G.W. Mills of New Zealand and many other reviewers from many countries. The contributions of all these members of the Small Hydro Subcommittee as well as other experts is gratefully recognized.

REFERENCES

1. K. Goldsmith. The Case for Small-scale Hydropower. *Water Power and Dam Construction.* United Kingdom, Reed Business Publishing Group, May 1991.
2. The World's Hydro Resources. *Water Power and Dam Construction.* United Kingdom, Reed Business Publishing Group, August 1992.
3. *1992 Handbook: Water Power and Dam Construction.* United Kingdom, Reed Business Publishing Group, 1992.
4. K. Goldsmith. Future Prospects of Hydropower. *Water Power and Dam Construction.* United Kingdom, Reed Business Publishing Group, August 1992.
5. K. Goldsmith, *Some Aspects of the Privatization of Small-scale Power Schemes.* Paper presented to Conference of the Electric Power Supply Industry (CEPSI) in Singapore, November 1990.
6. Small Hydropower on the Increase in China. *Modern Power Systems.* UK, January 1992.
7. *The Potential of Renewable Energy: An Interlaboratory White Paper.* Golden, CO, USA, US National Laboratories for US Department of Energy, 1990.

8. *Small Hydro Technology and Market Assessment* (draft). Ottawa, Canada, Alternative Energy Division, Energy, Mines and Resources, Canada, December 1991.
9. *Small Hydro Power: Water Power and Dam Construction.* United Kingdom, Reed Business Publishing Group, 1990.
10. M. Schomberg. *1992 Survey of Energy Resources.* London, UK, World Energy Council, 1992.
11. R. Price. *Effect of Government Incentives on Hydro Power in UK.* Harwell, UK, ETSU, Correspondence, August 1992.
12. UK Committee Calls For Quadrupling of Renewables Goal. *Newsletter of Independent Power Producers' Society of Ontario (IPPSO FACTO)*, Canada, Summer 1992.
13. T. Coe. Renewable Energy Would Gain From EC Tax On CO_2 Emissions. *International Solar Energy Intelligence Report*, USA, August 1992, pp. 127–128.
14. W. Rogers and J.P. Bourgeacq. Exporting the Hydro Experience. *Hydro Review*, Kansas City, MO, USA, HCI Publications, June 1992.
15. *Repowering Hydro: The Renewable Energy Technology for the 21st Century: Executive Summary – The US Perspective.* Proceedings of North American Hydroelectric Research and Development Forum, Kansas City, MO, USA, February 1992.

ANNEX

Terms of Reference

RENEWABLE ENERGY RESOURCES

For the purpose of the study reported in this book, renewable energy resources are defined as including solar, wind, oceanic, geothermal, tidal, mini-hydro, wood and biomass.

The intention of the study is to address growth opportunities and constraints to the full utilisation of renewable energy resources and to develop a global and long-term (15 to 20 years) perspective. Based on certain assumptions the study will attempt to present scenarios of the possible and desirable development of renewable resources and formulate recommendations for a forward-looking strategy at national and international levels.

The study will draw on existing information especially from recent WEC Technical/Study Committee reports, but the essential database for the study will result from a questionnaire to be circulated to all WEC Member Committees.

The study will be relevant to both developed and developing countries and will support the work of the WE Commission's *Energy for Tomorrow's World.*

INDEX